Logic, Epistemology, and the Unity of Science

Volume 37

Logic, Epistemology, and the Unity of Science aims to reconsider the question of the unity of science in light of recent developments in logic. At present, no single logical, semantical or methodological framework dominates the philosophy of science. However, the editors of this series believe that formal techniques like, for example, independence friendly logic, dialogical logics, multimodal logics, game theoretic semantics and linear logics, have the potential to cast new light on basic issues in the discussion of the unity of science.

This series provides a venue where philosophers and logicians can apply specific technical insights to fundamental philosophical problems. While the series is open to a wide variety of perspectives, including the study and analysis of argumentation and the critical discussion of the relationship between logic and the philosophy of science, the aim is to provide an integrated picture of the scientific enterprise in all its diversity.

More information about this series at http://www.springer.com/series/6936

Hourya Benis-Sinaceur · Marco Panza
Gabriel Sandu

Functions and Generality of Logic

Reflections on Dedekind's and Frege's
Logicisms

 Springer

Hourya Benis-Sinaceur
CNRS (IHPST, CNRS and University
   of Paris 1, Panthéon-Sorbonne)
Paris
France

Gabriel Sandu
Department of Philosophy
University of Helsinki
Helsinki
Finland

Marco Panza
CNRS (IHPST, CNRS and University
   of Paris 1, Panthéon-Sorbonne)
Paris
France

ISSN 2214-9775   ISSN 2214-9783   (electronic)
Logic, Epistemology, and the Unity of Science
ISBN 978-3-319-36782-8   ISBN 978-3-319-17109-8   (eBook)
DOI 10.1007/978-3-319-17109-8

Springer Cham Heidelberg New York Dordrecht London
© Springer International Publishing Switzerland 2015
Softcover reprint of the hardcover 1st edition 2015

Springer International Publishing AG Switzerland is part of Springer Science+Business Media
(www.springer.com)

Preface

The present book results from three papers. Each of them has been written independently of the others by one of us, but they share a common philosophical and historical background. All of them question a too easy reading of the origins of logicism, one which assimilates different views and purposes, both with one another and with more modern (but not necessarily more appropriate) conceptions. The common aim is to emphasise nuances and peculiarities among different ways of pursuing a program which only very broadly could be described as the reduction of (a part of) mathematics to logic. Though mainly devoted to discuss (some of) Dedekind's and Frege's views, they also deal with other conceptions somehow connected with these, in particular some endorsed by Lagrange, Cauchy, Weierstrass, Hilbert, Russell, Ramsey and Carnap.

The papers, or some of their previous versions, have somehow circulated within the scientific community, but have all remained unpublished up to now. We decided to put them together in a single volume, both because of their dealing with a common topic and because of their complementarity. They stem from shared standpoints and conceptions concerning the particular subject of enquiry, as well as on matters of philosophical and historical methodology, and their final versions, which we present here, ensue from many exchanges among us. But they pursue different specific aims. We hope they could jointly contribute to a better and more detailed picture of a crucial event in the development of philosophy of mathematics and logic. The common questions which our papers deal with and their different intents have been described in a newly written, coauthored introduction.

The *Institute d'Histoire et Philosophie des Sciences et des Techniques* in Paris (IHPST) has been the common context of our research. It is an intellectual home for all of us. Though written independently of each other, our papers have been prompted by a number of discussions we had among us, and with a large number of colleagues at the IHPST, at its seminars and workshops, but also, and possibly above all, during the everyday life at the Institute. To put it in another way: our book is the outcome of the rich intellectual dynamic made possible within the IHPST.

But it owes a great deal also to other influences, suggestions and comments. The list of all those who variously contributed to this and would deserve our acknowledgement would be too long. Let us thank some of them, as representative of all the others, namely: Andrew Arana, Mark van Atten, Michael Beaney, Jean-Pierre Belna, Francesca Boccuni, Méven Cadet, Stefania Centrone, Annalisa Coliva, Sorin Costreie, Michael Detlefsen, Jean Dhombres, Jacques Dubucs, Giovanni Ferraro, José Ferreirós, Sébastien Gandon, Jean Gayon, Jeremy Gray, Niccolò Guicciardini, Brice Halimi, Raclavsky Jiri, Joseph Johnson, Gregory Landini, Paolo Mancosu, Sebastiano Moruzzi, Alberto Naibo, Fabrice Pataut, Carlo Penco, Eva Picardi, Dag Prawitz, Shahid Rahman, Philippe de Rouilhan, Andrea Sereni, Stewart Shapiro, Dirk Schlimm, François Schmitz, Wilfried Sieg, Göran Sundholm, Jamie Tappenden, Luca Tranchini, Gabriele Usberti and Pierre Wagner.

Paris   Hourya Benis-Sinaceur
December 2014   Marco Panza
                                                                      Gabriel Sandu

Contents

Introduction

Logicism is usually presented as "the thesis that mathematics is reducible to logic", and is, then, "nothing but a part of logic". This is, at least, the way Carnap describes it in his influential 1931 paper ([41], p. 91, [10], p. 41). Though ascribing to Russell the role of "chief proponent" of it, Carnap also adds that "Frege was the first to espouse this view" (*ibid*).

Still, strictly speaking, Frege never argued for such a thesis. At most, he argued that arithmetic and real analysis are part of logic. But, also if it is so restricted, this thesis renders his view only very roughly. For what makes Frege's view distinctive is the way the inclusion relation between these mathematical theories and logic is conceived. And once this way is made clear, it also becomes clear that this relation does not depend, for him, on the mere possibility of a reduction of the former to the latter. Frege's point is, indeed, less that of showing how coming back from arithmetic and real analysis to logic, than that of developing logic enough so as to find natural and real numbers within it, and then show that what arithmetic and real analysis deal with is logical in nature.

Let us begin with arithmetic. In the *Vorwort* of *Grundgesetze*, he mentions the claim that "arithmetic is merely further developed logic [*weiter entwickelte Logik*]" ([97], *Vorwort*, p. VII, [110], p. VII$_1$), as the claim which he aims to argue for. This is only a rephrasing of the claim that Frege had taken himself to have established, though only informally, some years earlier, in the *Grundlagen*, namely that "Arithmetic is nothing but further pursued logic [*weiter ausgebildete Logik*], and every arithmetical statement is a law of logic, albeit a derived one" ([93], Sect. 87, [103], p. 99).[1] Frege's point seems, then, that "arithmetic is a branch [*Zweig*] of pure logic" ([97], *Einleitung*, pp. 1 and 3, [110], pp. 1$_1$ and 3$_1$) because the former results from an appropriate development (but not an extension), of the latter, that is,

---

[1] Here and later, from time to time, both in the present introduction and in the three following chapters, we feel free to slightly modify the English translations we quote, for sake of faithfulness to the original.

because "the simplest laws of cardinal number [*Anzahl*]^2 [...][are] derived by logical means alone" ([97], *Einleitung*, p. 1, [110], p. 1$_1$).

The crucial point of Frege's arithmetical logicism consists in fixing these logical means. This, in turn, entails identifying appropriate logical laws (or axioms, in modern terminology), deductive rules, and definitions from and according to which arithmetical truths follow. The purpose of the first and second parts of *Grundgesetze* ([97], Sects. I.1–II.54)^3 is precisely that of fixing these means and using them for deriving these truths.

In Frege's mind, what ensures the logical nature of these laws, rules and definitions is that the laws and rules are appropriately general, while the definitions are explicit and have recourse, in their *definiendum*, only to linguistic tools already introduced by previous analogous definitions, or directly belonging to the language in which the laws are stated. The appropriate generality of the laws and rules is, in turn, ensured by the fact that all they concern is (the values of) a small number of basic functions, defined by merely appealing to two basic objects, the True and the False (whose existence is taken for granted), and to the totalities of objects and of first- and second-level one-argument functions (so as to avoid any appeal to each of these totalities of functions for defining a function belonging to it). In other terms, these laws and rules are general because they merely pertain to (the values of) some basic functions, which are defined by relying on no device used for selecting some specific portions or elements of these totalities other than the True and the False. Now, defining these basic functions results in fixing a language to be used to form either names of values of these functions or of whatever other functions resulting from appropriately composing them,^{4} or general marks apt to "indeterminately indicate" [*unbestimmt andeuten*]" ([97], Sects. I.1, I.8, I.17, [110], pp. 5$_1$, 11$_1$, 31–32$_1$) these values. It follows, that, for Frege, the boundaries of logic are established by fixing a functional formal language, a small number of basic

---

^{2}We agree with Ebert and Rossberg in translating Frege's term 'Anzahl', when used in a technical context, with 'cardinal number', by conserving 'number' for his term 'Zahl' (cf. [110], "Translators' Introduction", p. xvi). A reason for using 'cardinal number', rather than 'natural number' is that Frege explicitly distinguishes (both in *Grundlagen* and in *Grundgesetze*) *endlich Anzahlen* from *unendliche* ones, namely finite cardinal numbers from infinite ones, among the latter of which he pays particular attention to the *Anzahl Endlos*, the cardinal number belonging to the concept ⌜*endliche Anzahl* ⌝ (cf. [93], Sects. 84–86 and [97], *Vorwort*, p. 5, and Sects. I.122–157). Notice, moreover, that, in *Grundlagen*, Frege also uses twice (Sects. 19 and 43) the term 'natürliche Zahl' (to be mandatorily translated with 'natural number')—in the latter case, merely in a quote from Schröder, but in the former by speaking on his own behalf—and many times (Sects. 76–79, 81–84, 104, and 108) the term 'natürlichen Zahlenreihe' (to be mandatorily translated with 'series of natural numbers' or 'natural numbers series'). Though in *Grundgesetze* (Sects. I.43–46, I.66, I.88, I.100, I.104, etc.), this last term is replaced with 'Anzahlenreihe' (to be translated with 'series of cardinal numbers' or 'cardinal numbers series'), it seems, then, that Frege takes a natural number to be a finite cardinal one.

^{3}We shall come back later on the third part.

^{4}To be more precise, Frege does not admit a direct composition of functions. According to him, functions are rather composed indirectly, so to say, by composing the names of their values (cf. [37], pp. 29-30). We shall avoid here to insist on this subtleties.

truths stated in this language, and a small number of rules used to draw truths stated in this language from other such truths. To put it briefly, when he speaks of logic, Frege is referring to a well-identified and (in his mind) appropriately established formal system, and when he claims that arithmetic is a branch of logic, he is implying that arithmetical truths are nothing but theorems of this system.

As we shall see pretty soon, this is not as trivial as it may appear at first glance. But it is still compatible with a conception of arithmetical logicism as a reductionistic program. To see what makes Frege's arithmetical logicism much more than that, one has to consider another distinctive and essential aspect of it. This depends on Frege's considering that values of functions (of whatever level) are objects, and that "objects stand opposed to functions", to the effect that "everything that is not a function" is an object ([97], Sects. I.2, [110], p. 7_1). Insofar as functions are, for him, unsaturated, this entails that cardinal and, *a fortiori*, natural numbers could not but be objects, for him. It follows that Frege's arithmetical logicism involves the thesis that natural numbers are objects, namely logical objects—objects whose intrinsic nature is made manifest by explicit definitions stated in the language of logic—and arithmetical truths are truths about these objects. But, insofar as it seems quite clear that natural numbers cannot be the True and the False, arguing for this thesis requires admitting that the language of logic is enough for defining some objects other than the True and the False.

The problem arises, then: how can such other objects be defined through this language, provided that it merely results from defining the basic functions of logic, and this is done by merely appealing to the True, the False and to the totalities of objects and first- and second-level one-argument functions? The answer depends (and could not but depend) on Frege's countenance, among his basic functions, of a function having values other than the True and the False. This is the case of the value-ranges function: a second-level one-argument function taking first-level one-argument functions (without any restriction), and giving value-ranges. Still, given the defining on the way basic functions are defined, taking such a function as a basic one entails renouncing restrictions mentioned above it explicitly, and, then, admitting of an implicit definition for it. Frege's infamous Basic Law V provides such a non-explicit definition: it implicitly defines value-ranges by stating an identity condition for value-ranges of first-level one-argument functions, that is, by asserting, as it is well known, that the value-range of a first-level one-argument function $\Phi(\xi)$ is the same as that of a first-level one-argument function $\Psi(\xi)$ if and only if the value of $\Phi(\xi)$ is the same as that of $\Psi(\xi)$ for whatever argument, which in Frege's formal language is expressed thus: $(\dot{\varepsilon}f(\varepsilon)=\dot{\alpha}g(\alpha))=(\underline{\quad\mathfrak{a}\quad}f(\mathfrak{a})=g(\mathfrak{a}))$, where '$f$' and '$g$' are marks used to indeterminately indicate first-level one-argument functions.

Frege was perfectly aware that, by admitting of such an implicit definition, he was derogating from the strict criterion of logicality that any other ingredient of his system meets. In the *Vorwort* of the *Grundgesetze* he recognises, indeed, that "a dispute" concerning the logical nature of this system "can arise [...] only concerning [...] Basic Law of value-ranges (V)" ([97], *Vorwort*, p. VII, [110], p. VII_1). Still, according to Frege, without this Law, and without value-ranges, there could not be other logical objects but the True and the False, and arithmetical logicism would, then, not be viable. This is what he openly claims in his tentative reply to Russell's paradox: "[…] even now I do not see how arithmetic can be founded scientifically, how the numbers can be apprehended as logical objects and brought under consideration, if it is not—at least conditionally—permissible to pass from a concept to its extension" ([97], *Nachwort*, p. 253, [110], p. 253_2).[5] Hence, for Frege, calling Basic Law V into question was not just calling into question his "approach to a foundation in particular, but rather the very possibility of any logical foundation of arithmetic" (*ibid*.). For, Frege seems to argue, if the value-range function is to be dismissed, what other logical function having other values than the True and the False is permissible? And if no such function may be permissible, how can natural numbers be logical objects? And if natural numbers are not logical objects, how can arithmetic be a branch of logic?

We know today that an alternative route for arithmetical logicism—allegedly understood, if not in the same way, at least in a way close to Frege's—has been suggested ([205], [118]). Still, it is clear that this route also depends on the admission of a basic function, namely the cardinal-number function, which, while being taken to be a logical function, is required to have as its possible values some particular objects whose existence is not a necessary condition for the admissibility of the relevant system of logic.

This is, in Frege's original terminology, a second-level one-argument function, like Frege's value-range one. And it is, like this latter function again, defined by a principle, namely Hume's principle, working as an axiom of the relevant system, and taken as an implicit definition. But, differently from Frege's value-range function, the cardinal-number function is not second-level insofar as its arguments are taken to be first-level functions. These arguments are rather taken to be concepts no more intended as functions from the totality of objects to the True and the False, but rather as the items designated by monadic first-order predicates.[6] The cardinal-number function is, thus, a total function, like the value-range one, only insofar as a previous restriction is, so to say, incorporated in the logical system its definition depends on: a restriction that makes the predicate variables of this system range only over concepts, rather than over items so generally conceived as to render the larger variety of Frege's first-level one-argument functions. This goes together with the fact that the values of the cardinal-number function are *ipso facto* cardinal numbers, rather than more general items among which cardinal and, more specifically, natural numbers, are selected with the help of appropriate explicit definitions (which might suggest that this function is not general enough to count as logical in Frege's sense).

It is not our purpose, here, to discuss neologicism. Touching upon it is only meant to emphasise the main difficulty with Frege's logicism, by showing that,

---

[5]Remember that a concept is, in Frege's terminology, a one-argument function whose values are either the True or the False, and its extension is nothing but its value-range.

[6]This entails that taking the cardinal-number function as a second-level function is imprecise, strictly speaking: this is, rather, a second-order function.

mutatis mutandis, it is still a crucial difficulty for its modern consistent version. This is the difficulty of fixing objects to be identified with natural numbers by having recourse only to means recognised as logical.

The way we have presented this difficulty hides a decisive aspect of it, however. This aspect only appears when it is made clear that, for Frege (as well as for the neologicists), nothing could be taken to be an object if it were not also taken to exist (in the only rightful sense in which anything can be taken to exist, both for him and for them), and no statement could be taken to be a truth if the singular terms and the first-order quantified variables included in it (if any) were not respectively taken to be names of, or to vary over existing individuals. The difficulty does not only consist, then, in defining natural numbers by having recourse only to means recognised as logical, but in doing it so as to ensure that these numbers exist, that is, that the (non-atomic) term that provides the *definiendum* of the explicit definition of each of them denotes an existing individual, and the (non-atomic) formula that provides the *definiendum* of the explicit definition of the property of being a natural number is satisfied by some (namely a countable infinity) of existing individuals (which means, in Frege's formalism, that the explicit definition of the first-level concept ⌜natural number⌝ designates a function whose value is the True for some, namely for countably many, arguments).

There is no room here for discussing the reason for neologicists to claim that their definitions comply with this condition. What is relevant is that, for Frege, no independent existence proof is needed for this purpose, since, for him, the relevant explicit definitions are so shaped as to ensure by themselves that this condition obtains. In other words, according to Frege, his explicit definition of each natural number directly exhibits an object to be identified with this number, while his definition of the property of being a natural number directly manifests that there are these numbers and which objects they are. This means that, according to Frege, these explicit definitions directly manifest that "there are logical objects" and that "the objects of arithmetic [i.e. the natural numbers] are such" ([97], Sect. II.147, [110], p. 149$_2$).

For real numbers, Frege does not seem to have thought that something like this would have been achievable, instead. Since, though he closed his informal exposition of the way he was planning to define these numbers by claiming that in this way he would have succeeded "in defining the real number purely arithmetically or logically as a ratio of magnitudes that are demonstrably there" ([97], Sect. II.164, [110], p. 162$_2$), his plan explicitly calls for an existence proof of domains of magnitudes going far beyond the simple inspection of the definition of these domains, and consisting, rather, in the independent exhibition of a particular domain of magnitudes generated from natural numbers. Frege actually fulfilled only a part of his plan: in the third part of *Grundgesetze* ([97], Sects. II.55–II.245), after having discussed and questioned several (informal) definitions of real numbers (*ibid*. II.55–II.155) and having exposed his plan (*ibid*. II.155–II.164), he proceeds to formally defining domains of magnitudes (*ibid*. II.165–II.245) and to prove some crucial properties of them, by leaving to a never appeared third volume of his treatise the accomplishment of the remaining part of the plan, including the existence proof of such domains, and the definition of real numbers as ratios over them.

Let $D(\xi)$ be the first-level concept of domains of magnitudes, namely the concept under which an object falls if and only if it is a domain of magnitudes, which means that $D(s)$ is the True if and only if s is such a domain. Frege's formal definition of domains of magnitudes consists in stating an identity like '$D(s) = \mathcal{D}(s)$', where '$\mathcal{D}(s)$' stands for an appropriate (non-atomic) formula (*ibid.* II.173–174 and II.197).[7] In modern terminology, this means that domains of magnitudes are explicitly defined as the objects that satisfy this formula. And this formula is such that s satisfies it (which means, in Frege's terminology, that $\mathcal{D}(s)$ is the True) if and only if s is the extension of another first-level concept $\mathcal{M}(\xi)$, under which an object falls, in turn, if and only if it is the extension of a first-level binary relation that, if taken together with all the other extensions of a first-level binary relation that fall under this very concept, forms a certain structure.[8] This means that domains of magnitudes are explicitly defined as extensions of first-level concepts under which fall the extensions of some first-level binary relations that form, when taken all together, a certain structure.

This definition is stated within the same functional formal language in which natural numbers are defined. Still, Frege openly claims (*ibid.* II.164) that it does not ensure that there are objects that stand to each other in some binary relations whose extensions, when taken all together, meet the relevant structural condition, and are many enough for the ratios over them to be identified with the real numbers. And, he argues, if there were no such objects, real numbers could not be defined as ratios over domains of magnitudes. The existence proof of domains of magnitudes envisaged by Frege should have consisted in showing how, by starting from natural numbers and by appropriately operating on them, one can get enough—i.e., continuous many—other suitable objects. It is not necessary to enter the details of the way Frege planned to conduct this proof, in order to understand that he could not have imagined that the relevant objects could be directly exhibited by explicit definitions, as he held to have done for natural numbers. This, together with the fact that he held that his definition of domains of magnitudes does not secure, by itself, the existence of appropriate such domains, is enough for concluding that real numbers could not have been taken by Frege as logical objects in the same sense as natural numbers. Hence, his logicism about real numbers, once completely expounded in agreement with his plan, could not have appeared similar in nature to his arithmetical logicism.

These short and quite general remarks should be enough to make clear that Frege's logicism is quite complex a thesis, or better that it consists of two distinct quite complex theses, respectively, pertaining to natural and real numbers that are

---

[7] As a matter of fact, this formula is not openly written by Frege, but it is easily deducible by other formulas which he openly writes.

[8] Remember that for Frege a binary first-level relation is a first-level two-arguments function whose values are either the True or the False.

only very partially and broadly rendered by the simple claim that arithmetic and real analysis are part of logic. It follows that it is not enough for someone to be credited with the same foundational program as Frege's that he made this same claim. Only a careful comparative scrutiny of the way this claim is justified and explained could allow one to evaluate whether this claim is an expression of logicism in the same sense as Frege's.

A case in point is that of Dedekind. From the very beginning of the *Vorwort* to the first edition of *Was sind und was sollen die Zahlen?*, he explicitly identifies the "simplest science" with "that part [*Theil*] of logic which deals with the theory of numbers [*Lehre von den Zahlen*]", and refers to it as to "arithmetic (algebra, analysis) [*Arithmetik (Algebra, Analysis)*]" ([49], p. VII, [53], p. 14). There is little doubt that what Dedekind means here with 'theory of numbers' is much more than the theory of natural numbers, and also includes real analysis. This is not only suggested by the parenthesis following the term 'Arithmetik', but also by the possessive pronouns 'its [*ihr*]' in what one finds some lines below: "it is only through the purely logical construction of the science of numbers and in its acquiring the continuous number-realm that we are prepared accurately to investigate our notions of space and time" (*ibid.*)[9]. It seems then quite clear that Dedekind here is endorsing the claim that both arithmetic and real analysis are part of logic. Still, the way this claim is justified with respect to arithmetic, as well as the tacit extension of it to real analysis, delineate a quite different conception than Frege's.

Dedekind's main point is that "the number-concept [*Zahlbegriff*] [is] entirely independent of the conceptions or intuitions of space and time", being rather "an immediate result from the laws of thought", since "what is done in [looking for] the number of a set [*Zahl der Menge*] or the number of some things [*Anzahl von Dingen*]" depends on "the ability of the mind to relate [*beziehen*] things to things, to let a thing correspond [*entsprechen*] to a thing, or to represent [*abzubilden*] a thing by a thing, an ability without which no thinking is possible" ([49], p. VIII, [53], p. 14).

The reference, here, is to the crucial role played, in Dedekind's definition of natural numbers, by the notion of a "mapping [*Abbildung*]". In his terminology: a "thing [*Ding*]" is "any object of our thought [*Gegenstand unseres Denkens*]" ([49], Sect. 1, [53], p. 21); a "system [*System*]" is that which "different things [...] constitute [*bilden*]" when they are "considered from a common point of view [*unter einem gemeinsamen Gesichtspuncte aufgefasst*]" and are, then, "associated in the mind [*im Geiste zusammengestellt*]" ([49], Sect. 2, [53], p. 21); the "elements [*Elemente*]" of a system are the things that constitute such a system (*ibid.*); and a "mapping [*Abbildung*][...] of a system *S* [...][is] a law [*Gesetz*] according to which

---

[9]This is also confirmed by the reference, which Dedekind makes a few lines below, to his supplement XI to Lejeune Dirichlet's *Vorlesugen über Zahlentheorie*, which is devoted to the theory of finite algebraic numbers ([66], pp. 434–626, esp. p. 470, footnote).

a determinate thing pertains to [*gehört zu*] every determinate element [...] of *S*"([49], Sect. 21, [53], p. 24).

His point is, then, that if these notions are appropriately used together, they provide enough conceptual tools to define the natural numbers and for setting up a theory of them. The way this is done in Dedekind's treatise is, however, essentially informal. The above explanations are everything Dedekind's exposition relies on in order to make these notions operate and carry out the required definition of natural numbers and the corresponding theorems. Hence, Dedekind's theory of natural numbers is in no way embedded, like Frege's, within a well-identified formal system. If it is logic, or a part of logic, it is, then, not because it is part of such a system, but because of the intellectual abilities these notions and our handling them hinge on. In other words, what is taken to be logical, in Dedekind's arithmetical logicism, is not a formal system and its ingredients, but some notions, informally explained, and our intellectual ability to handle them. Moreover, the relation of being part of, which relates arithmetic to logic, is not conceived as a relation of inclusion of a system into a system, but rather as a relation depending on the sufficiency of this ability for realising the relevant task.

Concerning real numbers, things are even clearer. Since Dedekind's extension of his arithmetical logicism to real analysis merely depends on his remark that the "creation [*Schöpfungen*] of [...] negative, fractional, irrational [...] numbers is always accomplished by reduction to the earlier concepts [...] without the introduction of foreign conceptions", as thoroughly shown, for irrational numbers, in *Stetigkeit und Irrationale Zahlen*, and suggested in section III of this very treatise, for the other numbers ([49], p. X, [53], p.15, [47]). The point, here, is not only that the definition of irrational numbers offered in *Stetigkeit* is as informal as that of natural numbers offered in *Was sind*, and that in the former treatise Dedekind advances no thesis assimilable to some sort of logicism, but also, and above all, that Dedekind seems to consider useless to show explicitly how the logical intellectual ability the theory of natural numbers depends on is also enough to pass from these numbers to real ones. All that is relevant, for him, is a generic appeal to the possibility of a conceptual reduction. There is nothing, then, like what justifies Frege's logicism for real numbers, namely a further development of the same formal system in which the definition of natural numbers is embedded, resulting in an independent formal definition of the former numbers.

But as essential as these differences might appear, they are far from being the only ones. Another, possibly even more essential one, already shines through Dedekind's speaking of creation of negative, fractional and irrational numbers by reduction to earlier concepts. It depends on Frege's and Dedekind's respective conceptions of the very nature of natural and real numbers. For Frege, they are objects, that is, individuals existing as such, and they are logical not insofar as their existence depends on logic, but rather insofar logic is enough for defining them (and, at least in the case of natural numbers, for ensuring their existence). This is not at all Dedekind's view. For him, they are, rather, "free creations [*frei Schöpfungen*] of the human mind" to be used for "apprehending more easily and more sharply the difference of things" ([49], pp. VII–VIII, [53], p.14), and these creations are logical insofar as logic, understood as we have said above, is enough for fixing their concept, which is all what is needed to create them. Since, for Dedekind, the only sense in which a number can be said to be an object depends of its being a thing, namely, as we have already said, an object of our thought, and "a thing is completely determined [*vollständig bestimmt*] by all that can be affirmed or thought concerning it" ([49], Sect. 1, [53], p. 21). The more difficult task of Frege's logicism, consisting in offering a justification of the existence of natural and real numbers is, then, simply dismissed by Dedekind, which merely implies that it is nonsensical.

Though the first edition of *Was sind* appeared four years after Frege's *Grundlagen*, no mention of the latter is made in the former. A short mention is made, instead, in the *Vorwort* of the second edition ([50], p. XVII, [53], p. 19). Though declaring of having become acquainted with Frege's treatise only after the publication of his own, and bringing to the reader's mind the difference of his and Frege's "view on the essence of number [*Wesen der Zahl*]", Dedekind emphasises Frege's standing "upon the same ground" with him. Had be been aware of the way Frege's view is spelt out in the *Grundgesetze*, whose first volume just appeared in the same year as the second edition of Dedekind's treatise, the latter would have probably advanced a different judgement, since the main differences among the two approaches are much more evident when *Was sind* is compared with *Grundgesetze* (as we have done above), rather than with *Grundlagen*, where Frege's logicism is, indeed, merely sketched out informally.

It was, then, up to Frege to emphasise the differences of his and Dedekind's approaches ([97], *Vorwort*, p. VII–VIII, [110], pp. VII₁–VIII₁):

My purpose demands some divergences from what is common in mathematics. This will be especially striking if one compares Mr Dedekind's essay, *Was sind und was sollen die Zahlen?*, the most thorough study I have seen in recent times concerning the foundations of arithmetic. It pursues, in much less space, the laws of arithmetic to a much higher level than here. This concision is achieved, of course, only because much is not in fact proven at all. Often, Mr. Dedekind merely states that a proof follows from such and such statement; he uses dots, as in '$\mathfrak{M}(A,B,C\ldots)$'; nowhere in his essay do we find a list of the logical or other laws he takes as basic; and even if it were there, one would have no chance to verify whether in fact no other laws were used, since, for this, the proofs would have to be not merely indicated but carried out gaplessly. Mr Dedekind too is of the opinion that the theory of numbers is a part of logic; but his essay barely contributes to the confirmation of this opinion since his use of the expressions 'system', 'a thing belongs to a thing' are neither customary in logic nor reducible to something acknowledged as logical.

If the most part of this quote keeps our attention to the first difference we have remarked above, namely the fact that Dedekind's presentation is informal, to the effect that his logicism is in no way the thesis that the relevant truths are theorems of a formal system taken as logic, the last remark points in a different direction: what makes Dedekind's alleged logic not be logic at all is its involving set-theoretic notions.

As Frege makes clear in the *Einleitung* of his treatise, by coming back to the same point, the complaint, here, is not merely with Dedekind's using the term 'system' with the "same intention [*Absisht*]" with which others were using, in the same years, the term 'set [*Menge*]' ([97], *Einleitung*, p. 1, [110], p. 1_l), or with his considering systems to be things to which other things belong.[10] After all, Frege himself will largely use a set-theoretic term as 'class [*Klasse*]' in the second volume of *Grundgesetze*, and will have no reticence to say that a class "contains [*umfassen*]" objects, and that an object "belongs [*angehört*]" to a class ([97], Sect. II.164 and II.173, for example). Still, for Frege, a class is the "extension of a concept" ([97], Sect. II.99 and II.161), that is, a value-range. This is exactly the point. For Dedekind, he argues, "every element [...] of a system [...] can itself be regarded as a system", and, "since in this case element and system coincide, it here becomes very perspicuous that, according to Dedekind, the elements are what properly makes out the system [*die Elemente den eigentlichen Bestand des Systemes ausmachen*]" ([97], *Einleitung*, p. 2, [110], p. 2_l), which is, in fact, inappropriate. Since, "every time a system has to be specified", even Dedekind, despite his "lack of insight", cannot but mention "the properties a thing must have in order to belong to the system, i.e. he defines a concept in virtue of its characteristic marks", and so makes it appear that "it is the characteristic marks that make out the concept, rather than the objects falling under it" ([97], *Einleitung*, p. 3, [110], p. 3_l; cf. also the quote from Frege's undated letter to Peano quoted in Sect. 3.5.5 of chapter 3, below). In other terms: if specifying a system reduces to defining a concept, what makes out the system cannot but be the same as what makes out the concept, and this is not given by the extension (or value-range) of this concept, but by what makes an object be part of such an extension. So, it is not Dedekind's use of set-theoretic notions that Frege is questioning, but his taking them as primitive notions, his not clearly submitting them to the more general notion of a concept, or, even, to the still more general notion of a function. Logic, Frege seems to mean, can include consideration of sets, but only insofar this is part of a more general study of the objects-concepts relations, of the falling of objects under concepts: dealing with the notions of a set as a primitive notion, and merely conceiving of a set as an aggregate of things is *ipso facto* departing from logic.

This is, possibly, the deeper difference between Frege's and Dedekind's logicism, and it sheds new light also on the differences between Dedekind's notion of a mapping and Frege's notion of a function. For the former, a mapping is something that "relates things to things" or makes "a thing correspond to a thing", or a thing "represent" a thing, and only appears, then, when appropriate things, which we independently take to be there as such, are related. For the latter, a function is that which an object is a value of, and it appears any time we refer to a certain object, through the way we refer to it, with the only exception of the primitive reference to the True and the False, which is necessary for defining the basic functions of logic. This exception, however, does not render our reference to the True and the False always independent of any function. Logic is, rather, for Frege, the study of our

---

[10]As a matter of fact, in *Was sind*, one finds only one occurrence, in Sect. 34, of the verb 'to belong to [*gehören zu*]' or its cognates used in this sense; possibly Frege's was here referring to *Stetigkeit*, where this use is much more frequent, namely in Sects. IV–V.

referring to the True and the False as values of particular sorts of functions, namely concepts and relations, and it is our referring to the True or to the False as the value of a certain concept or relation, for a certain object or a certain pair of objects as arguments, that makes this object fall or not fall under this concept, or the objects composing this pair are or are not related by this relation. The notions of a concept and a relation, rather than those of a system and its mappings are then, as Frege explicitly says, "the foundation stones [*Grundsteine*] on which [...][he] build [...] [his] construction" ([97], *Einleitung*, p. 3, [110], p. p. 3_1).

We have, thus, arrived at the two basic, strictly connected, topics the present book is devoted to, namely the way Dedekind's logicism differs from Frege's, and the conception of a function Frege's logicism is grounded on.

The first chapter, by Hourya Benis Sinaceur, is specifically devoted to the former topic, and goes much further than we could have done here in accounting for the specific features of Dedekind's logicism, and of his conception of logic, in connection, of course, with his conception of systems and mappings.

The second chapter, by Marco Panza, provides a historically situated account of Frege's notion of a function, by insisting on the intensional nature of this notion, and then questioning a widespread tendency to ascribe to Frege a Platonist view about functions, and, more specifically, concepts and relations.

This question is related to a lively discussion that took place after the publication, in 1992, of a paper by Jakko Hintikka and Gabriel Sandu ([129]), where the ascription to Frege of a notion assimilable to the modern, essentially extensional and set-theoretic notion of an arbitrary function is questioned. In the third chapter, Gabriel Sandu comes back on this discussion, by offering new arguments for his views, based on the consideration of Ramsey's reaction to Russell's logicism.

The intimate connection among these topics should be clear from the previous considerations. But it becomes even more evident when it is observed that the two features of Frege's logicism that are mainly responsible for the difference between this logicism and Dedekind's, namely Frege's conceiving of logic as a formal system, and his grounding it on a general notion of a function—as opposed, respectively, to Dedekind's taking as logical some informal notions and our intellectual ability to handle them, and to his conceiving of systems as aggregates of things that we take to be there as such—are strictly connected, in turn. For Frege's notion of a function is fashioned to provide the ground on which this logical system is erected, that is, it is, essentially, a linguistic notion. While Frege conceives of objects as existing individuals, he does not assign existence to functions, as such. He rather conceives them as matrices to be used to form object-names, which is precisely what, for him, makes a function appear through the way we refer to an object, as we have said above.

This has a crucial consequence on the way the totalities of objects and functions, and, then, an arbitrary function, are conceived by him. The question is largely discussed in Panza's and Sandu's chapters, but a short remark is in order, here as well.

True, Frege was aware of Cantor's distinction among different sorts of infinity. He entitled Sects. 84–86 of *Grundlagen* "Infinite cardinal numbers [*Unendliche*

Anzahlen]"[11] (in the plural), by explicitly referring to Cantor's very recently appeared *Grundlagen einer allgemeinen Mannigfaltigkeitslehre* ([38]). Though, in them, he specifically deals, in fact, only with "the cardinal number which belongs to the concept ⌜finite cardinal number⌝", he significantly denotes it with '∞_1', by suggesting that other infinite cardinal numbers follow, namely ∞_2, ∞_3, etc. After having come back to this same number in the first volume of *Grundgesetze*, this time by calling it 'Endlos'[12] and denoting it with '∞' ([97], *Vorwort*, p. 5, and Sects. 122–157), he remarks, in the second volume (*ibid.*, II.164) that his envisaged definition of real numbers as ratios of magnitudes could not be carried out without having available an infinite "class of objects" having an infinity greater than *Endlos*, namely that of the cardinal number of the concept ⌜class of finite cardinal numbers⌝. Still, he seems to have realised neither that the objects that such a class would contain could not all have a (finite) name (even in a language including countably infinite atomic names), nor that they would have allowed to form enough subclasses as to assign an extension to an even greater infinity of unnameable concepts, and, more in general, a value-range to an even greater infinity of unnameable functions. In other words, he does not seem to have realised that allowing for uncountable classes of objects goes together with admitting unnameable objects, including the possible value-ranges of unnameable functions and, then, also unnameable functions, in extension.

A symptom of his unawareness is his using Latin letters to form universal statements. Within Frege's system, a particular statement is formed by the name of a value of a concept or relation (which is a name of a truth-value), preceded by a special sign of assertion, namely '⊢'. Such a statement is taken to assert, then, that the truth-value named by this name is the True. A general statement is formed in the same way, except for the replacement of the name of a value of a concept or relation with a Roman object-marker, namely an expression involving Roman letters ([97], Sect. 26, [110], p. 44). And it is taken to assert that the object-name that is obtained from such an object-marker, by replacing within it each Roman letter for objects with whatever object-name (which actually refers to an object), and any Roman letter for functions with whatever function-name (of the appropriate sort), is a name of the True ([97], Sects. I.5, I.8, I.17 and I.19). So, to give a simple example, the

general statement '$\underset{a=b}{\underset{\mid}{\vdash}}\begin{array}{l}f(b)\\ f(a)\end{array}$', namely theorem III*c* of Frege's system ([97],

Sect. I.50), is taken to assert that the truth-value named by the object-name got from

'$\underset{a=b}{\underset{\mid}{}}\begin{array}{l}f(b)\\ f(a)\end{array}$', by replacing '*a*' and '*b*' with whatever object-names (which actually

---

[11]Cf. footnote 2, above.

[12]Cf. again footnote 2, above.

refer to an object), and '$f(\xi)$' with whatever name of a one-argument first-level function, is the True. It is clear that Frege could have not conceived of a general statement in this way if he had not admitted that any possible object and function can somehow be named.[13]

There is no doubt that a notion of a function leaving no room for the idea of an extensionally arbitrary function, and a notion of a set based on such a notion of a function could not have provided a basis for the development of mathematics and mathematical logic that followed the pioneering works of Frege, Dedekind and Cantor, among others. Hence, despite the lack of intelligibility that the idea of an extensionally arbitrary function brings with it, and despite the promises of clarification that alternative notions, intensional, in nature, could generate or have generated, it is a fact that a foundational program based on these last notions could hardly have been suitable for accounting for this development, even if it had presented no other internal difficulty. Still, the question that the present book is devoted to is not whether Frege's or Dedekind's different forms of logicism were based on suitable conceptions. Our aim is rather that of keeping the reader's attention on some crucial aspects of these forms of logicism, so as to contribute to a better understanding of their nature, their motivations, and their mutual differences.

Hourya Benis-Sinaceur
Marco Panza
Gabriel Sandu

---

[13]Unless he were admitting, instead, that the relevant generality is merely a linguistic one, that is, one that concerns only objects and function that can be somehow named, which is quite implausible, indeed.

Chapter 1
Is Dedekind a Logicist? Why Does Such a Question Arise?

Hourya Benis-Sinaceur

1.1 Introduction

Logicism is generally presented as the philosophical thesis that arithmetic, and therefore all of mathematics, can be deduced from logic alone or can be reduced to logic. Logicism is prominently associated with Frege's and Russell's achievements, which are taken as paradigmatic realisations of its aim. In Carnap's terms ([41], p. 91; [10], p. 41; my italics):

> Logicism is the *thesis* that mathematics is *reducible* to logic, hence nothing but a part of logic. *Frege was the first* to espouse this view (1884). In their great work, *Principia Mathematica*, the English mathematicians A. N. Whitehead and B. Russell produced a systematisation of logic from which they constructed mathematics.

On the one hand, even though Dedekind writes, in the first preface to *Was sind und was sollen die Zahlen?* [49], that arithmetic is a part of logic, Carnap does not mention him. On the other hand, some recent investigations, especially the studies by Boolos, have shown that if one wants to take the logicism credo *à la lettre*, neither Frege's foundation of arithmetic nor Whitehead and Russell's presentation of arithmetic in *Principia Mathematica* can be really called 'logicist'.[1] Boolos' main argument is that the fundament of Frege's reduction, i.e. the definition of numerical identity in terms of the one-to-one correspondence, named today 'Hume's principle',

---

[1] Boolos' statement is more general (Boolos [23, pp. 216–217]): "Neither Frege nor Dedekind showed arithmetic to be part of logic. Nor did Russell. Nor did Zermelo or von Neumann. Nor did the author of *Tractatus* 6.02 or his follower Church. They merely shed light on it".

© Springer International Publishing Switzerland 2015
H. Benis-Sinaceur et al., *Functions and Generality of Logic*,
Logic, Epistemology, and the Unity of Science 37,
DOI 10.1007/978-3-319-17109-8_1

or 'HP', in short,[2] is demonstrably consistent ([205]; [132], p. 138; [33]; [21], p. 174; [22]),[3] but not purely logical [27].[4]

Thus the state of affairs with logicism is not straightforward, even if one only considers its great father, Frege. Frege's great inventions (the function-argument analysis, the distinctions between *Sinn* and *Bedeutung*, between concept and object, etc.) have given birth to a huge amount of comments, interpretations, and debates in logic ands philosophy of mathematics. To deal with Frege's heritage is far beyond the scope of this paper. I shall rather focus on Frege's logicist *project*, not on its failure and/or its modern amendments.

Dedekind's construction of natural numbers in [49] and, secondarily, of real numbers in *Stetigkeit und irrationale Zahlen* [47] are also often viewed as a "logicist foundation" of arithmetic and real analysis. This view is based on some assertions in [49], mainly in the first preface. Actually, Dedekind shares Frege's aim of substituting logical standards of rigour for intuitive imports from spatiotemporal experience into the deductive presentation of arithmetic.[5] But sharing this aim does not mean having the same fundamental goal, nor following the same path to reach it. I will highlight the dissimilarities between Dedekind's and Frege's *actual* ways of doing and thinking, and I will bring out the fact that "there are considerable differences in their accounts of our *knowledge* of the existence and infinity of natural numbers" ([63], p. 52, my italics).

Moreover, pairing Dedekind with Frege often implies a distorted assessment of Dedekind's own achievements in [47, 49]: by the yardstick of logic and logicism proper, Dedekind's *Was sind* appears less deep and less thorough than Frege's *Die Grundlagen der Arithmetik* [93], without speaking of his *Grundgesetze der*

---

[2]The name of this principle comes from Frege's quoting, in [93] Sect. 63, Hume's claim that "When two numbers are so combined as that the one has always an unit answering to every unit of the other, we pronounce them equal" (*Treatise*, Book I, Part iii, Sect. 1). Later, in Sect. 73, Frege argues that "the cardinal number which belongs to the concept F is identical with the cardinal number which belongs to the concept G if the concept F is equinumerous [*gleichzähling*] to the concept G" ([103], p. 80), which he tries, then, to prove. What is today usually called 'Hume's principle' is the conjunction of this implication and the inverse one, which is formally (in the language of second-order logic) rendered as follows:

$$\forall F, G \left[(\#F = \#G) \Leftrightarrow F \approx G \right].$$

where '$F \approx G$' is a second-order formula expressing the existence of a one-to-one correspondence between the objects falling under F and those falling under G. Second-order logic $+$ HP is called 'Frege's arithmetic' or 'FA', for short. On the differences between Hume's statement about numbers and Frege's application to concepts cf. [191], who rightly points out that the definition of equinumerosity in terms of one-to-one correspondence is due to Cantor.

[3]What is proved is that HP has models with countable infinite domains.

[4]In *Grundlagen*, before trying to prove the if-direction of HP (cf. footnote (2), above), Frege claims to have reduced one-to-one correspondence to "purely logical relationships" ([93], Sect. 72: "Hiermit haben wir die beiderseits eindeutige Zuordnung auf rein logische Verhältnisse zurückgeführt").

[5]For example, Gödel ([114], p. 127; my italics) writes that the vicious circle principle "makes impredicative definitions impossible and thereby destroys the derivation of mathematics from logic, *effected by Dedekind and Frege*, and a good deal of modern mathematics itself".

Arithmetik [97], which set out the content of *Grundlagen* in the frame of the formal language introduced in his earlier *Begriffsschrift* [92]. For instance, Burgess thinks that what is missing in Dedekind's foundation of arithmetic is "any rigourous or even plausible derivation" of his axioms from something like HP ([28], introduction to part II, p. 141).[6] Such a derivation of arithmetic in second-order logic, which is outlined in the *Grundgestze* (by making an eliminable appeal to extensions of concepts) and worked out by Wright ([205], Chap. 4) and Boolos [23], is usually called 'Frege's theorem'. Thus, Burgess's criticism comes down to saying that Dedekind did not prove Frege's theorem,[7] which is hardly surprising, since Dedekind's aim was not to *derive* his axioms from something else but to *lay them* down as primitive, and to derive from them the definition of natural numbers and arithmetical operations.

Burgess considers that *deriving* mathematical induction was central in Frege's and Russell's attempts to provide a logical foundation for mathematics, as it also is a central goal when, following Zermelo, mathematics is developed in a set-theoretic framework ([28], introduction to part III, p. 345). But, Dedekind's purpose was *not to derive* mathematical induction from a *demonstrably* purely logical principle. Dedekind derives mathematical induction from his concept of chain ([49], Sect. 59), which is construed out of a *System S* and an *Abbildung*[8] φ with domain and codomain S.[9] Dedekind takes those concepts as resulting from two grounding operations, namely making a *System* from a multiplicity, and representing a thing by a thing, which are both taken to be "logical" in the sense that: (i) they are more general than the usual arithmetic operations, and (ii) they constitute the creative power of the mind. Besides that and above all else, Dedekind's first concern is not the question "How many?", therefore not the definition of cardinal numbers and not numerical identity (equinumerosity). Dedekind assumes Leibniz's indiscernible principle,[10] making the most of the idea of substitutability rather than of the narrower idea of equational identity.

Dedekind's philosophical assumptions are less explicit and much less systematic than those of Frege; one might then be tempted to interpret them as *germs* of Frege's definite standpoint. Such a view can even be supported by the fact that Frege knew Dedekind's essays on numbers. We have at least three pieces of evidence of this fact.

---

[6]Before Burgess, also Parsons [153] and Heck [119] made the same claim.

[7]Burgess relies on the fact that from HP one can prove the existence of Dedekind infinite sets (a set A is Dedekind infinite when there is a bijection φ from A onto a proper subset of A; for getting Dedekind infinite sets we need the infinity axiom + an equivalent of the axiom of choice).

[8]I keep untranslated Dedekind's terms 'System', which corresponds to our 'set', and 'Abbildung' (and cognates), which do not correspond to our mapping. Dedekind does not define an *Abbildung* by its graph as is defined a mapping in current set theory. The first translation made in [171, 209], namely 'representation', seems much better to me. In the light of Emmy Nœther's developments and of Category Theory, the *operation* of representing a thing by a thing is a morphism.

[9]Cf. [171], p. 247: "It is a most remarkable fact that Dedekind's previous assumptions suffice to demonstrate this theorem".

[10]Cf. [49], Sect. 1, [53], p. 21: "A thing is completely determined by all that can be affirmed or thought concerning it. A thing a is the same as b (identical with b), and b the same as a, when all that can be thought concerning a can also be thought concerning b, and when all that is true of b can also be thought of a".

(1) Frege quotes many times in *Grundlagen* ([93], Sects. 12, 26, 34, etc.) Lipschitz's *Lehrbuch der Analysis* [143], which was written down after harsh discussions with Dedekind on the definition of the "continuity" of the domain of the real numbers, and where Lipschitz introduces the operations on limits following Sect. 7 of *Stetigkeit*.

(2) In the first volume of *Grundgesetze*, Frege largely discusses some of Dedekind's fundamental views advanced in *Was sind*, both in the *Vorwort* and the *Einleitung* ([97], *Vorwort*, pp. VII–VIII and *Einleitung*, pp. 1–3), and he apparently takes from Dedekind's essay the term 'Abbildung' (*ibid., Vorwort*, pp. V and XI), and the verb 'abbilden' (*ibid., Vorwort*, p. XVI, and Sects. 39, 40, then pervasively from Sect. 53), who do note occur in *Begriffsschrift*,[11] and only occur in *Grundlagen* in relation to Schröder's *Lehrbuch der Arithmetik und Algebra* [177], in a quite different sense than that it has in *Was sind*.[12]

(3) In the second volume of *Grundgesetze*, ([97], Sects. II.138–140 and II.145–147) Frege examines thoroughly Dedekind's definition of real numbers. According to Heck's comment ([119], p. 598), Theorem 263 of *Grundgesetze* ([97], Sect. I.157) involves four conditions similar to Dedekind's conditions α, β, γ, δ of *Was sind* ([49], Sect. 71; cf. Sect. 1.4.1, below). According to Heck, in proving this theorem Frege proves, in fact, that the former conditions "determine a structure isomorphic to the natural numbers" which implies that these conditions works as "axioms for arithmetic which are different from, though closely related to, those due to Dedekind and Peano" ([119], p. 598).[13] Better, this is quite close to theorems 126 and 132 of *Was sind*.

It is difficult to say how much Frege was inspired by his reading of *Was sind*, but one thing seems clear: Frege does not isolate there the basic *general logical* laws, but the basic especial laws for arithmetic; thus, he strays far from *Grundlagen* and comes close to an axiomatic conception.

All these facts hint that, from 1879 onwards, Frege is familiar with Dedekind's works. By contrast, Dedekind comments very briefly on *Grundlagen* in the preface of the second edition of *Was sind* [50] and focuses on mathematical induction, the very arithmetical method. It is also striking that after he became aware of the logical paradox involved in his construction and having had it pointed out several times to him by Cantor, he did not even try to amend anything. No attempt like that of the *Nachwort* of the second volume of *Grundgesetze*, written after the discovery of Russell's paradox can be found in Dedekind's work. In the *Vorwort* to the third

---

[11] In *Begriffsschrift*, Frege rather uses the terms 'Function' and 'Verfahren' ([92], Sects. 24–31). 'Abbilder' occurs once (*ibid.*, Sect. 13): "Diese Regeln und die Gesetze deren Abbilder sie sind, können in der Begriffsschrift deshalb nicht ausgedrückt werden, weil sie ihr zu Grunde liegen".

[12] In *Grundlagen*, 'Verfahren' is replaced by 'Beziehung' ([93], Sect. 76ff.). Frege also uses 'eindeutige Zuordnung' (Sect. 62), 'beiderseits eindeutige Zuordnung' (Sect. 72) and 'beiderseits eindeutige Beziehung' (Sects. 78.5, 84) for expressing one-to-one relation, which is, in Dedekind's terms, 'ähnliche Abbildung'. For Frege (*ibid.*, Sect. 70): "Der Beziehungsbegriff gehört also wie der einfache der reinen Logik an. Es kommt hier nicht der besondere Inhalt der Beziehung in Betracht, sondern allein die logische Form" (Sect. 70).

[13] On this matter, cf. Sect. 3.3 of Chap. 3, below.

edition of *Was sind* [54], Dedekind affirms that "his trust in the internal harmony of our logic is not shaken" and he expresses his conviction that some means will certainly be found out in order to ground rigorously "the creative force thanks to which our mind creates out of some determinate elements a new element which is the *System* of them". 'Our logic' does not obviously refer here to logic *simpliciter*. So, to say that Dedekind's *logical* achievements are less thorough than those of Frege is commonplace. But are logical investigations, taken as such, one of Dedekind's concerns? Obviously not.

While a comparison between Dedekind and Frege is very instructive [7, 191], it seems to me inadequate to assess the contribution of the first by standards provided by the second (nor the reverse, but that is rarely the case). Even if their works are classified together "for good reasons" ([156], p. 339, footnote 6), we have to consider Dedekind's own way of viewing logic as a ground for arithmetic. Isn't it strange to value this way through questions that it was not designed to take on? If one keeps uniquely to Frege's conception, one cannot understand why Peano built on Dedekind's rather than on Frege's axioms; and one comes to misjudge Dedekind's main contribution, which was *mathematical* and quickly played a leading part in mathematical advances.[14] For instance, some outstanding interpreters of Frege's work disregard the fundamental difference between Dedekind's foundation of real numbers and that of Cantor or Weierstrass. However, the former is *not* based on the concepts of limit and convergence, just inversely Dedekind shows how to derive the concept of limit, and thus the usual theorems of real analysis, from the purely arithmetical definition of the concept of real number (cf. my presentation of *Was sind* in [56]). This beautiful result and the underlying account of real numbers exemplify how much Dedekind's view of arithmetic within the mathematical body and in respect of human understanding differs from Frege's one. This is just what I want to show.

Before coming to it, a quick survey of the literature making a comparison between Dedekind and Frege is in order.

There are many discussions on whether one could or could not assimilate Dedekind's foundational views with Frege's central thesis about the derivation of mathematics from pure logic. I will mention here only some of them.

Philip Kitcher begins his [133] with the remark that, since "our current understanding of what philosophy of mathematics might to be is so dominated by Frege's view of the field [...] Dedekind appears to us a lesser Frege, a man who groped toward some Fregean insights but who only saw dimly what Frege saw clearly" (*ibid.*, p. 299). Kitcher is describing here the general opinion of philosophers who know better Frege's work; Kitcher himself is at pains to rightly distinguish Dedekind's from Frege's philosophy. However, he finally argues that Dedekind developed "a version of the logicist thesis" (*ibid.*, p. 312), which already hints to sizeable differences between Frege and Dedekind.

---

[14]In 1877, Felix Klein writes to Dedekind ([67], p. 221): "Nothing that you have created in your solitary reflection went unheeded in the long term, everything decisive at its time has intervened in the development of mathematics, where it fructifies in a hundredfold way".

For Howard Stein, whose appraisal of Dedekind's work is illuminating, "Dedekind is a very important precursor of Hilbert as well as of logicism" ([185], p. 239). While Dedekind is certainly a precursor of Hilbert, he is *not*, in my opinion, a precursor of Frege, whereas Hilbert has certainly inherited from Dedekind, as well as from Frege and Russell-Whitehead, for conceiving of the foundations of arithmetic and logic simultaneously [128], since it later became obvious that no logic can be built up without presupposing arithmetic in some form or another (cf. for example Poincarés's papers on the "nature of arithmetical reasoning" and on relations of mathematics and logique: [160, 161]).

Tait, who is annoyed by the tendency to enhance Frege's "superior clarity of thought and powers of conceptual analysis", supports the view of the logicism of Dedekind, since, according to him, "Dedekind and Frege seem to agree on a conception of logic as comprising the most general truths, which do not concern any special subject matter" ([190], p. 313). He takes at face value Dedekind's assertion that arithmetic is a part of logic and attributes unquestionably Dedekind's *Abbilden* -ability to logic.[15]

Ferreirós also describes Dedekind's position as amounting to a logicist foundation ([91], esp. Chap. VII), though he has more recently argued that Dedekind's views count as a sort of "structural logicism".[16]

More recently, William Demopoulos and Peter Clark have entitled their Chap. 5 of Shapiro's *Handbook of philosophy of mathematics and logic* "The logicism of Frege, Dedekind, and Russell" [63].

For his part, Shapiro [181] develops as a neo-Fregean programme a technical treatment of real analysis using Dedekind's cuts instead of Frege's HP. This shows the possibility of developing a Dedekindian logicism, which is naturally not to be found in Dedekind's works.

In another direction, Erich Reck's [166], considers different plausible interpretations of Dedekind's assertions, and calls Dedekind's specific position 'logical structuralism'.

The question is, then: Is Dedekind a logicist? This question is justified because a clear-cut answer is not obvious. Speaking of Dedekind's logicism, everybody feels it necessary to qualify the term 'logicism', instead of using it purely and simply. And as one might expect, Kitcher, Stein, Tait, Ferreirós, Shapiro and Clark and Demopoulos, among others, understand 'logicism' in significantly different ways. Hence, my first task will be that of fixing an accurate definition of the term 'logicism', at least in its original meaning. Then, I shall explain to what extent Dedekind's contribution can be held as logicist in a loose sense, but I shall also insist on some fundamental differences with Frege.

---

[15]Tait translates 'Abbildung' by function: "I am very sympathetic with the view that the notion of function is a logical notion: a warrant for $\forall x \varphi (x)$ must be a function assigning to each b in the range of x a warrant to $\varphi (b)$, and a warrant for a proposition $A \rightarrow B$ is a function that assigns to each warrant of A a warrant of B. So the primitive truths of the logic of $\forall$ and $\rightarrow$ are truths about functions" ([190], p. 314).

[16]This view is presented in an unpublished paper, "On Dedekind logicism", which the author wrote after his being aware of my own view. Ferreirós was so kind as to send a copy of his paper to me.

In my opinion, there was no Frege-Dedekind tradition at the time of those two authors's works. As a testimony of my claim I quote footnote 5 in the *Vorwort* of *Grundgesetze* ([97], *Vorwort*, p. XI; Frege [110], p. XI$_1$):

> One searches in vain for my *Grundlagen der Arithmetik* in the *Jahrbuch über die Fortschritte der Mathematik*. Researchers in the same area, Mr Dedekind, Mr Otto Stolz, Mr von Helmholtz seem not to be acquainted with my works. Kronecker does not mention them in his essay on the concept of number either.

Frege criticises harshly Stolz, von Helmholtz, Kronecker, and Dedekind; he repeatedly strongly differentiates himself from them. His own claims should be taken more into account, even if it remains true that Hilbert, Russell, Gödel, Tarski and their followers forged their own views partly by blending elements from Dedekind's and Frege's works.

The blending happened and happens in so many ways that it may sometimes be difficult to discriminate what originally belonged to Dedekind and what belonged to Frege. Nonetheless we get more fine-featured information on Dedekind and on Frege by differentiating rather than unifying them under one and the same perspective. The least benefit from that may consist, for example, in avoiding a confusion between Dedekind's procedure for getting his "abstract numbers" and what Russell calls the 'principle of abstraction', i.e. HP. Naming this principle 'HP' has the advantage not to use the word 'abstraction', that Frege understands as denoting Aristotelean abstraction, which he rejects as being a psychological process,[17] while he admits abstract objects as *logical* objects recognisable through a *logical* means, thus equating 'abstract' with 'logical'. Once made the distinction between Dedekind's process of abstraction and Frege's method of transforming an equivalence relation into an identity, one cannot merge one of Dedekind's "shadowy forms [*schattenhafte Gestalten*]" ([49], *Vorwort*, p. IX; [53], p. 15) with Frege's "definite number[s] [*bestimmte, angebbare Zahl*]" ([93], *Einleitung*, p. I; [103], p. XIII), which are individual self-subsistent logical objects.[18]

Starting from here, my study will be divided in four parts. In Sect. 1.2, I will recall briefly the main features of Frege's logicism and Carnap's characterisation of the logicist thesis. In Sect. 1.3 I will show in detail how by the same fundamental words such as 'logic', 'number', 'thought', 'pure thought', 'laws of thought', 'concept', 'object', and 'function', Dedekind and Frege express radically different conceptual views. In Sect. 1.4, I will be devoted to the status of definitions: Dedekind and Frege thought differently about this important issue. In Sect. 1.5, I will take stock of the meaning of the term 'concept' and I will reconsider the meaning of structuralism and logicism respectively. My claim is that these terms describe two related but distinct perspectives.

---

[17]Famously, the first of the three fundamental principles stated in *Grundlagen* goes as follows ([93], *Einleitung*, p. X; [103], p. xxii): "There must be a sharp separation of the psychological from the logical, the subjective from the objective".

[18]As pointed out by Boolos ([23], p. 214), Frege's proof that every natural number has a successor depends on the assumption that cardinal numbers are objects. And it is only if one supposes cardinal numbers not to be objects that HP looks analytic or obvious.

1.2 The Logicist Thesis

1.2.1 The "New Logic"

Frege's work being considered as the very root of logicism, I shall leave aside, for the sake of clarity, Russell's changing-over-time elaboration[19] of the logico-philosophical position upheld by Frege, at least up to the publication of the second volume of *Grundgesetze*, in 1903.

Indeed, logicism originates from Frege's *Begriffsschrift*, while some traits are already in Leibniz' views. As he shall later write in *Grundgesetze*, Frege's explicit goal was a "renewal of logic",[20] which consists in (1) replacing the traditional Aristotelian splitting subject-predicate by the function-argument analysis, and (2) inventing a formal language appropriate for expressing the very logic of "pure thought" and the relations [*Beziehungen*] of concepts. *Begriffsschrift* establishes a formal system of logic (which contains the essentials of first- and second-order quantification with identity) with specific symbols and definite rules[21] according to which derivations are carried out exclusively by virtue of the "logical form" of expressions. The relations between concepts[22] are analysed in terms of function-argument dichotomy[23]: "It is easy to see", writes Frege in the *Vorwort*, "how taking a content as a function of an argument gives rise to concept formation" ([92], *Vorwort*, p. VII; [125], p. 7). In part III, he presents a logical reconstruction of "a general theory of sequences" which offers a "more detailed analysis o the concepts of arithmetic and a deeper foundation for its theorems" ([92], *Vorwort*, p. VIII; [125], p. 9).

Such a formal system of logic, which allows purely logic derivations written down in a specific artificial language, whose primitive symbols and liminal statements are explicitly enunciated and whose rules of inference are listed from the start, constitutes the prerequisite for what Carnap calls "the new logic", by contrast with "the old logic" [40], which was considered as "closed and completed" by Kant. In addition to Frege,

---

[19]Concerning Russell's views cf., e.g., [24], [26], esp. p. 292.

[20]Cf. [97], *Vorwort*, p. XXVI, ([110], p. XVI$_1$): "And so may this book, even if belatedly, contribute to a renaissance of logic."

[21]Frege's system has two connectives, negation and the conditional, six axioms, the universal quantifier introduced in *Grundgesetze*, Sect. 11, under the name 'Generality', with three more axioms, and two rules: explicitly *modus ponens* and, implicitly, rule of substitution. Relations between concepts are ruled by logical inference. In *Ausführungen über Sinn und Bedeutung* ([106], vol. 1, p. 128; [107], p. 118), Frege holds that all relations between concepts can be reduced to the "fundamental logical relation [...] of an object's falling under a concept", and adds that if an object falls under a concept, it falls under all concepts with the same extension, so that, "in relation to inference, and where the laws of logic are concerned, concepts differ only as so far their extensions are different".

[22]Usually, concepts are taken to be predicates of judgements, like in Kant's *First Critic* (*Transcendental Analytic*, I, Chap. 1), but Frege sees predicates as mathematical functions. By contrast, Dedekind does not consider the notion of judgement, because he does not tackle the question of the truth.

[23]On Frege's notion of function, cf. Chap. 2 of the present book.

Carnap counts Peano and Schröder as founders of "the new logic", and he recalls the chief work of Whitehead and Russell as the great fundament that all their successors have completed or reshaped. Carnap does not include Dedekind among the founders of "the new logic" (*ibid.*, p. 14). If one admits that, roughly, logicism is the thesis that mathematics can be reduced to *formal* logic, then the very first reason not to count Dedekind as a logicist is given by the term 'formal': Dedekind is not working in a formal language, indeed. Moreover, when Carnap brings up the reduction of the concepts of mathematical analysis to arithmetical concepts and deals with the "logical analysis [*logischen Zerlegung*]"[24] of the concept of number (*ibid.*, pp. 15 and 20–21), he does not mention Dedekind's work at all. If "logical analysis" means the analysis of statements into their *logical constituents*, which replaces the grammatical splitting predicate-subject with that of function-argument, and extracts contentual information from the way of using words, as Frege proposes in *Begriffsschrift* and *Grundlagen*,[25] there is no doubt that Dedekind does not practise this kind of analysis. Nowhere Dedekind does try to characterise numbers by using number-words or making a judgement involving numbers. His extension of the mathematical concept of function through his specific use of 'Abbildung' does not result from a logical analysis *of the language* of "pure thought [das reine Denken]" (cf. the complete title of *Begriffsschrift*: [92]).

The ways Frege and Dedekind go from mathematical function to *Begriff* and *Abbildung*, respectively, show two dissimilar ways of generalisation: Frege substitutes mathematics for grammar in a logical analysis of language, taking the "linguistic turn" and introducing quantification; Dedekind takes a mode of thinking into account and presents it as logical inasmuch as it applies everywhere in mathematics. However, if 'logical analysis' designates a work which takes place before construing a system, yields the primitive concepts and "articulate [their] sense clearly"(as Frege states in "Logik in der Mathematik": [106], vol. I, p. 228; [107], p. 211), then the letter to Keferstein of February 27, 1890 ([125], pp. 98–103) shows that Dedekind recognises the necessity of analysis before the synthesis, though his understanding of these operations is akin to the analysis/synthesis distinction of the Ancient, rather than to the Kantian distinction of analytic and synthetic judgements, from which Frege starts in *Grundlagen*, and which he abandons later in *Grundgesetze*, at the benefit of a more Euclidean conception of the former distinction.

The total absence of Dedekind in Carnap's picture leads us to reappraise the logicist interpretation of Dedekind's work on numbers and to wonder *when* this

---

[24]The expression is used by Frege, e.g. in "Logik in der Mathematik" ([106], vol. 1, pp. 225–228; [107], pp. 208–211), possibly his lecture notes of a course attended by Carnap in the spring of 1914.
[25]Cf. [93], Sect. 46; [103], p. 59: "To throw light on the matter, it will help to consider number in the context of a judgement that brings its ordinary use." Cf. also [106], vol. 1, "Zahl", p. 284, [107], pp. 265: "What [...] is the number itself? [...] We may seek to discover something about the number itself from the use we make of numerals and number-words. Numerals and number-words are used, like names of objects, as proper names." And again: "In arithmetic a number-word makes its appearance in the singular as a proper name of an object of this science; it is not accompanied by the indefinite article, but is saturated" ([106], vol. 1, "Aufzeichnungen für Ludwig Darmstaedter", pp. 276; [104], pp. 256).

interpretation became, in retrospect, a banal issue in the philosophy of mathematics. Though historically quite interesting, this is not the question I want to tackle here. What is more relevant for my purpose is that Carnap was certainly right from its point of view, since it is a matter of fact that Dedekind's work does meet none of the benchmarks of "the new logic" set up by Carnap: a symbolic formulation, in which primitive logical symbols and rules of inference are to be explicitly stated first[26]; the theory of relations (De Morgan and Peirce are mentioned as precursors); Russell's theory of types, which allows avoiding paradoxes[27]; the tautological or analytical character of all the logical and, consequently, of all the mathematical truths (a generalisation of Frege's view on the analytical character of arithmetical truths).[28]

À la lettre no more Frege's work meets all these benchmarks. Still, it is unquestionably a first, decisive step on the route that led to a form of logicism meeting them. In what follows I will, then, take this Carnapian characterisation of logicism as a description of a sort of idealisation of Frege's logicist program.

1.2.2 Logicist Foundations of Mathematics

In the introduction of *Grundlagen*, Frege claims that he will "make clear that even an inference like that from n to $n + 1$, which on the face of it is peculiar to mathematics, is based on the general laws of logic" ([93], *Vorwort*, p. IV; [103], p. XVI). This is a claim that was already grounded on, and justified by his results in *Begriffsschrift*, part III, where mathematical induction appears as a special case of what logical Whitehead and Russell named 'ancestral of a relation'.[29] So mathematical induction is a species of logical inference. In *Grundlagen*, Sect. 87, Frege is less affirmative. He writes that he only "hopes" that his work has "made it *probable* that the laws of arithmetic are analytic judgements and consequently *a priori*", and that "Arithmetic is nothing but further pursued logic, and every arithmetical statement a law of logic, albeit a derived one" ([93], Sect. 87; [103], p. 99; my italics).[30] Later on, Frege provides us with what he thinks to be a confirmation of this hope. Indeed, in *Grundgesetze*, he asserts ([97],

---

[26]In the preface of *Grundgesetze*, Frege stresses that in Dedekind *Was Sind* "nowhere [...] do we find a list of the logical or other laws he takes as basic " ([103], *Vorwort*, p. VIII; [110], p. VIII₁).

[27]Symptomatically, Carnap does not refer to Zermelo's way of avoiding the paradoxes.

[28]Gödel distinguishes between 'tautological' and 'analytic' and points out that the elementary theory of integers is demonstrably non-analytic as a consequence of his incompleteness theorem ([114], p. 139, footnote 46).

[29]Frege's formulation is as follows ([92], Sect. 26; [125], p. 60): "If from the two propositions that every result of an application of the procedure f to x has property F and that property F is hereditary in the f-sequence, it can be inferred, whatever F may be, that y has property F, then I say: 'y follows x in the f-sequence', or 'x precedes y in the f-sequence'".

[30]Cf. also [93], Sect. 109; [103], pp. 118–119: "From all the preceding it thus emerged as a very probable conclusion that the truths of arithmetic are analytic and *a priori*; and we achieved an improvement of the view of Kant".

Einleitung, p. 1; [110], p. 1; my italics): "In my *Grundlagen der Arithmetik*, I aimed to make it *plausible* that arithmetic is a branch of logic and needs to rely neither on experience nor intuition as a basis for its proofs. In the present book this is now to be established by deduction of the simplest laws of cardinal number by logical means alone". But the problem with Basic Law V soon appears and the question arises whether the logicist programme can be fulfilled.

I do not want to address this question, nor to consider the solutions proposed by Frege, Russell, and the modern supporters of neologicism. I am rather limiting myself to recall what is generally taken to be *the* logicist programme in terms of Carnap's twofold characterisation, which up till now is mostly endorsed. Indeed, in Carnap 1931 ([41], pp. 91–92; [10], p. 41) two requirements are enunciated:

> 1. The *concepts* of mathematics can be derived from logical concepts through explicit definitions. 2. The *theorems* of mathematics can be derived from logical axioms through purely logical deduction?

Let us begin with the first requirement. After an outline of the logical material necessary and sufficient for deriving natural numbers from logical concepts, Carnap quotes Frege's definition of natural numbers as "logical attributes which belong [...] to concepts", and mentions Russell's and Whitehead's work which corroborates "the logical status of the natural numbers" ([41], p. 93; [10], p. 42). Later, Carnap considers the derivation of the other kinds of numbers, and only then he briefly exposes Dedekind's cuts—and not Frege's conception of real numbers, which is based on the concepts of magnitude, measure and ratio—before passing to Russell's own remodelling of Dedekind's definition of real numbers through cuts. Carnap indicates that this process runs up against the problem of impredicative definitions. And he insists on the fact that "the logicist does not establish the existence of structures which have the properties of the real numbers by laying down *axioms* or postulates; rather, through *explicit definitions*,[31] he produces logical constructions that have, by virtue of these definitions, the usual properties of the real numbers" ([41], p. 94; [10], p. 44; my italics). And he adds "As there are no 'creative definitions',[32] definition is not creation but only name-giving to something whose existence has already been established" (*ibid.*).

According to Frege indeed, the real task is not making postulates (or axioms or "formal definitions"), but showing that they are satisfied. In other words, freedom of contradiction in a concept is not a sufficient guarantee that something falls under it.[33] As he writes: "The fundamental logical relation is indeed that of an object's falling under a concept[34]: all relations between concepts can be reduced to this" ([106], vol. 1 "Ausführungen über Sinn und Bedeutung", p. 128; [107], p. 118). Now, defining is

---

[31] This expression does not belong to the vocabulary of *Grundlagen*.

[32] About Frege's discussion of "Die schöpferischen Definitionen", cf. [97], Sects. 139–147.

[33] Cf. [93], Sect. 109, [97], Sect. II.86–137, where Frege contrasts *"die formale Arithmetik"* (by Heine and Thomae) with *"die inhaltliche Arithmetik"*, and Sect. II.138–147, where he questions Dedekind's, Hankel's and Stolz's definitions of real numbers. Cf. also the Frege-Hilbert correspondence in [105].

[34] Frege insists that this relation of subsumption is distinct from the relation of inclusion.

fixing, determining *what* is named by a name or designated [*bezeichnet*] by a sign. Frege holds that a name is the name *of an object* and that a definition lays down what a sign/word expresses, i.e. it determines univocally the conceptual *content* of the sign/word. Thus, concerning the cardinal numbers, "it is not a matter simply of giving names, but of designating for itself the numerical content" ([93], Sect. 28; [103], p. 39).[35] It will appear to Frege, in the 1890s,[36] that 'the conceptual content' is twofold; it is "Sinn" and "Bedeutung".[37] In *Grundgesetze* Frege insists again on the principle that "alle rechtmässig gebildeten Zeichen etwas bedeuten sollen" ([97], *Vorwort*, p. XII; cf. also Sect. 28). A definition indicates the "connection between sign and what is designated [*Zusammenhang zwischen Zeichen und Bezeichnetem*]" ([97], *Vorwort*, p. XIII; [110], p. XIII$_1$). By a definition "*something* is marked out ins sharp relief and designated by a name" (*ibid.*; my italics). Thus "formal definitions", in which one rests content with introducing signs without making a link with some object, be it concrete or abstract—*Sinne* and *Bedeutungen* are abstract objects—are not accepted. Notice that in Frege's view definitions by axioms are not necessarily *formal* definitions; in fact, according to him, they are not *definitions* at all.

Carnap stresses the "constructivistic" character of Frege's conception of definitions and claims that "this 'constructivistic' method forms part of the very texture of logicism" ([41], p. 94: [10], p. 44). He makes, then, a link with intuitionism. Note, however, that: (i) Frege's constructive definitions do not result from a construction of the mind, based on the *a priori* insight of time—as claims Brouwer; they are, rather, grounded on timeless logical objects and logical methods of inference; (ii) Frege does not subscribe to the algorithmic constructivism vindicated by Kronecker, who holds that the positive integer numbers are *given* by God and, consequently, need no definition. We shall see below (Sect. 1.4.2) what Frege means exactly by the expression 'constructive definition [*aufbauende Definition*]' employed in 1914 ([106], vol. I, "Logik in der Mathematik", p. 227; [107], p. 210).

As to Dedekind, there is no more comment in Carnap's paper. Nevertheless we know that, for Dedekind, creating new mathematical concepts is more than fruitful ([49], *Vorwort*), and he is used to laying down a small number of necessary and sufficient conditions as explicit starting point of his deductions, and to take them as definitions ([49], Sects. 71, 73). We know also that Dedekind clearly rejects the constructivistic standards as contrary to actual infinities, especially Kronecker's

---

[35]In *Grundlagen*, Sect. 43, Frege criticises Schröder's supposed assimilation of the number with a sign. Note that the term 'Begriffsschrift' is used by Frege to designate "the conceptual content [*den begrifflichen Inhalt*]" ([92], Sect. 3; [125], p. 12), which is independent from the peculiar statement which expresses it.

[36]Cf. the letter to Husserl of May, 25th, 1891 ([106], vol. 2, pp. 96–98), [96], and "Ausführungen über Sinn und Bedeutung" ([106], vol. 1, pp. 128–136; [107], p. 118–125).

[37]The *Sinn* of a statement is a thought; its *Bedeutung* is its truth-value. And judgement "could be characterised as a transition from a thought to a truth-value" (letter to Husserl of May, 25th, 1891: [106], vol. 2, p. 97; [108], p. 64).

"limitations upon the free formation of concepts [*Begriffsbildung*] in mathematics" (*ibid.*, Sect. 1, footnote; [53], p. 21).[38]

Consider now the second of Carnap's requirements. Carnap understand it as the requirement that "every provable mathematical sentence [...][be] translatable into a sentence which contains only primitive logical symbols and which is provable in logic" ([41], p. 95: [10], p. 44). The verb 'to translate' does not match Frege's view that arithmetical truths are *derivable* from purely logical laws provided that the logical definition of the concept of natural number is stated. But Frege's view clashes with Gödel's first incompleteness theorem.

Boolos calls Carnap's requirements (1) and (2) 'the definability thesis' and 'the provability thesis', respectively ([24], p. 270).[39] He draws attention to a supplementary distinction, that must hold between statements which can be expressed in the language of pure logic and statements true by virtue of logic alone. A statement assumed to be true and expressed in logical terms is not necessary a logical truth. Already Russell noticed it with the example of the axiom of infinity, that "though it can be enunciated in logical terms, it cannot be asserted by logic to be true" ([172], pp. 202–203). Parsons notes that, since the structure of natural numbers is second-order definable, "the simple translation of the language of arithmetic into that of second-order logic has been offered as a basis for a defence of the view that arithmetic is a part of logic", giving birth to a "logicist eliminative program" ([156], pp. 312–313) illustrated by Putnam's if-thenism [163] and by Putnam's and Hodes' recourse to modal notions instead of abstract objects such as sets and numbers [132, 164]. According to Parsons, for logicism proper, it's not sufficient to exhibit a mapping which translates all arithmetic truths into logical truths.[40] Hence Boolos's distinction between truths of logic and truths expressed in the language of logic ([23], p. 211). Boolos adds that the definability thesis alone does not suffice to show the truths of mathematics to be logical truths and no one "counts as a full-fledged logicist who does not endorse the provability thesis as well as the definability thesis" ([24], p. 271).[41] By the yardstick of a "full-fledged logicism", at which Frege aims, Dedekind is definitely not a logicist. Yet I have to explain why he has been or might be interpreted as advocating a kind of logicism. Before proceeding to this, some more words on Carnap's views.

Carnap displays the difficulties of the logicist programme, in particular in the treatment of the real numbers. He mentions Ramsey's solution: accepting impredicative definitions with the presupposition that the totality of properties already exists

---

[38] Also Frege employs the term 'Begriffsbildung' in his paper on Boole's "logical calculus" ([106], vol. I, "Booles rechnende Logik und die Begriffsschrift", p. 14). This term was common at that time, indeed. But while Dedekind aims at creating new *specific* mathematical concepts, Frege aims at showing the general *logical method* grounding the *uniform* process of *concept formation*.

[39] For a more refined distinction between language-logicism, consequence-logicism and truth-logicism, cf. [126]: Frege's project was truth-logicism as far as mathematical truths can be proved merely "on the basis of general logic laws and definitions" (*ibid.*, p. 206).

[40] Anyway, such a mapping cannot exist since arithmetic is undecidable ([23], p. 208).

[41] Boolos makes this remark in order to state that Russell advocates the definability thesis but not the provability thesis.

before their definition. This conception is akin to the "belief in a platonic realm of ideas which exist in themselves, independently of if and how finite human beings are able to think them" ([41], p. 102; [10], p. 50). Thus the logical structure of the purposed system involves or might involve a philosophical stand on the nature of the things designated by the signs or singled out by the definitions of that system. Carnap thinks: (i) that Frege does not share this belief in "theological mathematics", since, for him, "only that may be taken to exist whose existence has been proved" (*ibid.*)[42]; (ii) that Ramsey's solution can be accepted without falling in his "conceptual absolutism" ([41], p. 103; [10], p. 50). Leaving aside Ramsey's conception and Carnap's empiricist fighting against the conceptual absolutism, I shall focus on Frege's clearly asserted philosophical assumptions and compare them, at least on some crucial points, with Dedekind's less systematically developed views.

1.3 Similar Claims, Different Fundamental Conceptions

Before coming to the proper subject of this section, let me recall that Frege has more or less strongly changed his mind[43] over time about such fundamental issues as the logical status of cardinal numbers and their identification with extensions of concepts, the distinction between aggregates and extensions, the sharp distinction between arithmetic and geometry, the sharing out of mathematical statements into synthetic and arithmetic statements, the division between *a priori* and empirical truths, the uselessness or the need of intuition in the deductive development of arithmetic, the radical difference between, on the one hand, definitions that fix definitely the sense of the signs used or introduced and those that determinate for all time the objects to which mathematical statements refer and, on the other hand, the so-called creative definitions that single out a few primitive statements from which theorems can be derived. If we also take into account the fact that Frege's final views conflict with his

---

[42]This is Carnap's biased rephrasing of Frege's following statements: "In mathematics a mere moral conviction, supported by a mass of successful applications, is not good enough. Proof is now demanded for many things that formerly passed as self-evident. [...] In all directions the same ideals have been seen at work—rigour of proof, precise delimitations of extent of validity, and as a means to this, sharp definitions of concepts" ([93], Sect. 1; [103], p. 1). Nevertheless Frege assumes that abstract objects, such as thoughts or senses or numbers or mathematical truths, have a changeless existence, different from that of the real [*wirklich*] world and that of the inner world of an experiencing subject.

[43]Bynum ([37], p. 281) stresses that in *Grundlagen* "Frege did not consider the introduction of extensions to be necessary, and indeed he felt some discomfort in identifying them with numbers". Bynum thinks that this discomfort pushed Frege to deal with that part of logic that is independent of set-theory, namely "fundamental logic". Hodes ([132], pp. 143–144) points out the difficulty in interpreting Frege's "Tagebucheintragungen über den Begriff der Zahl", dated to March 23th–25th, 1924 ([106], vol. I, pp. 282–283; [107], pp. 263–264). According to Dummett ([74], p. 161), Frege's early writings do not contain "a complete systematic theory of philosophical logic comparable to, and in competition with, that propounded by him from 1891 onwards". Parsons [154] shows how Frege came to reject extensions as being really objects along with logicism.

first beliefs, sometimes straight out, plus the fact that HP turned out to be a consistent but not purely logical (purely analytical) principle, we will agree wholeheartedly with Boolos's conclusion according to which Frege himself was not a logicist in the strict meaning of the word ([23], pp. 216–217).

Nevertheless, there is actually a set of claims that are thought to be characteristic of the logicist project, on which many outstanding scholars are still working. The main claim is that FA $=$ HP $+$ second-order logic makes a consistent system which allows for interpretations as arithmetic.[44] Then the double question will be: (i) Is Dedekind's characterisation of the natural numbers and of the real numbers a planned, if not a successful, logicist reduction, i.e. are natural numbers defined by him as logical objects? (ii) Does Dedekind believe that thoughts or truths are subsisting by themselves? The difficulty on answering these questions is so much that Dedekind changed his mind—as happens to any thinker—, but rather that he was more involved in mathematical practice than in philosophical or logical investigations. Therefore, the scattered remarks of a philosophical or logical nature, that he made in *Stetigkeit*, *Was Sind*, and the letters to Lipschitz, Weber ([55], vol. III, pp. 464–482 and 483–490) and Keferstein ([184], pp. 259–278), and in some other rare places, have to be embedded in his mathematical writings.

This does not prevent from discerning a number of opinions on which Dedekind and Frege were dissenting. I shall come to them in the next sections. Namely, Sect. 1.3.1 is devoted to the dichotomy reason/intuition; Sect. 1.3.2 to the meaning to be attributed to of 'thought', 'law of thought', 'logic' and 'proof'; and Sect. 1.3.3 to the notion of truth.

1.3.1 Reason Versus Intuition and the Foundations of Arithmetic

As it is well known, the impulse to the search for rigour in the second half of the nineteenth century was the arithmetisation of infinitesimal analysis. As Dedekind puts it, endorsing Dirichlet's view, "every theorem of algebra and higher analysis, no matter how remote, can be expressed as a theorem about natural numbers" ([49], *Vorwort*, p. XI; [53], p. 16). Therefore the task is to give a definition of the natural numbers as "self-subsistent objects" ([93], Sect. 55; [103], p. 68) or to single out a few essential or "inner" properties ([51], pp. 54–55) of the natural numbers. Both Frege and Dedekind vindicate the autonomy of arithmetic *vis-à-vis* any intuition and experience in Kant's sense, and take Kant's transcendental aesthetics as a target for their criticisms. Dedekind thinks that mathematics does not proceed by construction of concepts into intuition, and that mathematical theories do not develop out of observation of facts nor of any apprehension of spatiotemporal data. Reason, or

---

[44]The claim that every arithmetic truth is a theorem of the system is abandoned because it clashes with Gödel's first incompleteness theorem.

"pure laws of though" alone are at work not only in arithmetic—as holds Frege[45]—, but also in the whole body of pure mathematics. Even though Dedekind takes on Gauss's view on the priority of arithmetic over geometry and affirms the autonomy of the former from the latter, he, contrary to Frege,[46] does not endorse the opinion that geometry is rooted in intuition. According to his innovative views, there is no epistemological difference between arithmetic and geometry; as a deductive science geometry is shaped in a similar way as arithmetic. And the Cartesian correspondence between curves and equations shows the common structure between real numbers and real functions of real variables. Moreover, for Dedekind 'arithmetic' refers to the whole body of numbers, be they natural numbers or negative or rational or irrational or complex numbers. Dedekind's goal is to achieve in a uniform way the gradual numerical extension of natural numbers without any help of any non-numerical notion,[47] in particular without appeal to geometrical notions or to the notion of measurable magnitude. One determines a measure by a number, not the other way round.

In *Grundlagen*, Frege agrees with that[48]; but, there, his only concern is the logical reduction of cardinal numbers. And, indeed, he also claims ([93], Sect. 105; [103], pp. 114–115) that "with the definition of fractions, complex numbers and the rest, everything will in the end come down to the search for a judgeable content which can be transformed into an identity whose sides precisely are the numbers". And he continues: "In other words, what we must do is fix sense of a recognition-judgement for the case of these numbers [...], then the new numbers are given to us as extensions of concepts". This means that Frege does not accept the successive numerical extensions out of the natural numbers. This is made openly clear in the second volume of *Grundgesetze*, where Frege openly suggests to define them as "ratios of magnitudes [*Grössenverhältnisse*]" ([97], Sect. II.157; [110], p. 155$_2$), and not through successive numerical extensions out of the natural numbers, which would result in a "piecemeal [*stückweise*]" definitions, which Frege rejects ([97], Sect. II.57; [110], p. 70$_2$; cf. also what he says on this matter in "Logik in der Mathematik": [106], vol. I, pp. 261–262; [107], pp. 242–243). As a consequence, he takes real numbers to be completely separate from natural numbers: for him, these two sorts of numbers belong to two "completely separate domains"[49]: the natural ones are those which answer the question 'how many?'; the real ones, which he calls 'measuring numbers

---

[45]Cf. ([93], Sect. 105; [103], p. 115): "In arithmetic we are not concerned with objects which we come to known as something alien from without through the medium of the senses, but with objects given directly to our reason and, as its nearest kin, utterly transparent to it".

[46]Cf. [93], Sects. 13 and 64, [103], pp. 19–20 and 75: "In geometry [...] it is quite intelligible that general propositions should be derived from intuition"; "Everything geometrical must be given originally in intuition".

[47]One may recall Aristotle's refusal of "μετάβασις εἰς ἄλλο γένος".

[48]Cf. [93], Sect. 19, [103], p. 255: "At this point I should like straight away to oppose the attempt to think of number geometrically, as a ratio between lengths or surfaces".

[49]Cf. [97], Sect. II.157, [110], p. 155$_2$: " [...] Darum ist es nicht möglich, das Gebiet der Anzahlen zu dem der reellen Zahlen zu erweitern; es sind eben ganz getrennte Gebiete".

[*Maasszahlen*]' ([97], Sects. II.58, footnote; II.157–160, and II.162]; [110], pp. 71$_2$, 155$_2$–157$_2$, 159$_2$–160$_2$), are those that are used in measuring continuous magnitudes. Remark, however, that the measuring numbers are not to be reduced to geometry: 'Grössenverhältnis' does not refer, indeed, to a ratio between lengths or surfaces, or alike. As a measuring number, a real number applies to different kinds of continuous magnitudes: geometrical magnitudes, but also temperatures, time-intervals, masses, etc.

Such an abstract concept of magnitude is already present in Euclid's *Elements*, book V, which presents a general theory of proportions for any sort of magnitudes, a theory that is then applied to (plane) geometry in book VI. Moreover, though in books VII–IX, it is also question of proportions among numbers, Euclid did not aim at grounding the concept of number on that of magnitudes or ratio of magnitudes, since for him 'number' only refers to positive integer number greater than 1.

By contrast, Frege defines real numbers as ratios of magnitudes and he holds indeed that the notion of ratio of magnitudes, in general, is arithmetical ([97], Sect. II.158; [110], p. 156$_2$), but he maintains that the real numbers do not result from successive extensions of natural numbers; they need a definition of their own. As *Grössenverhältnis*, a real number is not itself the measured magnitude; it rather measures such a magnitude. And a magnitude has not, in turn, to be confused with the object which has it: for example a length has not to be confused with a segment that has this length. Such a segment is not a magnitude, for Frege; only its length is so. To say it in general, a magnitude is, for him, the extension of a binary relation in which the elements of appropriate domains (like segments) can stay. For short, Frege calls such an extension 'Relation', in German, while he uses 'Beziehung' for what we call 'relation', in English. If we translate the German 'Relation' with the English 'Relation' (by preserving 'relation' for translating 'Beziehung', as it is usually done), we can say, in Frege's jargon, that a magnitude is a Relation, and "domains of magnitudes are classes of Relations" ([97], Sect. II.162; [110], p. 160$_2$). Real numbers, then, are ratios between magnitudes, namely between Relations. But, for Frege, a ratio is, in turn, the extension of a relation. Hence, ratios between Relations, namely real numbers, are, for him, extensions of relations between Relations, that is, "Relations on Relations" (*ibid.*).

What matters here is that: (i) Frege stresses the independence of arithmetic from geometry ([97], Sect. II.158); (ii) he treats the notions of a relation and of an extension of a relations as purely logical notions. Frege thinks that defining real numbers as ratios of magnitudes results, then, in a logical reduction similar to that realised in *Grundlagen* for cardinal numbers. The problem, of course, is, once more, that Frege's extensions clash with Russell's paradox. It has been noted, however, that Frege's treatment of real numbers contains valuable insights into what would later be developed as groups with orderings [75], but in that line of thinking Dedekind has priority since, in *Was sind* (and even earlier), he takes the fundamental steps of considering algebraic ordered structures, especially ordered groups and ordered fields. Defining the real numbers (up to isomorphism) out of rational numbers alone is extending the algebraic totally ordered structure of the field $\mathbb{Q}$ to the field $\mathbb{R}$ or, in modern terms, embedding $\mathbb{Q}$ in $\mathbb{R}$, then identifying $\mathbb{Q}$ with its image in $\mathbb{R}$, which constitutes a step inconceivable in Frege's frame.[50]

In the final year of his life, Frege comes back to Kant's epistemology and writes ([106], vol. I, "Zahlen und Arithmetik", p. 297; [107], p. 277)[51]:

> [...] that the series of integer numbers should eventually come to an end is not just false: we find the idea absurd. So an *a priori* mode of cognition must be involved here. But this cognition does not have to flow from purely logical laws, as I originally assumed [...]. The more I have thought the matter over, the more convinced I have become that arithmetic and geometry have developed on the same basis—a geometrical one in fact—so that mathematics in its entirety is really geometry.

Frege is here again at odds with Dedekind's constant view that arithmetic is *the* root of mathematics and that *any* branch of mathematics, even geometry, is a purely deductive science in the sense that it singles out the primitive propositions expressing the essential properties on the basis of which (possibly) all the theorems of the science could be proved.

For Dedekind even in the *science* of space, intuition is misleading and useless: Dedekind is the first mathematician who states that continuity (connectedness) is not given to us by spatial intuition; according to him, we do not have really a visual or intuitive apprehension of the continuity of a geometric line drawn on the blackboard, we conceive of it as a property that "we attribute to the line" (or to space) by a convenient axiom ([47], Sect. III; [53], p. 5), which must be explicitly formulated as a primitive—non provable—principle (47], Sect. V.III), since "for a great part of the science of space the continuity of its configurations is not even a necessary condition" ([49], *Vorwort*, p. XII; [53], p. 16). Thus the continuity principle is *not necessarily true* in any geometrical space; it is not a logically true principle valid in any space, even less in any *System* of elements. Moreover, "if we knew for certain that space was discontinuous there would be nothing to prevent us, in case we so desired, from filling up its gaps, *in thought*, and thus making it continuous" ([47], Sect. III; [53], pp. 5–6; my italics). A mathematical space is a thought-entity; the distinction between arithmetic and geometry comes down not to the division between concept (or relation in Frege's sense) and intuition,[52] but to the distinction between two *mathematical*

---

[50]Such identifications are very usual in mathematical practice, but the philosophical question about how to conceive of, e.g., the identity of the rational 2 and the real 2 gives still rise to subtle discussions.

[51]Cf. also [106], vol. I, "Neuen Versuch der Grundlegung der Arithmetik", pp. 298–299; [107], pp. 278–279: "I have to abandon the view that arithmetic does not need to appeal to intuition either in its proofs, understanding by intuition the geometrical source of knowledge, that is, the source from which flow the axioms of geometry [...]. I distinguish the following sources of knowledge for mathematics and physics: (1) Sense perception; (2) The Geometrical Source of Knowledge; (3) The Logical Source of Knowledge. The last of these is involved when inferences are drawn, and thus is almost always involved. Yet it seems that this on its own cannot yield us any objects [...] [and] probably [...] cannot yield numbers either [...]".

[52]In *Grundlagen*, Sect. 13, Frege holds that points, lines and plane are not individuated as are the numbers.

concepts: that of number (*Zahl*)[53] and that of magnitude (*Grösse*), and in particular that of real number and that of continuous magnitude, the former being independent from the latter. For Dedekind real *numbers* are as much numbers as natural numbers[54] and if we wanted to define numbers as the result of measuring a magnitude by another of the same kind (*gleichartige*), we would fail in the case of complex numbers. Then "arithmetic must develop itself out of itself" ([47], Sect. III; [53], p. 5), assuming the radical difference between number and magnitude, not between natural and real numbers.

When Dedekind writes, in the preface of the first edition of *Was sind* ([49], *Vorwort*, p. VII]; [53], p. 14)—nothing like this is to be found already in *Stetigkeit*—, that arithmetic, algebra and analysis are "a part of logic", he clarifies the point as follows: (i) they are "totally independent of the intuitions of space and time",[55] and, hence, (ii) the concept of number "flows *immediately* from the pure laws of thought" (my italics), what, in Dedekind's view, means that numbers *together with numerical operations*[56] are rooted in the constitution of the mind or, as Dedekind writes to Keferstein (February 27, 1890), they are "subsumed under more general notions and under *activities*[57] [my italic] of the understanding [*Verstand*] *without which no thinking is possible*" ([125], p. 100), and finally, (iii) "the numbers are free creations of the human mind [*menschlicher Geist*]", so that the entire number-realm, from natural to complex numbers, is "created in our mind".

In Frege's view, (i) holds at least from the time of *Begriffsschrift* to that of *Grundgesetze*, the last posthumous writings on number being excluded; (ii) holds only if one understands 'thought' and 'the laws of thought' in a way significantly different from Dedekind's understanding—as it will appear more clearly below[58]; (iii) certainly

---

[53] Dedekind uses 'Zahl' or 'natürliche Zahl' to refer to finite ordinal numbers (*Ordinalzahlen*): [49], Definition 73; for 'Anzahl', he has the same use as Frege ([93], Sect. 4, footnote; cf. above, footnote (2) of the *Introduction*), since both use it to refer to cardinal numbers ([49], Definition 161).

[54] Of course, they don't form the same structure, even though the totally ordered semi-ring of natural numbers is embedded in the totally ordered field of real numbers.

[55] Cf. also the following passage (*idid.*): "In speaking of arithmetic (algebra, analysis) as a part of logic I mean to imply that I consider the number-concept entirely independent of the conceptions or intuitions of space and time, that I consider it an immediate flow from the pure laws of pure thought [die reine Denkegesetze]". Notice that Dedekind does not write "laws of logic", but "laws of thought".

[56] Cf. e.g. [47], Sect. I, [53], p. 2, (my italics): "I regard the whole of arithmetic as a necessary, or at least natural, consequence of the simplest arithmetic act, that of counting, and counting itself as nothing else than the successive creation of the infinite sequence of positive integers in which each individual is defined by the one immediately preceding [...]. The chain of these numbers forms in itself an exceedingly useful instrument for the human mind; it presents an inexhaustible wealth of remarkable laws *obtained by the introduction of the four fundamental operations of arithmetic*. Addition is the combination of any arbitrary repetition of the above-mentioned act into a singular act [...]".

[57] In a famous letter to Bessel (of April 9, 1830), Gauss writes that "the number is a pure product of our mind" ([67], p. 40). And in a fragment dated to 1882, Dedekind maintains that "Analysis in its entirety is a necessary consequence of the thought as such" (*ibid.*, p. 199).

[58] Remark also that Frege does not only avoid to emphasise the strict connection between defining numbers and defining operations on them, but considers the former as essentially independent of the

does not hold at all: being derivable from purely logical concepts, or even from a geometrical source as in Frege's final texts,[59] is incompatible, according to Frege, with being created by or in our mind, or with being an "object of our thinking [*Gegenstand unseres Denken*]" as Dedekind says of any mathematical thing in general ([49], Sect. 1; [53], p. 21). Whoever is familiar with Dedekind's writings knows that an "object of our thinking" is not really an object, but a concept, 'concept' being understood in the context of mathematical practice: mathematical progress comes from new concepts such as those of a *System*, a group, a cut, a chain, a field, a module, an ideal, a lattice, and so forth.[60]

1.3.2 Pure Thought, Objectivity, Logic, Proof

1.3.2.1 The Laws of Thought

The expression 'the laws of thought' is used both by Frege[61] and by Dedekind and for both of them this refers to *the* laws of the mind.[62] Dedekind and Frege take thoughts as objective, and, following Kant, they both agree on understanding 'objective' as 'based on reason'.[63] But Frege goes one step further and repudiates the Kantian division between things in themselves and phenomena: he understands objective to be something whose (i) existence and (ii) apprehension do not depend on our sensation, intuition, ideation or any "result of a mental process" ([93], Sect. 26; [103],

---

latter. In the *Vorwort* of *Grundgesetze*, he feels no embarrass in observing that his "investigation", namely that offered in the first volume of his treatise, "does not yet include the negative, fractional, irrational, and complex numbers, nor addition, multiplication, etc." ([97], *Vorwort*, p. V; [110], p. V_1).

[59]Cf. [106], vol. I, "Erkenntnisquellen der Mathematik und der mathematischen Naturwissenschaften" p. 294, and "Zahlen und Arithmetik", p. 297; [107], pp. 274 and 277.

[60]I shall come back in Sect. 1.5.1 to Dedekind's and Frege's different notions of a concept.

[61]Cf. [92], *Vorwort*, pp. III–IV; [125], p. 5: "The most reliable way of carrying out a proof, obviously, is to follow pure logic, a way that, disregarding the particular characteristics of objects, depends solely on those laws upon which all knowledge rests. […] I first had to ascertain how far one could proceed in arithmetic by means of inferences alone, with the sole support of those laws of thought that transcend all particulars". Cf. also [93], *Vorwort*, p. III, [103], p. XV: "Thought is in essentials the same everywhere: it is not true that there are different kinds of laws of thought to suit the different kinds of objects thought about".

[62]Cf. [92], Sect. 23, [125], p. 55: "[…]pure thought irrespective of any content given by the senses or even by an intuition *a priori*, can, solely from the content that results from its *own constitution*, bring forth judgements that at first sight appear to be possible only on the basis of some intuition" (my italics). Cf. also [102], p. 74, [109], pp. 368–369: "Neither logic nor mathematics has the task of investigating minds and contents of consciousness owned by individual men. Their task could perhaps be represented rather as the investigation of *the* mind; of *the* mind, not of minds". Dedekind would completely agree with this assertion.

[63]Cf. [93], Sect. 26, [103], p. 35: "What is objective […] is what is subject to laws, what can be conceived and judged, what is expressible in words".

pp. 33–36).[64] But, for Frege, only the apprehension, not the existence, of what is objective depends on *reason*, while for Dedekind apprehension *and existence* depend on reason, since 'objective' means the same as 'constitutive of the rational activity of the mind': 'activity' does not mean the same as 'subjectivity'; all what Dedekind want to mean by using this term is that the objects of our thinking are not external to the thinking. Thus, for Frege, arithmetical objects are "immediately given to reason", where 'immediately' is used to mean that giving these objects to the reason does require no mediation of the senses ([93], Sect. 105),[65] while Dedekind holds that natural numbers are not immediately *given* to reason but that they "*flow* immediately from the pure laws of thought". One single word makes a big difference.[66]

Frege remarks that "although like all other disciplines mathematics, too is carried out in thoughts, still thoughts are otherwise not the object of its investigations" ([101], pp. 425–426; [109], p. 336). This points out sharply the difference between mathematicians' and logicians' stands. No further comment is needed to stress that Dedekind and Frege are using the sames words—namely 'thought' and 'pure laws of thought'—but they give them significantly different meanings.

1.3.2.2 The Laws of Logic

Frege frequently uses with the same meaning the expressions 'the laws of thought' and 'the laws of logic" or 'the general laws of logic', and he comes to prefer the two latter expressions, for they make clear unambiguously that logic is not concerned with what one holds to be true, but with what is true ([106], vol. I, "Logik", pp. 158–160; [107], pp. 146–148).

Now the expression 'laws of logic' is to be found nowhere in *Stetigkeit* or *Was sind*. We find there 'logic' and 'logical' qualifying a foundation which rests upon more general and more primitive concepts than the concepts usually taken as primitive in arithmetic or analysis. Thus the concept of the numerical real domain comes first; on it depend the notions of limit, continuity or convergence of a real function of real variables[67]; hence the logical priority of arithmetic *vis-à-vis* geometry does not mean, as

---

[64]Cf. Dummett's comments in [70], pp. 123–125.

[65]Cf. also [97], Sect. II.74, [110], p. 86$_2$: "We can distinguish physical from logical objects, by which of course no exhaustive classification is intended to be given. The former are in the proper sense actual; the latter not so, though no less objective because of that. While they cannot act on our senses, nonetheless they are graspable by our logical faculties. Such logical objects include our cardinal numbers; and it is probable that the remaining numbers also belong here".

[66]Frege's assertion that "the validity of Dedekind's proofs [in *Was sind*, Sect. 66] rests on the assumption that thoughts obtain independently of our thinking" ([106], vol. I, "Logik", p. 147, footnote; [107], p. 136) does not hold: Dedekind takes thoughts to be objective but not to obtain independently of our thinking. Actually, Dedekind's number-realm ([49], *Vorwort*, p. VIII; [53], p. 14) does not exist independently of our thinking.

[67]Indeed, *Stetigkeit* shows that the Dedekindian "completeness" of the real numbers field implies logically its Cauchy's completeness, once one defines a distance (a metric) on the field.

Frege holds, that geometry depends on intuition,[68] it simply means, as Frege maintains too in his early period, that numbers and numerical operations have an intrinsic definition, with no appeal to geometrical notion. According to Dedekind, "logic" also allows showing that the continuity of line and space neither is an explicit or implicit assumption among Euclid's definitions, axioms or postulates nor can be logically derived from them. Furthermore, "logic" allows showing that one can "establish with rigourous logic the science of numbers" upon "the definition of the infinite" ([50], *Vorwort zur zweiten Auflage*, p. XVI; [53], p. 19), i.e. that one can conceive of natural numbers as definable in terms of a "similar *Abbildung*" (injective function) on an infinite domain of abstract, i.e. non interpreted, elements. Dedekind deals also with logical dependence or independence, not with logical laws ruling the dependence relation. Moreover, in Dedekind's view, the most fundamental law of thought, the law which provides "the unique and therefore absolutely indispensable foundation [...][for] the whole science of numbers" is "the *ability* of the mind to relate things to things, to let a thing correspond to a thing, or to represent a thing by a thing, an ability without which no thinking is possible" ([49], *Vorwort*, p. VIII; [53], p. 14; my italics).[69] The most fundamental law of thought is also the *Abbilden*-ability.

Correspondingly Frege holds that thoughts have to be analysed into the function-argument dichotomy. Yet the perspective opened by the *Abbilden*-ability is very different from the perspective opened by the function-argument analysis: the traditional concept of function is generalised in totally different ways and for different purposes. In the first case, the *Abbilden*-ability is a dynamic rational process resulting into mathematical innovations and progress, because it permits taking one thing for another playing the same role. What matters is not about identity but about *analogy*, which can hold across *different* domains. In the second case, the function-argument analysis affords a static frame for decomposing thoughts into their logical constituents in order to find out their truth-value. Frege's notion of function comes close to our notion of logical predicate with one, two or more places; Frege's notion of generality comes close to our universal quantification.

1.3.2.3 Thought and Truth

For Frege, the laws of thought are the laws of logic, and the laws of logic are the laws of truth. 'Thought' has indeed a special meaning: a thought is "something for which the question of truth can arise at all" ([102], p. 60; [109], p. 353); thus thoughts are objects of logic, they fall outside the realm of mathematics proper as well as the head-on study of truth in and for itself. Frege explains that the laws of thought are the normative laws of logic and that there is no need for specific laws for arithmetic, for "aggregative thought" as he calls it ([93], *Einleitung*, p. III; [103], p. 15). The laws

---

[68]Dedekind does not put an exclusive disjunction between logic and geometry, as does Frege in his early writings. He holds that the mathematical general concept of space differs from the Euclidean space, taken as intuitive until the nineteenth century, and from the physical sensible space.

[69]'Numbers' here does not merely refer to natural numbers; it rather refers to any kind of numbers.

of logic are the laws of being true, not of being taken to be true, the laws of thought are not the laws of thinking.[70] Thus, logic is not only the theory of inference, but also the theory of truth, whose tools are judgement and concepts.[71]

By contrast, Dedekind does not consider truth as the task of logic but as the goal of human scientific activity, and he believes that "arithmetising",[72] as he calls it, is a fundamental activity of human reason, which is applied to empirical tasks, but whose laws of operating are neither rooted in, nor grounded on the experience. It is mostly with respect to this rejection of experience and intuition as being the basis that the laws of arithmetic are "a part of logic".

1.3.2.4 Logicality

In the same vein as Boolos, we take logicality to have three aspects:

(*i*) logicality of proving: in arithmetic the chains of inferences by which one goes from principles to consequences must convey no ingredient foreign to arithmetic, in particular no intuitive or geometric ingredient, and must be logically free of gaps; that leads to three requirements: (a) making explicit the principles and excluding any tacit assumption; (b) listing the rules of inference that will be used; (c) showing that any transition in a chain of inferences can be analysed into simple deductive/logical steps;

---

[70]Cf. [97], *Vorwort*, pp. 15–16, [110], pp. XV_1–XVI_1: "[...] being true is different from being taken to be true, be it by one, be it by many, be it by all, and is in no way reducible to it. It is no contradiction that something is true that is universally held to be false. By logical laws I do not understand psychological laws of taking to be true, but laws of being true". Cf. also [106], vol. I, "Logik", p. 158; and 146: "If a man holds something to be true [...] he thereby acknowledges that there is such a thing as something's being true. But in that case it is surely probable that there will be laws of truth as well, and if there are, these must provide the norm for holding something to be true. And these will be the laws of logic proper ". And again [102], p. 59, [109], p. 352: " I assign to logic the task of discovering the laws of truth, not the laws of taking things to be true or of thinking".

[71]The notions of a judgement and a concept are taken on from "the old logic" in general and, in particular, from Kant, but Frege's notion of a concept is idiosyncratic and Frege's way of connecting concepts and judgements with the notion of truth is totally new. More precisely, Frege holds that "the theory of concepts and of judgement is only preparatory to the theory of inference", and that "the task of logic is to set up laws according to which a judgement is justified by others, irrespective of whether they are themselves true"; thus " the laws of logic can guarantee the truth of a judgement only insofar as our original grounds for making it, reside in judgements that are true" ([106], vol. I, "17 Kernsäter zur Logik", dated to 1906 or earlier, p. 175, sentences 14, 15, and 16; [107], p. 175). Boolos and Heck ([30], p. 333) point out that the following question may have occurred to Frege: "Can the notion of a truth of logic be explained otherwise than via the notion of provability?". Insofar as he did not have the notion of interpretation, Frege could not have got the notion of logical consequence.

[72]The epigraph on the first-page of *Was Sind* is this: "Ἀεὶ ὁ ἄντρωπος ἀριτμητίζει". I discuss the matter in [56], pp. 101–113. It seems to me wrong to interpret the whole essay *Was Sind* as a "transcendental deduction" in Kant's specific sense, as suggested by Mc Carty [145], or to cut radically any link between Kant and Dedekind, as suggested by Reck, instead (Reck 2003).

(*ii*) logical nature of the basic concepts and basic propositions that are assumed, or into which the arithmetic concepts and the arithmetic propositions respectively are translatable;

(*iii*) logicality of the truth of the basic propositions, which are not only truths but logical truths.

About aspect (ii) a discussion may arise concerning the question whether the fundamental concepts of Dedekind's reconstruction of arithmetic, viz. the concepts of a *System* and of *Abbildung*, are logical concepts really. I leave this discussion for Sect. 1.5.2, below. Here, let me consider aspect (i), instead.

This aspect concerns logic as a theory of inference. Though prominent in his striving for logical rigour,[73] it is not thematised for itself by Dedekind, as it is, instead, in Frege's *Begriffsschrift* and *Grundlagen*. Requirements (a) and (c) in (i) are globally shared by Dedekind and Frege, which take "pure thought" to meet them. In fact, they are satisfied by any deductive system, or, to use an expression which is appropriate for Dedekind and can be fully applied to *Grundgesetze*,[74] by any (arithmetical) *axiomatic* system. Logic is involved as much as in any Euclidean enterprise, as it were. Therefore it seems to me unnecessary to qualify Dedekind's standpoint as "*logical* structuralism", as Reck did ([167]; cf. Sect. 1.1, above, p. 6): any sort of structuralism aims at showing the logical relations between propositions through a deductive presentation, just as any logicism deals with (formal) axiomatic systems. There is indeed a common concern: the deductive concern. But, when Frege focuses on mathematical axiomatic systems, he brings to the fore the logical elements involved in them: axioms are *truths* and theorems are truths *inferred* from axioms in accordance with the logical laws of inference. And, according to Frege, (α) truths are absolute so that there is one unique axiomatic system for geometry, namely Euclid's system, and one unique system for arithmetic, namely the system of *Grundgesetze*; (β) "we cannot regard as definition the system of sentences in each of which there occur several of the expressions that need defining" ([106], vol. I, "Logik in der Mathematik", p. 229; [107], p. 212).[75] That means that a definition fixes the sense unambiguously: to a sign should be assigned, *via* a "constructive definition", one unique sense; a sign must not only indicate, but designate a determined object. A

---

[73] A place where Dedekind expresses his permanent concern for rigour is his Letter to Weber of November 8, 1878, where he exhorts him to "not renounce to use logic" in secondary school ([55], vol. III, p. 485).

[74] Cf. [97], *Vorwort*, p. VII, [110], p. VII$_1$: "The gaplessness of the chains of inferences contrives to bring to light each axiom, each presupposition, hypothesis, or whatever one may want to call that on which a proof rests; and thus we gain a basis for an assessment of the epistemological nature of the proven law". Frege discusses the nature of axioms in "Logik in der Mathematik", [106], vol. I, pp. 221–222.

[75] I leave out of consideration Dummett's remark, which Demopoulos renders as follows: "Frege's basic approach [to numbers] would have been problematic even if no inconsistency had been discovered since there is an unacceptable circularity in Frege's procedure: the abstraction principle which introduces the numbers contains an implicit first-order quantifier, so the numbers introduced on the left occur within the range of the variables bounded on the right in the explicit definition of one-one correspondence" ([60], p. 220).

consequence of this requirement is, for instance, that if the sign '2' is defined first as a *designans* of a certain natural number, it cannot be also defined as a *designans* of a real number, since these two numbers are different objects. Indeed: 'if the first definition is already complete and has drawn sharp boundaries, then the second definition either draws the same boundaries and is then to be rejected since its content should be proven as a theorem, or it draws different boundaries and thereby contradicts the first" ([97], Sect. II.58; [110], pp. 71_2–72_2). (α) and (β) say how much Frege diverges from the mathematical understanding of axiomatics.

The requirement (b) is satisfied by Frege, but not by Dedekind. This makes much more prominent the genuine logical aspect of deduction and brings to the light the constitution of a *formal logical* system, such as that of *Begriffsschrift* and of *Grundgesetze*.

1.3.2.5 Reference to Euclid

It is remarkable that both Dedekind and Frege refer to Euclid's system, pointing out what is lacking in it. But what is lacking according to Frege is not what is lacking according to Dedekind.

For his part, Dedekind proves that any Euclidean construction is feasible using only the *algebraic* real numbers, and, then, that a continuity axiom is not only explicitly lacking but even not necessary in Euclid's geometrical constructions ([49], *Vorwort*, pp. XII–XIV).[76] Moreover, Dedekind proves that Euclid's theory of proportions assumes implicitly only the Archimedean axiom, which is not sufficient to guarantee the continuity of the domain of "incommensurable magnitudes" (cf. the letters to Lipschitz mentioned in footnote (76), above). Thus, Dedekind makes explicit what is *logically* deducible from Euclid's assumptions and which *mathematical* supplement is needed for reasoning correctly on continuous magnitudes or on real numbers, real functions of real variables, etc. (though most people knows this major contribution only through Hilbert's *two* continuity axioms in *Die Grundlagen der Geometrie*, namely Archimedes's axiom V.1 and the linear completeness axiom V.2: [127], Sect. 8).

On the other hand, Frege shows what, according to him, goes beyond Euclid's ideal, and in the same way beyond Dedekind's achievement, namely the specification in advance of all methods of inference. That is to say that an *axiomatic* system, that makes explicit the deductive structure of a mathematical theory, is *not yet* a formally constructed *logical* system of that theory. Or, put differently, the logical aspect involved in an axiomatic system is not sufficient to fulfil the logicist requirements.

This differentiation between an axiomatic and a logicist requirement is the dividing line between Dedekind and Frege until the publication of *Grundgesetze*, and continues after to impact Frege's conception of proofs and definitions. Whereas Dedekind is attentive to defining everything which can be defined and to proving any

---

[76]Cf. also [67], Appendix XXXI, and the letters to Lipschitz of June 6th and July 27th, 1876 ([55], vol. III, pp. 468–479).

statement which is provable,[77] in order to obtain the most simple concepts and the very primitive statements, and hence to make clear the logical connections between mathematical propositions, he does not develop a systematic detailed reflection on logical inference in itself nor on what is or how must be an adequate definition or a correct proof. *Stetigkeit* and *Was sind* show *practically* that a definition is adequate when, starting from that definition, chains of logically correct inferences lead (i) to the definitions of usual arithmetical operations on the real numbers[78] and on the natural numbers respectively, and (ii) to the proof of propositions involving these operations, such as the proof of the upper bound theorem for the real numbers, the proof of the continuity of the rational operations extended to the real numbers, or the proof of mathematical induction on natural numbers. Like Frege, Dedekind did care about a "really scientific foundation for arithmetic" ([47], [Preface], p. 9; [53], p. 1) and logical rigour in his "presentation [*Darstellung*]" of the natural numbers. He insists on the long sequence of simple inferences constituting "the chains of reasoning on which the laws of numbers depend", assuming that the recognition of a mathematical truth "is never given by inner consciousness", but rather by a "step-by-step under-standing" ([49], *Vorwort*, p. IX; [53], p. 15). Our sequential understanding cannot but establish arithmetic laws progressively, by a long chain of inferences.

By contrast Frege has many comments on inferring and defining.[79] These com-ments are part of Frege's research on logic. The renewal of logic that Frege wants to achieve implies considering logic not only as giving a firm ground for arithmetic but also and mainly as a field on its own. Frege wrote a series of papers on the essence of logic in which he deals with the laws of truth and the laws of valid inference, with the definition of objects and the distinction between object and concept[80] and between sense and *Bedeutung*, with the sharp distinction between psychology and logic (i.e. in his terms, between thinking and thought) and the affinity between logic and ethic. As Dummett points out, in large parts of these papers one cannot find any reference to mathematics nor any mention of a mathematical example ([69], p. 96) For Frege logic applies everywhere, not only in the foundations of mathematics; it is *coextensive* with language, and the logical work is first a "struggle against language" (cf. [106], vol. I, "Erkenntnisquellen der Mathematik und der mathematischen Naturwissenschaften" p. 289; [107], p. 270). This is a decisive component of Frege's perspective on logic, a component which has triggered the linguistic turn, and which is totally absent from Dedekind's views.

---

[77]Cf. [49], *Vorwort*, p. VII, [53], p. 14: "In science nothing capable of proof ought to be accepted without proof". That's the very first sentence of *Was sind*.

[78]Dedekind insists upon the fact that we must define in its entirety (up to an isomorphism) the domain of the real numbers in order to have the possibility to define *in a general way* the operations on it: otherwise how could we know that, e.g. the result c of the addition or of the multiplication of some two individual real numbers a and b is again a real number? Letter to Lipschitz, June 10, 1876 ([55], vol. III, pp. 462–474).

[79]Definitions are pervasively treated in [97], Sects. I.26-33, II.55–65 and in "Logik in der Mathe-matik" ([106], vol. I, pp. 219–270; [107], p. 201–250).

[80]Remember the third fundamental principle stated in *Grundlagen*: "never to lose sight of the distinction between concept and object" ([93], *Einleitung*, p. X; [103], p. XXII).

1.3.3 More on Inference: Truths and Logical Truths

In the first volume of *Grundgesetze*, Frege criticises Dedekind's chains of inferences in *Was sind* ([97], *Vorwort*, pp. VII–VIII; [110], pp. VII$_1$–VIII$_1$; my italics)[81]:

> Mr Dedekind's essay, *Was sind und was sollen die Zahlen?*, the most thorough study I have seen in recent times concerning the foundations of arithmetic [...] pursues, in much less space, the laws of arithmetic to a much higher level than here.[82] This concision is achieved, of course, only because much is not in fact proven at all. [...] nowhere in his essay do we find a list of the logical or *other* laws he takes as basic [...].

In fact, in *Was sind*, Sect. 71, Dedekind *does list* the four basic laws that, reformulated in a formal language by Peano, are known as Dedekind-Peano axioms for the natural numbers; just they are clearly *not* logical, i.e. universally valid, laws and they are not rules but premises of inference within a specific mathematical domain. Thus *Was sind* does not match with aspect (iii) of logicality. And as Boolos remarks ([24], p. 270; my italics):

> It is evident that one who claims to have enumerated all the ideas and steps involved in mathematical reasoning need not imply that that reasoning is *logical* reasoning [...]; however justly, it might well be said that the Zermelo-Fraenkel set theory provides such an enumeration: to say so is, obviously, not to be committed to the view that its axioms are logical truths.

As we know, Hilbert has been very much impressed and influenced by the deductive style of *Was sind*, Peano took on Dedekind's axioms for arithmetic, and Zermelo's axioms are a completed and amended version of Dedekind's axioms. Thus, Dedekind furnished the explicit basis for the multi-sided development of the axiomatic approach, which has become a fundamental constituent of mathematical reasoning. It became so clear that 'axiomatic' does not coincide with 'logical' that Frege himself admits in 1914 the possibility of general laws that are specific to mathematics, that also are *not* laws of logic. He even recognises that "one can reduce a mode of inference that is peculiar to mathematics to a general law, if not a law of logic, then *one of mathematics* [...][a]nd from this law one can then draw consequences *in accordance with general logical laws*" ([106], vol. I, "Logik in der Mathematik" pp. 220; [107], pp. 203–204; my italics).[83] Frege's remark applies very well to Dedekind, who indeed does not investigate logical issues in and for themselves, but aims at showing how to "establish *with rigourous logic* the science of numbers" ([50], *Vorwort zur zweiten Auflage*, p. XVI; [53], p. 19), namely to lay

---

[81]Here is what Frege writes some line above: "Although it has already been announced many times that arithmetic is merely further developed logic, still this remains disputable as long as there occur transitions in the proofs which do not conform to acknowledged logical laws but rather seem to rest on intuitive knowledge. Only when these transitions are analysed into simple logical steps can one be convinced that nothing but logic forms the basis".

[82]Frege means Dedekind's definition of addition, multiplication, and so on.

[83]But this is only one first step; going further leads to construing the logical system lying at the bottom of mathematics.

down a deductive presentation of arithmetic that bans intuition, geometric notions, and any non-arithmetical notion.

Let's turn now to definitions.

1.4 On Definitions

1.4.1 Dedekind's Definition by Axioms

Dedekind claims that he creates new concepts and, in fact, he does that by stating explicitly new primitive "laws [*Gesetze*]" or "conditions [*Bedingungen*]" for characterising concepts "intrinsically [*wesentlich*]". These laws are used as a "definition [*Erklärung*]" from which theorems can be deduced ([49], Sects. 71 and 73, for the sequence of natural numbers; [47], Sect. V.IV, for the continuity of the domain of real numbers). Here is a point that Ferreirós ([91], p. 247) brought to light: Dedekind does not use the term 'axiom' except for his continuity principle and Cantor's axiom of continuity ([47], [Preface], p. 11, and Sect. 3); continuity is indeed not a necessary property of space whereas conditions α, β, γ, δ of *Was Sind*, Sect. 71 (cf. below) are essential properties of natural numbers; they are, as the letter to Keferstein mentioned above makes totally clear, necessary and sufficient conditions. A *System S* is simply infinite if and only if there exists a distinguished element $e \in S$ and a bijective *Abbildung* $\varphi : S \mapsto S - \{e\}$ such that induction holds, in Dedekind's terms:

$$\alpha. \ \varphi (S) \subset S$$
$$\beta. \ S = \{e\}_0$$
$$\gamma. \ e \notin \varphi (S)$$
$$\delta. \ \varphi : S \mapsto S \ \text{ is injective}$$

where $\{e\}_0$ is the chain of $\{e\}$, i.e. the least set containing $\{e\}$ and closed under φ.

Even though it may originate in Kant's view according to which there are no axioms in arithmetic, the lack of the term 'axiom' does not make *Was sind* a pre-axiomatic presentation. For, in Dedekind's time and before, 'essential property', 'law', and 'condition' were used as we use now 'axiom', they played the same role, and the axiomatic method was practised long before its codification by Hilbert. The conjunction of conditions α, β, γ, δ characterises the *structure* of the natural numbers as a simply infinite *System*: *any System S* of uninterpreted things (or "shadowy forms [*schattenhaften Gestalten*]": [49], *Vorwort*, p. IX; [53], p. 15) that satisfies the conditions α, β, γ, δ behaves *as the System N* of natural numbers; there are a distinguished element $e \in S$, and a bijective *Abbildung* $\varphi : S \mapsto S - \{e\}$, which makes the conditions α, γ, δ be satisfied,[84] such that induction holds, i.e. condition β

---

[84] As noted by Dedekind, this simple statement involves infinity of S.

is also satisfied. Any such S is isomorphic to N (categoricity theorem: [49], Sect. 132), which does not mean that we can take any S as *the* natural numbers.[85] Hence, if the word 'axiom' is lacking in *Was sind*, the thing itself is really there and makes the substance of the definition of the kind of structure instantiated by the natural numbers, which is a progression with Russell's term, an ω–sequence in our modern terminology.

Tait's provocative opinion according to which "Dedekind's view is not the so-called structuralist view" ([190], p. 317) means the following: the term 'structuralist' holds in the case where, when we are asserting an arithmetic proposition A we assert that A holds in every simple infinite *System*, whereas, according to Tait, Dedekind asserts A of the natural numbers themselves. Tait is fighting in particular Dummett's interpretation of Dedekind's numbers as structural objects, i.e. objects "that have no properties save those that derive from position in 'the' abstract simple infinite system (sequence of order type ω)" ([74], p. 295). Indeed Dedekind specifies a structure on ordinal numbers in terms of which they can be characterised categorically by the conditions α, β, γ, δ, and have all the properties derivable from the latter. According to Tait, defining a structure out of the numbers is a specifically logical operation. Hence Tait's view of Dedekind being a logicist. Tait ([190], pp. 316–317) judges that

> Dedekind's treatment is certainly superior [to Frege's one] in at least one respect: namely, by proving the categoricity of the second-order theory of a simple infinite system, he fixes the sense of arithmetic propositions independently of whether we can in some sense prove the existence of such a system; whereas not having isolated the axioms of a simple infinite system and proved categoricity, Frege's treatment of arithmetic propositions fails absolutely with the failure of his identification of them with equipollence classes of some system of objects.

1.4.2 Frege's Ontological Conception of Definitions of Objects

Frege wonders: "what definition is and what it can achieve [was Definieren ist und was dadurch erreicht werden kann]" ([97], *Vorwort*, pp. XIII; [110], pp. XXII$_1$). And he gives roughly two different answers. Here I consider the first one; I shall come to the second in Sect. 1.4.6.

This first question is advanced in *Grundlagen*, and is akin to Plato's[86] search about what is designated by some individual term—such as 'Socrates'—or by some general term—such as 'beauty' or 'science'—, and how this is. This is a search after the essence, after "das Wesen der Sache" ([97], *Einleitung*, p. 1; [110], p. 1$_1$).

---

[85]We should not forget that Dedekind's construct is based upon "a prior analysis of the sequence of natural numbers just as it presents itself in experience, so to speak, for our consideration" (Letter to Keferstein, February 27, 1890: [184], pp. 271–272; [125], p. 99). For a worthwhile discussion of this matter, cf. [156], pp. 306–311.

[86]Frege recalls the Socratic aphorism: "The first prerequisite for learning [...] is [...] the knowledge that we do not know" ([93], *Einleitung*, p. III; [103], p. XV). Cf. also [106], vol. I, "Logik in der Mathematik", p. 239; [107], p. 221.

Obviously 'Wesen' and 'wesentlich' do not have the same meaning for Frege as for Dedekind. Indeed Frege's very first question in *Grundlagen* is "what the number one is, or what does the symbol 1 mean [*bedeutet*]?" ([93], *Einleitung*, p. I; [103], p. XIII), and it is oriented towards a definition of *the* number one, namely it is a question about a definite particular *object*, whose *properties* are to be specified, and simultaneously about the meaning of the symbol (singular term) by which this object is designated.[87] Being, meaning and naming are all linked in Frege's study.

Now, to say what is a particular natural number, such as 1 or 2, prepares us to say what are natural numbers in general, i.e. what is the concept of natural number. In a Platonistic way, Frege examines first, in about 60 sections, answers given by his predecessors or contemporaries and rules them out showing the logical difficulties raised by each one. Frege shows how we ought not to start from five apples, three fingers or *the* moon to get respectively the number five, the number three and the number one, for numbers are neither physical things nor attributes of things. There is no direct route from physical things to arithmetical objects. We should rather start with the linguistic expressions 'five apples', 'three fingers' or '*the* moon', each expression taken in the context of a statement, a judgement, from which we will realise that (cardinal) numbers are ascribed only to concepts: "a statement of number is an assertion about a concept" ([93], Sect. 46; [103], p. 59). Frege shows how we pass from statements about a definite cardinal number, possibly zero, to this very same definite number as *arithmetical object*, i.e. how we pass from a certain multitude of objects to the concept of the cardinal number of *this* multitude, or, better, to the concept of the cardinal number of the concept F that identifies this multitude (that is, the concept F of which this multitude is the extension), and then to this very cardinal number, namely the extension of the concept ⌜equinumerous with F⌝. Since, the cardinal number of a first-level concept F is the extension of the second-level concept under which fall all and only those first-level concepts equinumerous with F.

The complete and final answer to the question 'How many?' cannot be given before we "fix the sense of a numerical identity" ([93], p. 73; [103], p. 73): the question about the definition of an object involves the search for a criterion of its identity. "If we are to use the sign 'a' to designate an object, we must have a criterion for deciding in all cases whether b is the same as a" ([93], Sect. 62; [103], p. 73). Thus the question is threefold: (i) what is the specific object denoted by a numerical sign 'a'?; (ii) how are cardinal numbers given to us? (iii) which criterion permits us to *recognise* that the sign 'a' denotes the same object as the sign 'b' in '$a = b$'[88]? Frege's aim is "to construct the content of a judgement which can be taken as an

---

[87]Cf. [106], vol. I, "Logik in der Mathematik", pp. 234 and 262; [107], pp. 216 and 243; my italics: "Is that [...] a science which proves sentences without knowing *what* it proves?"; "Definitions must be given once and for all".

[88]Frege summarises his method as follows ([106], vol. I, "Logik", p. 154; [107], p. 143): "The first and most important task is to set out clearly what the objects to be investigated are. Only if we do this shall we be able to recognise the same as the same: in logic too such acts of recognition probably constitute the fundamental discoveries".

identity such that each side of it is a number" ([93], Sect. 63; [103], p. 74), and to achieve this construction of the *identity* of cardinal numbers on the basis of a general concept of identity that does not hold only for (cardinal) numbers: from a principle about numerical identity, namely the principle stated by Hume (mentioned in footnote (2), above), Frege draws, firstly, an explicit definition of the cardinal number that belong to a concept ([93], Sect. 68; [103], pp. 79–80)[89]:

> [...] the cardinal number which belongs to the concept F is the extension of the concept $\ulcorner$ equinumerous to the concept $F \urcorner$

and, then, a definition of a logical relation ([93], Sect. 73; [103], p. 85)[90]:

> the cardinal number which belongs to the concept F is identical with the cardinal number which belongs to the concept G if the concept F is equinumerous to the concept G.

Cardinal numbers are what concepts share when they are mutually equinumerous, that is, when their extensions are one-to-one correspondent.

The question about the essence of numbers is also, as it was in the philosophical tradition, a question about quiddity and identity. But to answer it, as to cardinal numbers, Frege provides an explicit definition and an identity condition. This opens the way to the "new logic" and to logical philosophy, which would replace the traditional metaphysics. However, a logical difficulty arises from the treatment of cardinal numbers as objects that both fall under concepts and are associated with concepts as their numbers.[91] Moreover, also the epistemic problem is not solved.[92]

Dedekind's concept of number is radically different: as we saw, Dedekind does not focus on the individual cardinal numbers nor even on the concept of cardinal number,[93] therefore, (i) he does not aim at deducing the numerical equality, i.e. the *identity* of cardinal numbers, from one-to-one correspondence (among concepts), and (ii) he does consider neither the relations between proper names and individual numbers, nor that between general names and concepts. What concerns him is not what a singular natural number like, for example, 1, is (which depends, for Frege, on a concept under which fall so many *different* things, like *the* moon, *the* sun, *the* Pythagorean theorem, etc.), but rather a generalization of the function successor, which holds not only for natural numbers but also, possibly, for elements other than numbers. According to Dedekind, the linear total ordering that structures any progression, and from which one can derive a general form of mathematical induction ([49], Sect. 59) and the *recursive definition* (*ibid.*, Sect. 126) of the *operations* of addition, multiplication, difference, power, etc. as *ordinal* operations are simpler than, and so prior to the cardinality aspect, which he takes to be more intricate. As to the concept of equality, Dedekind takes on Leibniz's definition of substitutability,

---

[89]Cf. footnote (2), above.

[90]Cf., again, footnote (2), above.

[91]Cf. [27], p. 309. Also [153] displays some difficulties with the thesis that numbers are objects.

[92]Cf. the quote from the *Nachwort* of *Grundgesetze*, at the beginning of Sect. 1.4.3, below.

[93]This is one reason why "Dedekind would not have been happy with the suggestion that the existence of infinite systems be derived from Hume's principle" ([23], p. 216).

whereas Frege takes on Leibniz's principle of the identity of indiscernibles (for Dedekind 'a' can be replaced with 'b' provided that b has the same properties as a, and this can happens also if b is an object distinct from a; for Frege, no distinct objects can have the same proprieties, so that 'a' can be replaced with 'b' only if a is the same object as b).

1.4.3 Frege's Epistemology

In Frege's view, the search for a definition of the concept of natural number is tied with an *ontological assumption* and with an *epistemic task*: numbers, thoughts, truths are timeless self-subsistent objects,[94] and we have to apprehend them, for "if there are logical objects at all—and the objects of arithmetic are such—then there must also be a means to grasp them, to recognise them" ([97], Sect. II.147; [110], p. 149$_2$). Frege's final sentence of the Afterword to the second volume of the *Grundgesetze* (written to expound a tentative way-out to Russell's paradox) is this ([97], *Nachwort*, p. 265; [110], p. 265$_2$; my italics):

> This question may be viewed as the fundamental problem of arithmetic: how are we to *apprehend* logical objects, in particular, the numbers? What *justifies us* to *acknowledge* numbers as objects? Even if this problem is not solved to the extent that I thought it was when composing this volume, I do not doubt that the path to the solution is found.

The epistemological task is double: it is about access to what *there is* and about the *justification* of the judgement of recognition of what there is. The answer is twofold.

Firstly, access is through meaning in a linguistic context, "since it is only in the context of a statement that words have any meaning", so that the problem "becomes this: To define the sense of a statement in which a number-word occurs" ([93], Sect. 62; [103], p. 73).[95] Indeed, Frege thinks that we get the arithmetical objects

---

[94]The following passage could be interpreted as conflicting with Frege's ontological assumptions ([93], Sect.60; [103], pp. 72): "The self-subsistence which I am claiming for number is not to be taken to mean that a number-word designates something when removed from the context of a statements, but only to preclude the use of such words as predicates or attributes, which appreciably alters their meaning [*Bedeutung*]". But for Dummett ([69], pp. 83, 81) the context principle is "a thesis about reference, not just about sense", it is used "to justify regarding abstract terms as standing for genuine, objective objects"; and what conflicts with it is the doctrine that truth-values are objects.

[95]In retrospect Frege writes ([97], *Vorwort*, p. X; [110], p. X$_1$):

> Previously I distinguished two components in that whose external form is a declarative statement [*Behauptungssatz*]: 1) the acknowledgement of truth [this is the definition of a judgement, given in *Grundgesetze*, Sect. I.5], 2) the content, which is acknowledged as true. The content I called 'judgeable content [*beurtheilbarer Inhalt*]'. This now splits for me into what I call 'thought' and what I call 'truth-value'. This is a consequence of the distinction between the sense and the reference [*Bedeutung*] of a sign. In this instance, the thought is the sense of a statement and the truth-value is its reference. In addition, there is the acknowledgment that the truth-value is the True.

not through some kind of Kantian synthesis but through the logical analysis of arithmetical statements. What matters is always a statement about some specified cardinal number applied to some multitude of objects, whether these objects be concrete or not, real or not. Thus a *logical analysis* of the language is introduced[96] along with the context principle ("never to ask for the meaning of a word in isolation, but only in the context of a statement": [93], *Einleitung*, p. X; [103], p. XXII), and the radical separation between concept and object, in order to answer the ontological question: "What, then, are numbers themselves?" ([106], vol. I, "Zahl" p. 284; [107], p. 265), to which Frege answer this way: "We may seek to discover something about numbers themselves from the use we make of the numerals and number-words. Numeral and number-words are used, like names of objects, as proper names" (*ibid.*).[97] In Frege's view the linguistic turn is closely tied with an ontological commitment.

Secondly, we need a criterion for numerical identity, a criterion that decides with absolute certainty whether the object designated by a number-word *a* is *the same* as the object designated by the number-word *b*. The criterion cannot be but logical since the numerals refer to logical objects that we know by analytical judgements. Contrary to Kant, Frege holds that arithmetical judgements are analytical *a priori*[98] and, at the same time, that logic is fruitful as a tool for clarifying what is embedded in our mathematical discourse.[99] Logic alone affords the needed justification for the recognition of *what there is*.

---

[96]Cf. [106], vol. I, "Meine grundlegeden logischen Einsichten" p. 272; [107], p. 252: "Work in logic just is, to a large extent, a struggle with the logical defects of language".

[97]Cf. also also [106], vol. I, "Aufzeichnungen für Ludwig Dermstaedter", p. 276, [107], p. 256: "In arithmetic a number-word makes its appearance in the singular as a proper name of an object of this science; it is not accompanied by the indefinite article, but is saturated".

[98]Needless to say that 'analytical' in Kant's conception conforms to Aristotle's analysis of a proposition into subject and predicate (the predicate is contained in the subject). With the analysis into argument and function, Frege introduces a new sense of the adjective 'analytical' ([93], Sects. 3, 16–17). First, the analytical/synthetical, and *a priori*/ *a posteriori* distinctions "concern [...] not the content of the judgement, but the justification for making the judgement" (*ibid.*, Sect. 3; [103], p. 3). Second, for Frege, analysis is a *process* similar to chemists' decomposition; thus a truth resulting from an analysis (an analytical proposition) is *a posteriori*, at least in Kant's sense. But, in mathematics, justification is "finding [...][a] proof and [...] following it up right back to the primitive truths" (*ibid.*,; [103], p. 4). Now, "if, in carrying out this process, we come only on general logical laws and on definitions, then the truth is an analytic one", and "if [...][it] can be derived exclusively from general laws, which themselves neither need nor admit of proof, then the truth is *a priori*" (*ibid.*).

[99]Cf. [93], Sect. 17, where Frege expresses the innovative view that logic can provide us with substantive knowledge; if one can, writes Frege, show the inner link of arithmetic with logic, then "the prodigious development of arithmetical studies, with their multitudinous applications, will suffice to put an end to the widespread contempt of analytic judgements and to the legend of the sterility of pure logic" (*ibid.*; [103], p. 24). Cf. also [93], Sect. 91, [103], p. 104: "statements which extend our knowledge can have analytic judgements for their content". Frege's followers will dispute on the mathematical fruitfulness of logic: Poincaré and Wittgenstein will be against; Tarski, Abraham Robinson, Kreisel, Feferman, among others, will concretely show how logical analysis may be used as a tool for proving or discovering mathematical results.

1.4.4 Dedekind's Treppen-Verstand and Stückeweise Definitions

As seen above, Dedekind's aim is that of characterising *structurally* the essence of numerical continuity and of the natural numbers. Epistemologically speaking, Dedekind keeps to the critical line of Kant. He focuses, indeed, on the power of reason and the limits of the human understanding [*Verstand*] rather than on being, truth and the justification of our recognition of them. He does not tackle proper ontological questions, because he thinks they are out of the scope of science.[100] Dedekind takes his starting point neither in the physical world (fingers, apples, moon, sun, strokes on a sheet of paper, etc.) nor in language, namely in phrases or statements containing number-words. He considers straight away a scientific domain, namely elementary arithmetic, and asks what we are *doing* when we carry out elementary operations. And the answer comes down to excluding intuition, seeking for "*inner* " (structural) properties, and to promoting the step-by-step understanding [*Treppen-Verstand*], which is building *gradually* chains of inferences from primitive assumptions to deduced properties.

Dedekind may well be considered as a great pioneer of the epistemic turn realised by structuralism: primitive assumptions are not fixed once and for all (unlike Kant, Plato and Frege); they are fixed within a given system and they vary with the system. Definitions emerge first for a restricted domain, then they are gradually generalised, for example by embedding the initial domain into more comprehensive domains under preservation of the initial operations (but not necessarily of *all* properties of the initial operations). They are "*stückweise Definitions*", which Frege rejects. Moreover, the historical aspect of knowledge is taken into account, simply because mathematical invention cannot be separated from knowledge of the previous mathematical concepts and methods.[101] And it is not a matter of the psychological or sociological aspects

---

[100]Cf. [55], Vol. III, "Über die Einführung neurer Funktionen in der Mathematik", pp. 428–429 (my translation; Dedekind's italics): "The chief task of any science is striving to ground the *truth*, [...] towards which one can but go farther [without being capable with our step-by-step understanding to attain it]. But science itself, which represents the course of human knowledge, is open to an infinite variety of presentations [*Darstellungen*][...] it may be framed into different systems, because as human work it is submitted to arbitrariness and affected by all the imperfections of the human intellectual powers". By contrast, Frege thinks that the logical presentation of arithmetic is fundamentally unique.

[101]I think Dedekind would have agreed with Frege's following remark: "What is known as the history of concepts is really a history either of our knowledge of concepts or of the meanings of words [*Bedeutungen der Wörter*]. Often it is only after immense intellectual effort, which may have continued over centuries, that humanity al lest succeeds in achieving knowledge of a concept in its pure form, in stripping off the irrelevant accretions which veil it from the eyes of the mind" ([93], *Vorwort*, p. VII; [103], p. XIX). But Dedekind does not consider that the history of knowledge is psychology of knowledge; knowing historically mathematical notions may lead to "stripping off the irrelevant accretions" and to throwing light on ignored aspects of them.

of an invention,[102] it is a matter of the epistemic conditions of its emergence: its content has "inner" links with previously established results so that the shape of the whole structure is modified by it. As Dedekind writes ([55], Vol. III, "Über die Einführung neurer Funktionen in der Mathematik ", pp. 430; my italics): "progress in the development of any science *reacts* always again on the system thanks to which one tries to conceive of its organism, giving a new shape, and that is not only a historical fact, but it is also based upon an internal necessity".[103]

I have shown elsewhere ([11]; [56], p. 220; [12]) that Dedekind, through his influence on Jean Cavaillès [43], is the first contributor to our modern "conceptual history", whereas Frege originated the "conceptual analysis" practised by Gödel, Tarski, A. Robinson, Feferman and others. Frege recognises well that "the history of earlier discoveries is a useful study, as preparation for further research", but this "should not set up to usurp their place" ([93], *Einleitung*, p. VIII; [103], p. XX). Dedekind does not see a clash between the historical process and the logical rigour in substantial advances. Dedekind's "creation of concepts" points at the working mathematician and the newly introduced practice: defining new concepts encompassing many and various results, in accordance with logical laws, for a better systematisation of knowledge. The letters to Lipschitz show clearly that Dedekind aims at a renewal of Euclid's enterprise.

1.4.5 Frege's Criticism of Dedekind's Stückweise and Creative Definitions

Frege recognises that Dedekind's definitions are not "formal", since, in contrast with those of Thomae, Heine, Stolz or Hankel,[104] they do not apply to mere signs but to what signs express. Dedekind's arithmetic is "*inhaltlich*" ([97], Sect. II.138) and escapes "the mathematical sickness of our time, [...][i.e.] confusing sign with what is signified" ([106], vol. I, "Logische Mängel in der Mathematik" p. 172; [107], p. 158). But Dedekind's definitions are "*stückweise*" and "creative". Frege fights against "*das stückweise definieren* " because a definition must fit once and for all "the definiteness and fixity of the concepts and objects of mathematics" ([93], *Einleitung*, pp. V–VI; [103], pp. XXII–XVII). Moreover, Frege fights also against "creation", and this for two connected reasons.

---

[102]It is noteworthy that neither in *Grundlagen* nor in *Grundgesetze* Frege criticises Dedekind's way on grounds of psychologism. Dummett's psychologistic reading of Dedekind ([74], Chap. 2, "Frege and the paradox of analysis", p. 49) is very questionable.

[103]Frege is at odds with this dynamic view. Here is what he writes, instead ([106], vol. I, "Logik in der Mathematik" p. 261; [107], pp. 241–242): "We must always distinguish between history and system. In history we have development; a system is static. [...] what is once standing must remain, or else the whole system must be dismantled in order that a new one may be constructed".

[104]On the relations between Frege and Thomae and between Frege and Hankel, cf. [46, 157].

The first one is grounded on his questionable philosophical division of the world into two exclusive parts, the purely logical part and the rest, that can be physical or psychological. What is purely logical never changes; and it can only be discovered, not invented. Mathematical propositions are true forever, and they have been or can be proved because they are true, not the other way around. They are true even if we fully ignore them or do not recognise them as true; nevertheless, we have to recognise them as true and, in order to succeed in this task, logic is the only appropriate means, because it alone allows to recognise and justify truths.

The second unquestionable argument is logical: he points out that a mathematical definition does not create anything whose existence has not been proved beforehand. But one may wonder whether Frege himself should not have, in *Grundlagen*, proved the existence of the finite cardinal numbers before defining them. In fact, he just *assumes* these numbers to be logical, self-subsistent objects; hence he credits them with a timeless existence in a "third realm". Is this to say that, if the logical reduction succeeds, then we should conclude that the question of existence is also just reduced to making precise in what sense logical objects exist, or rather that the question of existence persists, being only pushed to the level of extensions of concepts, as Russell will show?

Until 1903 Frege faces neither the first nor the second question, because of two strong ontological assumptions: (i) he has no doubt that logical objects exist independently from space, time and cognitive acts, and (ii) he believes that the numbers "are *immediately given* to reason" ([93], Sect. 105; [6], p. 126). We do not have to prove the existence of something whose existence is immediately given to us; this is why the definition of cardinal numbers in *Grundlagen* presupposes from the outset the existence of these numbers and provides rather a logical criterion for their identity.

In *Grundgesetze*, Sect. II.143, Frege relates creative definitions to Otto Stolz, and he states that a mathematician should, before performing a creative act, prove that the *properties* that he will attribute to the object he wants to create do not mutually contradict, which he/she can only prove by proving that there exists an object that has all the properties in question. And if he can do it, then he does not need to create such an object. This criticism points out a difficulty for purely formal theories, i.e. in Frege's sense, theories for which no model is known in advance.

Frege is right, and, indeed, Dedekind is not formalist in this sense; he is speaking not of creating objects but of creating concepts that bring to the light the inner structure of a family of *Systeme* of objects. He writes to Keferstein that the fundamental properties of natural numbers, namely their meeting conditions α, β, γ, δ stated in Sect. 1.4.1, above, must be mutually compatible, independent from each other, and sufficient for deriving all arithmetic theorems (cf. the quote from this latter in Sect. 1.5.2.2, below). But he does seek to demonstrably show neither the compatibility and the independence of these properties, nor the coincidence between arithmetic truths and theorems derivable from these conditions. The reason is given by Dedekind himself: he found out those properties "after protracted labour, based upon a prior analysis of the sequence of natural numbers just as it presents itself, in experience, so to speak, for our consideration" ([184], pp. 271–272; [125], p. 99).[105] In modern terms, Dedekind construed the theory looking at a model whose consistency is therefore beyond doubt. From the point of view of a working mathematician, this suffices to avoid the problem of the possible vacuity of arithmetical statements posed by Parsons [156]. But at Dedekind's times, this was not as clear as today, and, in any case, the philosophical question of the mode of reality of mathematical entities still remains.

Anyway, Dedekind already feels in the late 1880s the need to prove the existence of infinite *Systeme*, on the basis of which the whole domain of numbers lays, and he attempts to build a proof in *Was sind*, Sect. 66.[106] Such a proof might have been felt necessary since *actual* infinite systems constitute a mathematical object different in nature from the given sequence of natural numbers. Dedekind addresses, regarding actual infinities, the existential/ontological question, which he generally leaves untouched, but the proof fails.

Frege's alternative solution, namely "to transform the generality of an equality into an equality" between logical objects,[107] comes up against the existence of extensions of concepts.[108]

---

[105]Cf. footnote (85), above.

[106]Something like the theorem proved here is lacking from the first draft ([67], Appendix LVI).

[107]Cf. [97], Sect. II.147 and II.157, respectively, [110], p. 149_2 and 155_2:

> If there are logical objects at all—and the objects of arithmetic are such—then there must also be a means to grasp them, to recognise them. The basic law of logicd which permits the transformation of the generality of an equality into an equality serves for this purpose. Without such a means, a scientific foundation of arithmetic would be impossible. For us it serves the purposes that other mathematicians intend to achieve by the creation of new numbers. […] In any case, our creation, if one wishes so to call it, is not unconstrained and arbitrary, but rather the way of proceeding, and its permissibility, is settled once and for all. And with this, all the difficulties and concerns that otherwise put into question the logical possibility of creation vanish; and by means of our value-ranges we may hope to achieve everything that these other approaches fall short of.

> We have been reminded of our transformation of the generality of an equality into an equality of value-ranges that promises to accomplish what the creative definitions of other mathematicians are not capable of.

What Frege is evoking here is Basic Law five, which in a modern notation can be rephrased as follows:

$$[ValueRange\,(f) = ValueRange\,(g)] \Leftrightarrow \forall x\,[f\,(x) = g\,(x)].$$

Since this makes the generality of an equality (or identity), '$\forall x\,[f\,(x) = g\,(x)]$', equivalent to an equality (or identity) of value-ranges, '$[ValueRange\,(f) = ValueRange\,(g)]$'. The generality is, of course, expressed by the universal quantifier.

[108]It has been remarked that, in *Grundlagen*, Frege makes no use of extensions once HP is derived (in Sect. 73). By contrast, extensions (or more generally value-ranges) are used throughout *Grundgesetze*. However [119] shows that they are eliminable except in the proof of HP.

The failure both of Dedekind's proof and of Frege's Basic Law V led Russell and Zermelo to admit an axiom of infinity along with the arithmetical axioms, which finally comes down to accept "creative definitions" (whatever ontological status may be so ascribed to the introduced entities).

1.4.6 Frege's Technical Conception of Definitions

The second answer to the question advanced at the beginning of Sect. 1.4.2 is rather more technical than ontological. It hinges on the constriction of a formal system whose primitive signs stand for logical objects and functions, and whose primitive laws are assumed to be purely logical. But even in this technical sense, for which "definition is really only concerned with signs" ([106], vol. I, "Logik in der Mathe-matik", pp. 224; [107], p. 208), a definition fixes *once and for all* the sense of a sign, since the logical system to be construed is *unique*.[109]

In "Logik in der Mathematik" ([106], vol. I, pp. 227–229; [107], pp. 210–211), Frege distinguishes two different cases.

The first concerns definitions proper or "definitions *tout court*". These are "con-structive [*aufbauende*] definitions", since we "construct a sense out of its constituents and introduce a sign to express this sense". A definition *tout court* is, then, "an arbi-trary stipulation which confers a sense on a simple sign [*the definiendum*] which previously had none", a sense which has "to be expressed by a complex sign [*the definiens*] whose sense results from the way in which it is put together". Despite its being an arbitrary stipulation, once it is made, a definition in this sense must remain the same everywhere in *the* system, since this is unique. Moreover, we can dispense with the newly introduced, abbreviating, sign, and keep the *definiens*. Thus, from a logical point of view, argues Frege, definition is quite inessential. If so, the ques-tion arises immediately: why did Frege invest so much care to define, explicitly and contextually, the concept of a cardinal number?[110]

The second case concerns what Frege calls 'analysing definitions [*zerlegende Def-initionen*]'.[111] These follow the reverse procedure; they consist of a logical analysis of the sense of a long-established sign (or concept-word), which provides a complex expression that, provided that the analysis is correct, has the same sense as such a long-established sign. But how can one recognise that the analysis is correct? Indeed, the sameness of sense is open to question, and, Frege says, it can be grasped only when it is self-evident and can be "recognised by an immediate insight", to the effect

---

[109]This is the essential reason why Frege does not have the notion of logical consequence, let us say from a set S of logical formulas to a logical formula A. He does not consider a formula under a range of interpretations.

[110]Cf., e.g., [74], Chap. 2, "Frege and the paradox of analysis", pp. 17–52. Note in passing that Frege does not use the adjective 'explicit' to qualify explicit definitions (since, for him, any suitable definition is explicit).

[111]Cf. footnote (112), below.

that "what we should here like to call a definition is really to be regarded as an axiom",[112] and "it is really only relative to a particular system that one can speak of something as an axiom".[113]

We are far from the view of *Grundlagen* according to which a definition, not an intuition, must capture the very essence of a thing ("what [...] numbers themselves [are]"). The intervening intuition and the relativity of axioms are two reasons for Frege's rejecting analysing definitions as being not definitions proper. However, if it is right that axioms generally result from an analysis of the received sense of some mathematical signs, the new senses yielded by the stipulation of the axioms obtained by analysis cannot be the same as the previous ones; they have to be *new* ones. Why does Frege want that sense be preserved from a long-established sign up to a new axiom? Because Frege just cannot accept that senses—that is to say, thoughts or thought-constituents—may evolve. What is evolving, according to him, is only our knowledge of them, and this happens through elucidation [*Erläuterung*], which make clearer a sense that existed before but was grasped only in an unclear or partial way. Frege proposes regarding logical analysis "only as a preparatory work which does not itself make any appearance" ([106], vol. I, "Logik in der Mathematik", p. 228; [107], pp. 211)[114] in the system to be constructed from the ground up on the basis of a proper definition, namely a constructive definition.

This conception is partly close to Dedekind's brief genealogical description in the letter to Keferstein mentioned above. Dedekind splits the mathematical work into analysis and synthesis, endorsing the sense given to these two terms by the Ancients. A long-standing analysis of the pre-theoretic sequence of natural numbers allowed the axioms for the synthetic presentation offered in *Was sind* to be found. Contrary to Frege, Dedekind makes no radical difference between axioms and definitions:

---

[112] That's the paradox of (logical) analysis, which results from an immediate insight and yields an axiom instead of giving an identity each member of which is a logical object. Regarding worries caused by the expression '*zerlegende Definition*', and its relations with "analytische Wahrheiten" and "analytische Grundsätzen", which Frege deals with in *Grundlagen* ([93], Sects. 3–4)—where we do not find 'analytische Definition', but rather 'Auflösung der Begriffe', for 'conceptual analysis'—, cf. [73], Chap. 2. Dummett renders 'zerlegende Definition' with 'analytic definition', in according with the translation of [107]. But, as rightly observed by Beaney ([6], p. 316, footnote 10), Frege's *zerlegende Definitionen* are not *analytisch* in the Kantian sense. According to Beaney (*ibid.*), "where a definition is 'analytic', then it must be understood as either a 'constructive definition' or an 'axiom'" (I suppose that he takes constructive definitions to include the definition of individual cardinal numbers in *Grundlagen* and *Grundgesetze*). But if "analytic" definitions may be "axioms", the task remains to explain why Frege continues, as late as in 1914, to reject axioms as (implicit) definitions. After all, following Frege's terminology, there are not only "logical concepts" and "logical objects", but also "basic laws", which might be taken as logical axioms.

[113] This last sentence occurs some pages earlier: [106], vol. I, "Logik in der Mathematics", p. 206), [107], p. 206. Still, the truth of a statement that might count as an axiom is *not* relative. Compare with Dedekind's view according to which "Drehen und Wenden der Definitionen, den aufgefundenen Gesetzen oder Wahrheiten zuliebe, in denen sie eine Rolle spielen, bildet die grösste Kunst des Systematikers" ([55], vol. III, p. 430). Yet in mathematics this turning and shifting leaves no room for arbitrariness.

[114] A similar point is made few line below: "The effect of [...] logical analysis [...] will be precisely this—to articulate the sense clearly".

as observed above, his four axioms for the natural numbers (namely the conditions α, β, γ, and δ stated in Sect. 1.4.1, above) work as definitions ([49], Sects. 71 and 73); similarly the continuity axiom is stated as the fourth basic law for defining the real numbers ([47], Sect. 5). What Dedekind calls 'axiom' is, for him, a defining condition ([126], p. 537), while Frege wants it to be a basic logical truth. Dedekind does not encounter Frege's problem concerning the coincidence of the result of analysis with our pre-analytic conception; he readily admits that a reader of *Was sind* "will scarcely recognise in the shadowy forms which [...][he] bring[s] before him his numbers which all his life long have accompanied him as faithful and familiar friends" ([49], *Vorwort*, p. IX; [53], p. 15). The shadowy forms, not the familiar numbers, are a free creation of the human mind. Practice will provide them with familiarity and some kind of substance.

1.4.7 Frege's and Dedekind's Philosophical Assumptions

For both, Dedekind and Frege, mathematical or rational thought are objective in the sense given above (Sects. 1.3.2.1–1.3.2.4). But for Dedekind, mathematical thinking is a creative and evolving activity, whereas for Frege, paradoxically, 'thought' has nothing to do with 'thinking', since it does not have to be thought at all.

For the latter, a thought is the sense [*Sinn*] of a statement [*Satz*]; a statement expresses a thought, which is permanently either true or false (*tertium non datur*). So the *Bedeutung* of a statement is its truth-value, in a way parallel to that which assigns to a name its bearer as its *Bedeutung*. According to Dummett ([69], p. 87)[115] "to know the sense is to know the condition for the expression to have a given reference", in the same way as knowing the sense of a name is knowing a mode of presentation of its referent. "I begin"—Frege writes—"by giving pride of place to the content of the word 'true', and then immediately go on to introduce a thought as that to which the question 'Is it true?' is in principle applicable" ([106], vol. I, "Aufzeichnungen für Ludwig Darmstaedter", p. 273; [107], pp. 253). The *Begriffsschrift* was invented in order to make easier the control of the validity of proofs and went together with the presentation of logic as a theory of *inference*. From the 1890s onwards logic will appear as a theory of truth. Truth becomes the central affair of logic, its very aim [116]: the laws of logic are the laws of the True and the False, and what True is,

---

[115]Cf. [153], for a discussion of this matter.

[116]Its aim or goal, not its essence, which is, rather, "the assertoric force with which a sentence is uttered" ([106], vol. I, "Meine grundlegenden logischen Einsichten", p. 272; [107], pp. 252). The following quote, from the beginning of "Der Gedanke", is even clearer ([102], pp. 58–59; [109], p. 351–352): "Just as 'beautiful' points the ways for aesthetics and 'good' for ethics, so do words like 'true' for logic. All sciences have truth as their goal; but logic is also concerned with it in a quite different way: it has much the same relation to truth as physics has to weight or heat. To discover truths is the task of all sciences; it falls to logic to discern the laws of truth. [...] I assign to logic the task of discovering the laws of truth, not the laws of taking things to be true or of thinking".

is indefinable.[117] Moreover, according to Frege, thoughts constitute a "third realm" of changeless entities, and "the work of science does not consist in creation, but in the discovery of true thoughts"; more in general, "in thinking we do not produce thoughts, we grasp them" ([102], pp. 69 and 74; [109], pp. 363 and 368). Again, to grasp a thought is the same as knowing the conditions for it to be true. Hence, we need to separate the content from the act of thinking, and provide the content, namely the thought, with a criterion of identity independent from the subject's mental life. The most Frege can concede is that thoughts have a kind of actuality "quite different from the actuality of things" and that "their action is brought about by a performance of the thinker", and yet the thinker does not create thoughts, nor can he react on them, he just "must take them as they are" ([102], pp. 77; [109], pp. 371–372).[118]

There is absolutely no ambiguity: Frege's universalistic conception of logic and truth is backed up with a ontological realism, which, unsurprisingly, goes so far as to finally admit a logical intuition intervening in grasping logical objects, recognising their logical identity, and making a judgement about their being true or false. It would be wrong to conceive of grasping, recognising, judging as our acting on thoughts. It is rather the case that "it may be possible to speak of thoughts as acting on us" ([106], vol. I, "Logik", p. 150; [107], p. 138).[119] Dummett notes that Frege's realism, the "myth of the third realm" [71],[120] is certainly not "a logical precondition" of his major achievements in logic ([69], p. 80); yet it is a philosophical assumption, which Frege maintains and even reinforces until the last years of his life: his permanent concern is to isolate the logical from any psychological process and to separate the sense (thought) from its linguistic expression. Carnap ([41], p. 102; [10], p. 50) wrongly exempts Frege from holding the "absolutist conception" and the "theological mathematics" that he attributes to Ramsey, probably just for providing an ancestor to his own empirical logicism.

What may be said, instead, concerning Dedekind's philosophical assumptions? Dedekind is definitely *not* a realist: he promotes actual infinities but does not think them to exist independently of our thinking. In accordance with Kantian optimistic rationalism, mathematical concepts are created *and* objective, they are abstract but they are not genuine self-subsistent objects—which is the distinguishing mark of

---

[117]Cf. [106], vol. I, "17 Kernsätze zur Logik", p. 189, sentence 7, [107], p. 174: "What true is, I hold to be indefinable". Cf. also [106], vol. I, "Logik", pp. 139–140, [107], pp. 128–129: a remarkable foreinsight of Tarski's undefinability theorem.

[118]Cf. also [106], vol. I, "Logik", p. 149; [107], p. 137: "The metaphors that underlie the expression we use when we speak of grasping a thought, of conceiving, laying hold of, seizing, understanding, of *capere, percipere, comprehendere, intelligere*, put the matter in essentially the right perspective. What is grasped, taken hold of, is already there and all we do is take possession of it".

[119]Compare with Gödel's more affirmative opinion about the axioms of set-theory, which "force themselves upon us as being true." ([113], p. 268).

[120]That our current understanding of mathematical realism, which originates from Bolzano's "*Sätze an sich* [16] and Frege's "third realm", does not fit with Plato's account of the being of mathematical objects is soundly argued by Tait [189, 190] and McLarty [147], whose conclusion is that "Plato was not a mathematical Platonist" (*ibid.*, p. 120). Hence my discriminant use of 'realism' and 'Platonism'.

Frege's *logical objects*. Now, can Dedekind be taken as a non-realist logicist? In other words, how can one think of Dedekind's structuralism as being a form of logicism?

1.5 *System* and *Abbildung*: Structuralism and/or Logicism

To answer these questions, I tackle now the outstanding question: are the fundamental concepts involved in Dedekind's reconstruction of arithmetic, viz. the concepts of *System* and *Abbildung*, really concepts of logic?

1.5.1 Concept

When he does not use it as a synonym for the vague term 'notion', Dedekind understands 'concept [*Begriff*]' as such to refer to a domain (a *System* in his terminology) together with appropriate operations on it, that is to say a structure, in our modern language. Dedekind, as most of his contemporaries or later followers like Hilbert, Emmy Nœther, B.L. van der Warden, Emil Artin, etc., does not use the term 'structure', that has been most popularised by Bourbaki. And yet Dedekind is the mathematician with whom structuralism originates, even if he provides nothing as a theory of abstract structures.

Already in 1854, Dedekind uses 'System' and 'systematising' for 'structure' and 'structuring' respectively.[121] Indeed, 'systematising' indicates the action of isolating primitive assumptions from their logical consequences. Later, in *Was sind*, 'System' is used with the same meaning as 'domain of uninterpreted elements'. What is so called affords, then, the basis for defining general operations whose instantiation results in the definition of the finite ordinal numbers and of the operations on them: +, ×, etc. A *System* results from considering "things [...] from a common point of view", and it is extensionally conceived; in our current terminology, it is a set. Here is Dedekind's definition ([49], Sect. 2; [53], p. 21):

> It very frequently happens that different things, $a, b, c, \ldots$ for some reason can be considered from a common point of view, can be associated in the mind, and we say that they form a *System S*; we call the things $a, b, c, \ldots$ 'elements of the *System S*', they are *contained* in S; conversely, S *consists* of these elements. Such a system S (an aggregate, a manifold, a totality) as an object of our thought is likewise a thing; it is completely determined when with respect to every thing it is determined whether it is an element of S or not.

Still, Dedekind's reference to Euclid's *Elements*, as well as to Galois and Riemann, among others "*Systematikers*" ([55], vol. III, p. 430), his strong interest in the

---

[121]Cf. [55], vol. III, p. 428: "Die weitere Entwicklung einer jeden Wissenschat immer wieder auf das System, durch welches man ihren Organismus zu erfassen sucht, neubildend zurück wirkt, ist nicht allein eine historische Tatsache, sondern beruht auch auf einer innern Notwendigkeit".

deductive character of a theory, and his use of the method of analysis and synthesis leave no doubt about his promoting a structuralist mathematical practice ([156], pp. 306–311; [176]; [183]; [148]). Logic is a necessary tool for this promotion.

But mathematical substance is also indispensable. Dedekind's historical concern is a precondition of the search for firm grounds, and both attitudes, grounding *and transforming* the mathematical substance, are tightly bound with a close eye on mathematical practice and its history.[122] As learnt from looking at the history, Dedekind writes, "the greatest and most fruitful advances in mathematics and other sciences have invariably been made by the creation and introduction of new concepts" ([49], *Vorwort*, p. XI; [53], p. 16). New concepts are conceived of as new modes of determination [*Art der Bestimmung*], or modes of presentation [*Darstellung*] that meet a higher standard of logical rigour and respect the hierarchy which places arithmetic at the head of the whole mathematical body.

One can say that what Dedekind calls 'concept' is close to one aspect of what Frege calls 'sense', namely to the aspect that result from taking sense to be a way in which a *Bedeutung* is given, as opposed to the other aspect of Fregean senses consisting in their being themselves logical objects offered to our possible grasping.[123] But on the one hand, Dedekind's concern is far from elaborating the ontological and logical status of concepts. He rather endorses the Kantian conception of a concept as being the human power to organise and unify things, a thing being "every object of our thought" ([49], Sect. 1; [53], p. 21), i.e. a thought-object. A concept has no existence independent from our mind; it is not there before the mind *creates* it.[124] It is a *tool* used for grounding and generalising mathematical methods and for opening new perspectives for the mathematical *activity* as well. On the other hand, Dedekind does not elaborate upon the distinction he makes sometimes between object and concept[125] and, in particular, he naturally does not see a concept as a step in the identification of an individual object, let alone as a step in the determination of the truth-value of thoughts related to the identifiable object. As above-said, Dedekind does not tackle the question of mathematical, less alone of logical, truth.

As it is well known, Frege makes, instead, a very specific use of 'concept'. First understood, in *Begriffsschrift*, in opposition to objects, as functions of one argument resulting from the decomposition of judgeable contents, concepts appear in *Grundlagen* to be that which cardinal numbers belong to, and are defined in *Grundgesetze* as functions having truth-values (taken as being two particular objects, in turn) as

---

[122]Cf. [55], vol. III, p. 428 (my italics): "Diese Vorlesung hat nicht etwa [...] die Einführung einer bestimmten Klasse neuer Funktionen in die Mathematik, sondern vielmehr *die Art und Weise* [my italics] zum Gegenstande, wie in der *fortschreitenden Entwicklung* [my italics] dieser Wissenschaft neue Funktionen, oder, wie man ebensowohl sagen kann, neue *Operationen* [Dedekind's italics] zu der Kette der bisherigen hinzugefügt werden".

[123]For a criticism of Frege's twofold conception of sense, cf. [72], pp. 276–281.

[124]This makes Dedekind's concepts close to middle ages universals (contrary to Boolos' suggestion, in [29], p. 149, it's not so easy to give the same ontological status to Plato's Forms and to universals).

[125]Cf. the letters to Lipschitz of July 27, 1876 and to Weber of January 24, 1888 ([55], vol. III, pp. 474–479 and 488–490).

their values. The use of concepts in forming judgements (or statements) manifests, then, the step from the level of sense to the level of *Bedeutung*,[126] then from concepts themselves to their extensions. Frege regards the passage from a concept to its extension as the only way of establishing the existence of an object on logical grounds.[127] While objects are regarded as belonging to different kinds, like thoughts, truth-values, and value ranges (among which there are numbers), concepts are uniformly understood as one-argument functions (of different levels),[128] and are, then, taken to be essentially "unsaturated" (just as are unsaturated mathematical functions), and then prior to their extensions. Hence the radical distinction between concepts and objects, understood as completely different sorts of entities.[129] A concept is nevertheless something objective ([93], Sect. 47), better concepts and objects are on the same level of objectuality ([106], letter to Husserl of May, 25th, 1891, p. 97).

As it is well known, such a radical distinction between concepts and objects involves difficulties from the logical point of view ([171], Appendix A), and also from the point of view of mathematical practice. Not surprisingly, Dedekind does not share Frege's requirement of distinguishing *radically* and once and for all concepts from objects. This does not mean that Dedekind does not make any distinction between 'object' and 'concept'. He uses the first term to refer to individual objects, and the second to refer to whole domains equipped with some operations and laws governing them. For instance, Dedekind writes to Lipschitz ([55], vol. III, letter to Lipschitz of July 27th, 1876, p. 475) that he intended not to invent a "new object for mathematical research", or some previously unknown irrational numbers, but rather to define at once the *complete* domain of irrational numbers and the concept of irrational number, without considering the individual numbers that fall under this concept, which is the same as defining the algebraic ordered structure of the domain, by listing a small number of properties that it is required to satisfy. From a *logical* point of view, we are in a second-order language, and, from *Frege's logical point of view*, we are dealing only with objects, namely with the real numbers, on the one side, and with their domain (which is the extension of a certain concept), on the other. A concept *qua* structure, as understood by Dedekind, may be dealt (by mathematicians) as an object resulting from a process that Husserl will later call 'thematizing activity [*Thematik*]' ([130], Sects. 8–11; [131], pp. 33–47). I don't wish to enter here into discussing what structures and objects are. I just want to seize the irreducible difference between

---

[126]Cf. [96], p. 35, [104], p. 65: "Judgements can be regarded as advances from a thought to a truth-value", Cf. also [106], vol. I, "Ausführungen über Sinn und Bedeutung", p. 133, [107], p. 122: "The laws of logic are first and foremost laws in the realm of *Bedeutungen* and only relate indirectly to sense".

[127]Cf. [97], *Nachwort*, p. 253, [110], p. 253$_2$: "Even now, I do not see how arithmetic can be founded scientifically, how the numbers can be apprehended as logical objects and brought under consideration, if it is not—at least conditionally—permissible to pass from a concept to its extension".

[128]A concept under which objects fall is a concept of first level, a concept under which concepts of first level fall is a concept of second level, etc.

[129]Cf. the third fundamental principle of *Grundlagen*: "Never to lose sight of the distinction between concept and object" ([93], *Einleitung*, p. X; [103], pp. XXII).

Dedekind's and Frege's use of the words 'concept' and 'object'. Frege's numbers are individual logical objects; Dedekind's numbers instantiate an uninterpreted (abstract) structure, which is itself "the first object of the *science of numbers or arithmetic*".[130]

Frege's sophisticated notion of concept is connected with that of truth-value, which is the decisive step for a semantic theory, and which is absent from Dedekind's works (as is a syntactic theory also absent).[131] In *Grundlagen* Sect. 74, Frege writes [93], Sect. 74; [103], pp. 87–88):

> On my use of the word 'concept', 'a falls under the concept *F*' is the general form of a judgeable content which deals an object *a* and permits of the insertion for *a* of anything whatever.

According to his mature theory, a concept is a function whose values are not judgeable contents, but truth-values. In Frege's schema given in the letter to Husserl of May, 25th, 1891 ([106], p. 97), one sees the permanent correlation between sense and *Bedeutung* and the analogical role of statements, proper names and concept-words: the sense of a statement is a thought and its *Bedeutung* is a truth-value; for a proper name, the sense is correlated with an object, which is the *Bedeutung* of such a name; the *Bedeutung* of a concept-word is the concept itself[132] as distinguished from its extension (constituted by the objects—possibly none—falling under it; Frege is clearly considering here only first-level concepts). Objects, truth-values, concepts are all the reference of expressions of different logical types; truth-values and concepts are abstract objects, i.e. logical objects. Dummett sees the equation of *Bedeutung* with semantic value as the first stone for constructing a compositional semantic theory: he assigns a reference to the constituent parts of a statement so that the statement is true or false in accordance with the semantic value of its components.[133]

---

[130]Cf. [49], Sect. 73, [53], pp. 33–34) (Dedekind's italics): "The relations or laws which are derived entirely from the conditions α, β, γ, δ in (71) [cf. Sect. 1.4.1 above] and therefore are always the same in all ordered simply infinite systems, whatever names may happen to be given to the individual elements […], form the first object of the *science of numbers or arithmetic*".

[131]I do not want to say that a semantic point of view is absent in Dedekind's work. What I take to be absent is a *semantic theory*, that is, a theory of truth. Likewise, I do not want to say that Dedekind has no syntactic views. What I want to say is that he has no *syntactic theory*, that is, no theory of inference.

[132]Cf. [74], Chap. 10, p. 235: "We can make no sense, for example, of the thesis that the content of a statement of number consists in predicating something of a concept unless we view the concept as being the *reference* of the concept-word".

[133]Dummett ([74], Chap. 9, p. 215) distinguishes between thesis (T) that truth-values are the references [*Bedeutungen*] of statements and thesis (O) that truth-values are objects. According to him, (O) is "objectionable", but (T) is not.

1.5.2 System *and* Abbildung: *the Search of Generality*

Now, not retrospectively within our current set-theoretic frame, but from Dedekind's own point of view, are the concepts of *System* and and that of *Abbildung* concepts of logic?

A first answer is easy: for Dedekind, these concepts result from fundamental *operations* of the understanding, which are more *general* than the numerical operations proper. This type of generality explains the applicability of these operations not only to arithmetic, but also to other mathematical branches and elsewhere. Once we have brought into light the structure of totally ordered simply infinite (countable) *Systeme*, we can transfer this structure, for example, to the domain of algebraic numbers and algebraic functions, as it is actually done by Dedekind and Weber [57]. There is no doubt that Dedekind views the ascent from arithmetic proper to general "arithmetising", that is, from natural numbers as such to arithmetical structures, as a logical ascent—and this is probably one reason why Tait takes him as a logicist. But Dedekind conceives of logic as the structure of the operative mind and not as something that the mind recognises as being independent of itself. Frege comes close to that only once, when he appeals to "the logical disposition of man" ([106], vol. I, "Erkenntnisquellen der Mathematik und der mathematischen Naturwissenschaften", dated to 1924/1925, p. 288; [107], p. 269).

1.5.2.1 Dedekind's Conception of the Operative Mind and Cantor's Paradox

Indeed, for Dedekind, a *System* results from "the creative power [*Schöpferkraft*] of the mind to create out of determinate elements a new determinate element which is their *System*" ([54], *Vorwort zur dritten Auflage*, p. XIII)[134]; this power is crucial for Dedekind who considers natural numbers as forming an autonomous *System*, and who introduces actual infinities. *Was sind* is grounded upon the "aggregative thought" so harshly criticised by Frege (cf. Sect. 1.3.2.3, above). A *System* is also named 'aggregate', 'manifold' or 'totality' by him (cf. Dedekind's definition, quoted in Sect. 1.5.1, above), and and Dedekind deals with "object[s] of our *thinking* " ([49], Sect. 1; [53], p. 21; my italics),[135] rather than with objects of thought, because he holds essential *passing* from things to a *System* of them, a procedure informally used by Dedekind a long time before its explicit setting.

---

[134]Taking this time 'concept' in its common meaning, Frege writes similarly ([93], Sect. 48; [103], p. 61): "The concept has a power of collecting together far superior to the unifying power of synthetic apperception".

[135]Beman's English translation has 'thought' instead of 'thinking', but I take 'thinking' to be more appropriate here. Here is the German text: "Im folgenden verstehe ich unter einem Ding jeden Gegenstand unseres Denkens". But notice that *Denkens* is not something subjective, for Dedekind, as it is for Frege. Then, in Sect. 2, Dedekind writes: "Ein […] System *S* (oder ein Inbegriff, eine Mannigfaltigkeit, eine Gesamtheit) ist als Gegenstand unseres Denkens ebenfalls ein Ding".

Frege rejects this procedure as being outside logic, although according to Dedekind it consists in forming a concept, the latter being considered as an object of our thinking , a thought-object (cf., again, Dedekind's definition, quoted in Sect. 1.5.1, above). Dedekind is convinced that the fundamental operation ⌜set of⌝ must be preserved anyway, and that a logical solution will certainly be found for the logical flaw emerging from its use ([54], *Vorwort zur dritten Auflage*), because sets are linked with the most fundamental operation, the *Abbilden*-ability.[136] Dedekind did not try himself to overcome the difficulty; he was not even willing to face up to it.

As it is well known, Cantor communicated, several times, what we call 'Cantor's paradox' to Dedekind. Felix Bernstein reported that, in winter 1896/97, Cantor wrote to Dedekind about the set of all things and asked him to take a position on this default of his construction ([55], vol. III, p. 449). Later, in his famous letter to Dedekind of August 3rd, 1899, Cantor came back to his distinction between consistent and non-consistent multiplicities, and, in a successive letter of August 28th of the same year, he asked Dedekind for a discussion in a face-to-face meeting.[137] But Dedekind resisted to the idea that there might be infinities that cannot be *actual* or cannot be brought to constitute a *consistent System*. Here is what he replies to Cantor ([67], p. 261, Dedekind's letter to Cantor of August 29th, 1899; the insertion in double brackets is from Dugac):

> [...] zur Diskussion Ihrer Mittheilung bin ich noch lang nicht reif [...] obgleich ich ihren Brief vom 3. August mehrere Male durchgelesen habe, mir über ihre Eintheilung der Inbegriffe in konsistente und inkonsistente [Vielheiten] hoch nicht klar geworden bin; ich weiss nicht, was Sie mit dem "Zusammensein aller Elemente einer Vielheit" und mit dem Gegentheil davon meinen.

Finally, on September 4th 1899, after having meet him, Dedekind confessed that Cantor did give the *"coup de grace "* to his error ([142], p. 54). Nonetheless, he did not try at all to search for the means to neutralise the paradox.

In the brief preface to the third edition of *Was sind* [54], at a time when Frege's *Grundgesetze*, Russell's *Principles of mathematics* [171], Hilbert's first paper on "Foundations of logic and arithmetic" [128], Zermelo's first "Investigations on the foundations of set theory" [211], and the first volume of Whitehead Russell's *Principia Mathematica* [206] were already published, Dedekind wrote that he did not doubt of the intrinsic value of his mathematical foundation, leaving to others the task

---

[136] Gödel thinks too that set-theoretical paradoxes are "a very serious problem, not for mathematics, however, but for logic and epistemology" ([113], p. 268, footnote 40). He also already points out the analogy between the "naïve" use of the concept of set, understood as the generating of unities out of manifolds, and Kant's categories of pure understanding.

[137] The two letters have been published together as a single letter in [39], p. 443–447. Here what Cantor writes in the first of them (*ibid.*, p. 443): "Eine Vielheit kann nämlich so beschaffen sein, dass die Annahme eines 'Zusammenseins' *aller* ihrer Elemente auf einen Widerspruch führt, so das es unmöglich ist, die Vielheit al eine Einheit, als ein 'fertiges Ding' aufzufassen. Solche Vielheiten nenne ich *absolut unendliche* oder *inkonsistente Vielheiten*. Wie man sich leicht überzeugt, ist, z.B. der 'Inbegriff alles Denkbaren' eine solche Vielheit". Cf. [155] for detailed comments. The request for a face-to-face discussion is included in the last part of the letter, which is omitted in the mentioned edition: cf. [67], p. 260.

to amend the logical flaw. Surprisingly, he did not even mention Zermelo's amendment through the *"Aussonderungsaxiom"* and the assumption of an infinity axiom. He probably considered not *his* job to take on logical aspects of his set-theoretical axiomatisation of integers, while the mathematical part of the construction has been well done.[138] Dedekind views *Systeme* as "logical" operations of the mind, but not so far as to tackle the logical difficulty involved in their unrestrictive use. Indeed a mathematical theory carried out in a logically inconsistent system need not to be ruined by the inconsistency.

1.5.2.2 Dedekindian Abstraction

Tait explains the procedure he calls 'Dedekind's abstraction' as follows ([189], p. 369, footnote 12):

> For example, in set theory we construct the system $\langle \omega, \phi, \sigma \rangle$ of finite von Neuman ordinals, where $\sigma x = x \cup \{x\}$. We may now abstract from the particular nature of these ordinals to obtain the system $\mathcal{N}$ of natural numbers. In other words, we introduce $\mathcal{N}$ together with an isomorphism between the two systems. In the same way we can introduce the continuum, for example, by Dedekind abstraction from the system of Dedekind cuts.

One may say that, in Tait's analysis, 'abstraction' has, for Dedekind, just the same meaning as 'idealisation' in Husserl's terminology. Indeed, Tait sets in contrast Platonistic idealisation with Aristotelean abstraction from sensible things, and takes modern mathematics to be "inalterably Platonistic" in a sense faithful to Plato's writings, whereas, he says, the use of 'Platonism' to refer to the view that mathematical objects "really" exist does not fit Plato's theories ([200], p. 304).[139]

Dedekind's question is not what *are* numbers themselves, but "what *is done* in counting" ([49], *Vorwort*, p. VIII; [53], p. 14; my italics).[140] From analysing the *actual* process of counting within the particular model provided by natural numbers, we are lead to consider ordinals, i.e. counting-numbers. According to Dedekind, this

---

[138]Parsons ([155], p. 526) uses modality in order to save the idea that *any* multiplicity of objects constitutes a set: "The idea that any available objects can be formed into a set is, I believe, correct, provided that it is expressed abstractly enough, so that 'availability' has neither the force of existence at a particular *time* nor of giveness to the human mind, and formation is not thought of as an action or Husserlian *Akt*. What we need to do is to replace the language of time and activity by the more bloodless language of potentiality and actuality". By the way, Dedekind's operation or activity of mind follows a Kantian line.

[139]I have mentioned above, in footnote 120, McLarty's elaboration on the distinction between Plato's original theories and our modern use of 'Platonism' and cognates, together with my discriminant use of 'realism' and 'Platonism'.

[140]Cf. also [47], Sect. 1, [53], p. 2, (my italics): "I regard the whole of arithmetic as a necessary, or at least natural, consequence of the simplest arithmetic act, that of counting, and counting itself as nothing else than the successive creation of the infinite series of positive integers in which each individual is defined by the one immediately preceding; the simplest act is passing from an already-formed individual to the consecutive one to be formed. [...] Addition is the combination of any arbitrary repetition of the above-mentioned simplest act into a single act; from it in a similar way arises multiplication".

is just what has provided the starting point of his own definition of natural numbers. Here is how he states the basic questions this definition depends on, in his letter of February 27th 1890 to Keferstein ([184], p. 272; [125], pp. 99–100):

> What are the mutually independent fundamental properties of the sequence N, that is, those properties that are not derivable from one another but from which all others follow? And how should we divest these properties of their specifically arithmetic character so that they are subsumed under more general notions and under activities of the understanding *without* which no thinking is possible at all but *with* which a foundation is provided for the reliability and completeness of proofs and for the construction of consistent notions and definitions?

Thus, for him, the "logical process of building up the science of numbers" ([49], *Vorwort*, p. VIII; [53], p. 14) on the basis of the counting-practice does not depend on what these numbers are, but it rather depends on finding out the primitive mutually independent and consistent properties of their sequence, and from divesting them of their "specifically arithmetic character". This "divesting" is quite close to Plato's and Husserl's idealisation; it is not a psychological process. One should also quote in its entirety Sect. 73 of *Was sind*, already partially quoted in footnote (130), above, in order to make this clear ([49], *Vorwort*, Sect. 73; [53], pp. 33–34):

> If in the consideration of a simple infinite system N set in order by an *Abbildung* φ we entirely neglect the special character of the elements, simply retaining their distinguishability and taking into account only the relations to one another in which they are placed by the order-setting *Abbildung* φ, then are these elements called 'natural numbers' or 'ordinal numbers' or simply 'numbers', and the base-element 1 is called 'the base-number' of the *number-series* N. With reference to this freeing the elements from any other content (abstraction) we are justified in calling numbers a free creation of the human mind. The relations or laws which are derived entirely from the conditions $\alpha, \beta, \gamma, \delta$ in (71) [cf. Sect. 1.4.1 above] and therefore are always the same in all ordered simple infinite systems, whatever names may happen to be given to the individual elements (compare 134), form the first object of the *science of numbers or arithmetic*.

In order to prevent a psychologistic misunderstanding, like that of Dummett, for example (cf. p. 33, above), Tait [191] presents Dedekind's abstraction as a typical logical abstractive procedure, contrasting its logical nature with Aristotelean abstraction's going from empirical data to mathematical objects. Tait sees Dedekind as a logicist rather than a structuralist, arguing that what is essential is that propositions about the abstract objects translate into propositions about the things from which they are abstracted, so that the truth of the former depends on the truth of the latter.[141] However, what is at stake is really the abstract structure itself, i.e. the total ordering imposed by an injective *Abbildung* φ, which makes a simple infinite *System S* the chain of a distinguished singleton $\{e\} \subset S$ (i.e. the least set containing $\{e\}$ and closed under φ), rather than the abstract (i.e. logical, in Frege's view) character of the natural numbers themselves. Dedekind's *abstract* numbers are uninterpreted elements, not logical objects. "The *science* of numbers" depends only on the theory of simple infinite sets, with axioms $\alpha, \beta, \gamma, \delta$, and not on the choice of any particular such *System*.

---

[141]Parsons [156] discusses Tait's view and offers a deep analysis of the notion of mathematical structure.

Dedekind explains that this is why his ordinal numbers, "the abstract elements of *the* simple infinite *System*" are "new individuals to be created [*neu zu schaffenden Individuen*]" ([55], vol. III, letter to Weber of January 24th, 1888, pp. 489–490; my italics). In a similar way, he also takes an irrational number to be something new, created and represented by, but not identical to, the corresponding cut.[142]

Frege could not have accepted, in contrast, that one take as primitive so unspecified properties of the numbers, because the statements of these properties do not have a determinate sense as long as we leave uninterpreted the shadowy elements; they are not complete statements, hence they cannot express thoughts which may be judged once and for all true or false. In Frege's view, Dedekind's way to obtain generality is not logical, since logic is the theory of truth (along with the theory of inference).

1.5.2.3 *Abbildung* and one-to-one Correspondence: Dedekind Versus Frege

Digging for the foundations of arithmetic, both Dedekind and Frege come to recognise, but in very different ways, the essential role of the one-to-one correspondence. The differences are those between a structural practice, which seeks no entity beyond the intrastructural relations, and the logicist view, which looks for the truth-conditions of identity statements among number-names and replaces postulation of objects with explicit definition in terms of extensions of concepts.

In *Was sind*, the primitive relation on *Systeme* is inclusion, which is expressed in terms of an *ähnliche Abbildung* from a *System S* into *S* itself. An *Abbildung* results from the "ability [*Fähigkeit*] of the mind to relate things to things, to let a thing correspond to a thing, or to represent [*abbilden*] a thing by a thing" ([49], *Vorwort*, p. VIII; [53], p. 14). *Abbildung* is representation or correspondence in general; it may be "*ähnlich*", that is one-to-one (or injective), and more particularly, one-to-one from a *System S* onto its image (or bijective). Is saying that *Abbildung* is a very general operation of the mind the same as saying that it is a purely logical notion? An affirmative answer (like that advanced by Ferreirós [91], p. 229) would find partial justification in the expressions Dedekind uses in the prefaces to the first and second editions of *Was sind*. But for him, logic is logic of the operative mind. Moreover, as a matter of fact, Dedekind introduces first, in the 1850s, the notions of *System* and *Abbildung* in his algebraic works and in his theory of algebraic numbers,[143] without

---

[142]Compare with Benacerraf's view (advanced in [8]), according to which no set-theoretic representation should be taken as defining natural numbers. As I recalled above (42), Dedekind had written to Lipschitz that he did not want to invent some previously unknown irrational numbers. However, there is no contradiction between this early view (1876) and the one communicated to Weber in 1888, because the abstract elements of the theory of simple infinite sets—the "shadowy forms"—and the real numbers produced by cuts are not identical with the familiar numbers as they were commonly used.

[143]Cf., e.g., the note "Aus den Gruppen-Studien", dated to 1855–58 ([55], vol. III, pp. 439–446), where Emmy Nœther found the germ of his own "*Homomorphiesatzes*" (*ibid.*, p. 446; [149]), and the famous "Xth Supplément" to the second edition of Dirichlet's lectures on the theory of numbers

speaking of logic at all. In addition to that, even in *Was sind* the practical perspective is not really logical, since *Systeme* and *Abbildungen* are not defined as or derived from logical notions proper, such as it is the case for Frege's concepts and relations.

Whereas Frege uses the one-to-one correspondence for defining equinumerosity by passing from arithmetical notions to purely logical ones, Dedekind uses the one-to-one correspondence both for discriminating infinite *Systeme* from finite ones, and for characterising simple infinite *Systeme* (that is, in modern terminology, countable totally ordered sets): a *System* S is infinite if there is an *ähnliche Abbildung* (injection)

$$\varphi : S \mapsto T \quad \text{with} \quad T \subset S,$$

such that $\varphi^{-1}(T) = S$ (that is, φ is one-to-one from S to T, with $T \subset S$), and it is finite if it is not infinite (this definition is now classic—modulo a replacement of Dedekind's terminology with the current set-theoretic one). And S is simply infinite if there is an *ähnliche Abbildung* $\varphi : S \mapsto S$, such that the φ-chain of a singleton $\{e\} \subset S$ is S itself, and $\{e\} \not\subset \varphi(S)$.[144] Here S is a domain of uninterpreted (abstract) elements; for getting the natural numbers *System* N, it is enough to specify the singleton $\{e\}$, by taking it to be $\{1\}$. The structure of N is that of any simply infinite *System* (any progression). Moreover, an *ähnliche Abbildung* φ defined on a domain S into a domain T is not identified with the set of ordered pairs $(x, \varphi(x))$ for $x \in S$ and $\varphi(x) \in T$. It may, then, serve to connect not only different domains having the same structure, but also different structures which may be themselves satisfied by different domains. What matters is the one-to-one correspondence, not its domain and codomain. That will become very clear with Emmy Nœther's general homomorphism theorems (of groups, rings, modules, algebras: [149, 150]), the roots of which their author found in Dedekind.

On the other side, Frege considers the one-to-one correspondence neither between cardinal numbers as finite multitudes, nor between sets, but between objects falling under some concepts, since he defines cardinal numbers as things belonging to concepts. Thus, any such a correspondence is, in fact, for him, a binary relation among objects, but it induces a second-level relation among concepts, the relation of being "equinumerous", which, as such, belongs to pure logic ([93], Sects. 70–72), and is, then, characterised independently of any sort of numbers. This characterisation is comparable to Dedekind's definition of simply infinite *Systeme* independently of numbers. It allows Frege to define equality between cardinal numbers in logic. He does it in three steps:

(*i*) Two concepts F and G are said to be equinumerous, which could be denoted by '$F \approx G$', if there is a correspondence Φ that correlates one-to-one the objects falling under F with the objects falling under G;

([66], pp. 434–626), where Dedekind defines the field structure and develops his general theory of ideals using set-theoretical operations.

[144]The φ-chain of a singleton $\{e\}$, namely $\varphi_0(\{e\})$, in Dedekind's notation, is the intersection of all chains $K \subset S$ such that $\{e\} \subset K$ ($K \subset S$ is a φ-chain if $\varphi(K) \subset K$).

(*ii*) The cardinal number *n* belonging to *F* is identified with the extension of the second-level concept ⌜equinumerous with *F*⌝;

(*iii*) The cardinal number which belongs to *F* is taken to be equal to the cardinal number which belongs to *G*, if the concept *F* is equinumerous to the concept *G*.

With all this at hand, Frege is then able to define the individual natural numbers ([93], Sects. 74, 77, and 82; [103], pp. 87, 90, and 95):

0 is the cardinal number belonging to the concept ⌜not identical with itself⌝;

1 is the cardinal number belonging to the concept ⌜ identical with 0⌝;

...

For any natural number *a*, the number that follows in the series of natural numbers directly after *a* is the cardinal number belonging to the concept ⌜member of the series of natural numbers ending with *a*⌝.

This last concept requires a definition, of course, which Frege also provides by appealing to the concept of a cardinal number belonging to a concept *F*, but, of course, not to the concept of a natural number ([93] Sects. 76, 79, 81; [103], pp. 89, 92, 94).[145] Finally ([93] Sect. 83; [103], p. 96), Frege defines the concept of a (finite) natural number by stating that *n* is such a number if (and only if) it is a member of

---

[145] In short, the definition is as follows. Let the statement

'*n* follows in the series of natural numbers directly after *m*'

mean the same as the statement

'there exists a concept *F*, and an object *x* falling under it, such that the cardinal number belonging to the concept *F* is *n* and the cardinal number belonging to the concept ⌜falling under *F* but not identical with *x*⌝ is *m*'.

Let us say, for short, that *n* stands in the relation SUCC with *m* if (and only if) *n* follows in the series of natural numbers directly after *m* (this notation is not Frege's, which firstly gives his definitions for a generic φ-series, where φ is a one-to-one correspondence whatsoever, then particularises them them by replacing 'φ-series' with 'series of natural numbers'). Let the statement

'*y* follows in the series of natural numbers after *x*'

mean the same as the statement

'if every object to which *x* stands in the relation SUCC falls under the concept *F*, and if from the proposition that *d* falls under the concept *F* it follows universally, whatever *d* may be, that every object to which *d* stands in the relation SUCC falls under the concept *F*, then *y* falls under the concept *F*, whatever concept F may be'.

(in modern terminology this provides a definition of the strong ancestral relation of SUCC). Let, finally, the statement

'*n* is a member of the series of natural numbers ending with *a*'

mean the same as the statement

'*a* follows in the series of natural numbers after *n* or *a* is the same as *n*'.

the series of natural numbers beginning with 0 (provided that the statement '*a* is a member of the series of natural numbers beginning with *n*' mean the same as the statement '*a* follows in the series of natural numbers after *n* or *a* is the same as *n*': [93] Sect. 81; [103], p. 94).[146]

Thus Frege uses concepts and relations, where Dedekind uses *Systeme*. For Frege, a one-to-one correspondence is a binary relation; for Dedekind it is a special kind of *Abbildung*, namely an injection. From Frege's concepts and relations, quantification theory has been developed; from Dedekind's *Systeme* and *Abbildungen*, set theory has been developed, together with a "working structuralism" ([148], p. 360). Now, does the latter stand against the former? This is the question I shall try to answer in the following section.

1.5.3 Dedekind's Chains and Frege's Following in a *φ*-Sequence

In the preface to the second edition of *Was sind*, Dedekind writes ([50], *Vorwort zur zweiten Auflage*, p. XVII; [53], p. 19) that he had not read *Grundlagen* until 1889, and recognises in retrospect "very close points of contact" between his and Frege's works, especially between his notion of chain and Frege's notion of following in a *φ*-sequence, presented both in *Begriffsschrift* ([92], part III, Sects. 23–31) and in *Grundlagen* ([93] Sects. 79–84).[147] Speaking of "points of contact", as for two geometric curves, Dedekind points to two different paths. He adds that the "the positiveness with which [...][Frege] speaks of the inference from *n* to *n* + 1 [...] shows plainly that here he stands upon the same ground with me". On the same ground indeed,[148] but at diametrically opposite poles.

For, from the outset, Frege's goal is to show that "even an inference like that from *n* to *n* + 1, which, on the face of it, is peculiar to mathematics, is based on general laws of logic, and that there is no need of special laws for aggregative thought" ([93], *Einleitung*, p. IV; [103], p. XVI).[149] Thus, if the goal is achieved, every arithmetical theorem becomes "a logical law" and "calculation becomes deduction", as Frege writes in *Grundlagen* ([93], Sect. 87; [103], p. 99), in accordance with the programme already stated in *Begriffsschrift* ([92], *Vorwort*, p. IV; [125], p. 5): "I first had to ascertain how far one could proceed in arithmetic by means of inferences alone, with the sole support of those laws of thought that transcend all particulars. My initial

---

[146]Cf. footnote (145), above.

[147]Cf. footnote (145), above.

[148]For the formal similarities between Dedekind's chains and Frege's following in a *φ*-sequence, cf., e.g. [63], pp. 140 and 141.

[149]Cf. also what Frege writes in Sect. 45 ([93], Sect. 45; [103], p. 58): "The terms 'multitude', 'set' and 'plurality', are unsuitable, owing to their vagueness, for use in defining number". 'Vagueness' obviously means the same as 'multivocity'.

step was to attempt to reduce the concept of ordering in a sequence to that of *logical* inference [*Folge*],[150] so as to proceed from there to the the concept of number".

Quite the reverse, Dedekind's achievement in *Stetigkeit* shows that numerical total ordering $<$ is, with the identity as well, the primitive relation, which permits passing from rational numbers to real numbers without any notion of magnitude. In *Was sind* the inference from n to $n + 1$ is reduced to an order-setting *Abbildung*,[151] injective but not surjective, defined on an infinite *System* whose elements are not necessarily numbers.[152] The performed order is inclusion of sets, namely $\subset$. For defining the general notion of a chain ([49], Sect. 37), one needs only a domain S of undetermined elements and an *Abbildung* φ with domain and codomain S; then, $K \subset S$ is a φ-chain of S if and only if $\varphi(K) \subset K$. The intersection of all chains of S containing A, i.e. the chain of A (*ibid.*, Sect. 44), permits to get the general theorem of complete induction (*ibid.*, Sect. 59), which, applied to the case where $S = N$ and φ is injective, provides the arithmetical induction (*ibid.*, Sect. 80) and the theorem of definition by induction, or finite recursion (*ibid.*, Sect. 126). Sure, in agreement with Frege's conception of logic, Dedekind's notion of a chain is not logical: it would be so only if one counted the notions of a *System* and an *Abbildung*, as well as that of inclusion, to be logical in turn, which is certainly not Frege's view. In Dedekind's view those notions belong to a new mathematical discipline, which is general arithmetic. In opposition to Frege's logicist *reduction*, Dedekind saw in his *generalisation* the source of new mathematical developments, and, indeed, this was an important source for the emergence of set theory[153] and axiomatics. Dedekind's immediate followers axiomatised set theory, interpreting *Abbildungen* as mappings, while modern category-theorists rightly understand *Abbildungen* as morphisms, namely arrows connecting structures. Despite his evocation of logic, Dedekind's interpretation of Leibniz's 'calculemus!' does not aim at constructing the calculus of reasoning, but at showing that reasoning is fundamentally arithmetic generalised to undetermined elements. What matters is the *structural generalisation*, not the logical reduction, of arithmetic. The nuance is quite significant: Dedekind's logic of the mind is something else than Frege's logic conceived of as a formalisation of the notions of inference and truth.

Actually, Dedekind *and* Frege agree to disagree not only on their respective conceptions of logic, but also on their conceptions of the nature of numbers. Frege writes ([97], *Vorwort*, p. VIII; [110]; p. VIII$_1$):

> Mr Dedekind too is of the opinion that the theory of numbers is a part of logic; but his
> essay barely contributes to the confirmation of this opinion since his use of the expressions

---

[150]Translating 'Folge' by 'consequence', in accordance with our current usage, would be misleading since Frege actually deals with inference.

[151]For the logical similarity with the ancestral of a relation (defined as in footnote (145), above), and for setting on a par Dedekind's introduction of the real numbers, as corresponding uniquely to cuts, and Frege's introduction of extensions, as objects corresponding uniquely to concepts, cf. [25], pp. 249–254.

[152]Once again, Dedekind's generality through *Abbildung* differs from Frege's logical generality obtained by the ancestral of a relation.

[153]For the respective roles of Dedekind and Cantor in the emergence of this theory, cf. [90].

'system', 'a thing belongs to a thing' are neither customary in logic nor reducible to something acknowledged as logical.

Thus in Frege's opinion, set and membership are not of logical nature; Dedekind's reduction to *Systeme* is not a logical reduction. In other words, set theory is not logic. One cannot but admire Frege's perspicuous eye: indeed, in a language with the membership relation as a *primitive term*, such as first-order ZF, no mathematical relation is logical.[154] The question 'What is logic?' might still be disputable, yet Frege's judgement confirms that Dedekind and Frege answer it differently,[155] and that lead to think that the demarcation between set theory and logic still is open to question.

1.6 Conclusions

Some final remarks, now.

1. My first point is that a bias is introduced by isolating Dedekind's two essays on numbers from the rest of his production, and that things become even worst if one points to *Was sind* alone, since this is the only piece where Dedekind regards arithmetic as a part of logic.[156] Dedekind's mathematical production concerns a wide range of mathematical topics, including geometry, infinitesimal analysis, arithmetic, algebra, topology, and in every domain he inaugurated a structural way to do mathematics, without necessarily making a link with logical concerns. For instance, it appears from his theory of ideals that it is possible to isolate the "inner properties" of the concept to be defined (or created), namely the concept of an ideal—i.e. spelling the necessary and sufficient conditions that an ideal is required to meet, from which theorems on ideals can be deduced—in close connection with the substantive mathematical results at hand (especially Gauss's and Kummer's on number theory), and without calling for logic [4, 76].[157] Moreover, his work contains *neither* a specifically logical development nor a body of articulated philosophical views. Nevertheless, his scattered remarks about the nature of number or the essence of

---

[154]Tarski [193] shows that the answer to the question: 'Is mathematics reducible to logic?' depends on the choice of the language. On the assumption of Tarski's criterion of logicality, according to which logicality is set-theoretically defined through invariance under permutations of the domain of individuals, the answer is affirmative in Russell's simple type theory, but negative in Zermelo's first-order system. His paper prompted a rich discussion on the very nature of logic which is still open: Tarski's criterion is accepted as a necessary but not sufficient condition for defining logicality in semantic terms [146], but it is criticised for reducing logic to set-theory [87].

[155]Gödel stands on the side of Frege: he distinguishes between sets or classes, on the one hand, and concepts, on the other hand, and aims too at establishing a "theory of concepts" ([114]; [113]; [199], pp. 297–299 and 309–312; [200], Chap. 8).

[156]But cf. also Dedekind's use of the term 'Systemlehre der Logic' in a paper dating back to 1897, and mentioned by Ferreirós ([52], Sect. 4; [91], pp. 225–226).

[157]Dedekind highlights the inner link between his concept of cut and his concept of ideal, e.g. in the introduction of [48], on which cf. [67], pp. 65–72.

continuity form a coherent picture, which pertains to knowledge and epistemology rather than to ontology.[158] Dedekind endorses the Kantian split between epistemology and ontology, whereas Frege renews the ancient connection between the two.

2. Dedekind writes that Frege stands on the same ground with himself. Logic is this common ground. But one cannot give to 'logic' and 'logical', under Dedekind's pen, the same meaning as these words have for Frege. In particular Dedekind's logic has anything to do neither with analysis of language, nor with a theory of inference, nor with a theory of truth. So, if one persists in taking *à la lettre* Dedekind's claim that arithmetic is a part of logic, one should make precise that the building tools he uses to substantiate this claim, viz. the notion of a *System* and an *Abbildung*, together with that of inclusion, are not logical in Frege's sense of the word. For there is room in Frege's logical realm neither for the aggregative thought, nor for the kind of generality prompted by Dedekind's *Abbildung*: the operative mind representing a thing by a thing allows substituting the second for the first not because the two are identical, but just because they play the same role in determined conditions. Dedekind's *Abbildung* conflicts straight out with Frege's concern for objectual identity. As Dedekind's work shows everywhere, identity is much less fruitful than *analogy* (representation in Dedekind's wording), which affords not an *objectual* but a *functional* identity. The undetermination of the elements is, from a mathematical point of view, not a flaw *vis-à-vis* the question of truth, but the condition for bringing structures to the light, which may apply to domains of different elements and, thus, facilitate substantial interactions and substantial developments of mathematical stuff.

3. For Dedekind, what matters are not the numbers themselves, but their structure. Arithmetic is fundamental not only because numbers are applied everywhere, but because, by following the arithmetical laws, we can calculate with things which are not numbers. What matters is not what can be said of numbers in themselves, but how the four conditions α, β, γ, δ, brought to the light in Sect. of *Was sind* (cf. Sect. 1.4.1, above) are satisfied. And this is why we can say that arithmetic is a formal structure of our experience ([133], p. 328). The "logic of the mind" is arithmetic taken generally. As Dedekind writes, "every *thinking* man, even if he is not clearly aware of that, is an arithmetic-man, an arithmetician" ([67], p. 315; my italics), for thinking is representing a thing by a thing, relating a thing with a thing. Hence, I understand Dedekind's claim that arithmetic is a part of logic as meaning that arithmetic *also* affords a rational (logical) norm of thinking.

4. Dedekind's construction in *Was sind* shows, in fact, that Dedekind assimilates logic to set theory—rather than the reverse—, what Frege refused for good reasons ([153], Sects. VI–VII). However, the demarcation between logic and set theory stressed by Frege is still in debate. In a semantic set-theoretic approach, different invariance criteria across structures are proposed for capturing logical notions, for example by Tarski [193], Sher [182], McGee [146], Van Benthem [13, 14], Feferman [87], and Bonnay [17]. In a syntactical proof-theoretic approach, logicality is defined in terms of some set of basic inference patterns, for example by Gentzen

---

[158]Stein ([186], p. 247) rightly points out that Dedekind's work is "quite free of the preoccupations with 'ontology' that so dominated Frege, and had so fascinated later philosophers".

[112], Prawitz [162], Martin-Löf [144], Feferman [88], and others. Worth mentioning are also the game-theoretic approach and the computational approach, among others. Logicality may also be understood in a "holistic" way, concerning a whole language. Different stances are taken as to the status of the second- and higher-order quantification. Hence, logicians vindicate the autonomy and priority of logic over set theory, but they diverge on what is logic. Parsons ([153], pp. 165–167) and Feferman ([86], p. 45) defend Quine's view according to which second- and higher-order quantification is nothing else but "set theory in sheep's clothing" on the ground that the meaning of such quantification depends on which sets exist ([196], pp. 66–68; [86], p. 22). According to Parsons, "the justification for not assimilating high-order logic to set theory would have to be an ontological theory like Frege's theory of concepts as fundamentally different from objects, because 'unsaturated' ", and even in that case "high-order logic is more comparable to set-theory than to first-order logic" ([153], p. 166). Feferman argues that, in contrast with operations of second-order logic, operations of first-order logic with equality have the same meaning independently of the domain of individuals over which they are applied ([86], pp. 38 and 45). By contrast, Boolos (e.g. [18–20]); Resnik (e.g. [167]); Shapiro (e.g. [179, 180]), and others refuse to regard the line between first- and second-order logic as the line between logic and mathematics and they take second- and higher-order quantification to be genuine logic. Boolos's following judgement is noteworthy, for example: "Of special interest in Dedekind's work [...] is the use of what Quine would regard as set-theory and what I [...] would call logic" ([25], p. 254). Actually, as van Benthem pinpoints, there are many intuitive aspects of logicality, and no one of the various formal characterizations exhausts the notion [198]. Yet such a liberalism does not kill the search after criteria for logical notions as universal notions independent from what there is, and, in particular, criteria not definable in set-theoretic terms [87]. I will recall here the structural analogy between proofs and programmes expressed by the Curry-Howard isomorphism, which states a structural correspondence between formulas and types. In face of it, logic and arithmetic appear to be two faces of the same process.

5. A last point. It is risky to cut a piece of work from its practical and historical context. I would strongly speak in favour of what is now named the 'practical turn' in philosophy of science and also in favour of "the historical turn" in analytic philosophy. I think we gain a more accurate view on mathematical or logical concepts and methods when we start with mathematical or logical practices and build philosophical reflections from the practical ground. I think also we gain a better philosophical analysis and appraisal of a piece of work when we do not ignore its historical entrenchment.

Chapter 2
From Lagrange to Frege: Functions and Expressions

Marco Panza

2.1 Introduction

Part I of Frege's *Grundgesetze* is devoted to the "exposition [*Darlegung*]" of his formal system. It opens with the following claim ([97], Sect. I.1, p. 5; [110], p. 5₁):

> If the task is to give the original reference [*Bedeutung*] of the word 'function' in its mathematical usage, then it is easy to slip into calling a function of x any expression [*Ausdruck*] that is formed from 'x' and certain determinate numbers by means of the notations [*Bezeichnungen*] for sum, product, power, difference, etc. This is inappropriate [*unzutreffend*], since in this way a function is depicted [*hingestellt*] as an *expression*, as a combination of signs [*Verbindung von Zeichen*], and not as what is designated [*Bezeichnete*] thereby. One will therefore be tempted to say 'reference of an expression' instead of 'expression'.

Frege does not explicitly ascribe this inappropriateness to anyone, though he could have ascribed it to many.[1] One is Lagrange, who, a little less than one century earlier, defined functions as follows, both in the *Théorie des fonctions analytiques* and in the *Leçons sur le calcul des fonctions* ([134], Sect. 1, p. 1; [140], *Introduction*, p. 1; [137], p. 6; [138], p. 6; I quote from the *Théorie*; the analogous passage of the *Leçons* presents some inessential changes):

> One calls a 'function' of one or several quantities any calculational expression [*expression de calcul*] into which these quantities enter in any way whatsoever combined or not with other quantities which are regarded as having given and invariable values, whereas the quantities of the function may receive any possible value.

---

[1] Baker [5] has considered, but finally rejected, the idea that Frege could also have ascribed this inappropriateness to himself, by referring to Sects. 9–10 of his [92]. My purpose here is not to describe the evolution of Frege's views. I shall rather confine myself to considering his mature views, as they emerge from *Grundgesetze* or other contemporary works.

© Springer International Publishing Switzerland 2015
H. Benis-Sinaceur et al., *Functions and Generality of Logic*,
Logic, Epistemology, and the Unity of Science 37,
DOI 10.1007/978-3-319-17109-8_2

According to our modern view, this definition is based on an inadmissible conflation of syntactical items and their *designata*. Frege's warning against a definition like this appears, then, to be not merely motivated by a different approach, but rather mandated by conceptual clarity and rigour. Historically speaking, things are not so simple, however: if considered in context, Lagrange's definition reflects a quite precise conception of mathematics, which is, in some crucial respects, close to Frege's.

Like Frege's *Grundgesetze*, Lagrange's treatises also pursue a foundational program. Still, the former's program is not only crucially different from the latter's, it also depends on a different idea of what a foundation of mathematics should be like.[2] Despite both this contrast and that between warning and Lagrange's definition, the notion of a function plays similar roles in their respective programs. My purpose is to emphasise this similarity. In doing so, I hope to contribute to a better understanding of Frege's logicism, especially in relation to its crucial differences from a set-theoretic foundational perspective. This should also shed some light on a question raised by Hintikka and Sandu in a widely discussed paper [129], namely whether Frege should or should not be credited with the notion of an arbitrary function that underlies our standard interpretation of second-order logic.[3]

In Sect. 2.2, I shall recount Lagrange's notion of a function.[4] In Sect. 2.3, I shall advance some remarks on connected historical matters. This will provide an appropriate framework for discussing the role played by the notion of a function in Frege's *Grundgesetze*, to which Sect. 2.4 is devoted. Some concluding remarks will close the chapter.

2.2 Lagrange's Notion of a Function

Lagrange's treatises aim at offering a non-infinitesimalist interpretation of the differential formalism. For this purpose, the calculus is embedded in a general theory of functions, often termed 'algebraic analysis'.[5] Though this theory originated with Euler's *Introductio* [77], Lagrange suggests a new way of integrating the calculus within it, quite different from that suggested by Euler himself in his *Institutiones* [80]. This rests on a more general conception of mathematics, that, though close to Euler's, significantly differs from it.[6]

---

[2] Broadly speaking, a foundation of mathematics aims at a reorganisation of mathematics according to a suitable order. For Frege, such an order ought both to reflect an objective order of truths ([93], Sect. 2) and to provide mathematics, especially arithmetic and real analysis, with an epistemically sound basis (*ibid.*, Sect. 3). For Lagrange, it should rather obey an ideal of purity (as I and my coauthor G. Ferraro have largely argued for in [89]).

[3] Cf. Sect. 3.1 of Chap. 3, below.

[4] A more comprehensive discussion is offered in [89].

[5] This is suggested by the complete title of the *Théorie*: 'Théorie des fonctions analytiques contenant les principes du calcul différentiel [...] réduits à l'analyse algébrique des quantités finies'.

[6] For a discussion of Euler's conception, I refer the reader to [152].

The basic idea is that of recasting all mathematics within algebraic analysis, understood as a general theory of functions. Hence, functions are not merely conceived as objects to be studied by a branch of mathematics; they are taken to be what all mathematics is about.

This idea is structurally similar to that which underlies the set-theoretic foundational program. But, while this program is often understood ontologically—i.e. it is taken to entail the requirement that all mathematical objects be ultimately identified with appropriate sets—Lagrange does not argue that all that mathematics deals with has to be ultimately identified with functions. For him, mathematics is the science of quantities, as classically maintained, and not every quantity is ultimately a function: this is not so for numbers or geometrical and mechanical magnitudes, which are endowed with a specific, irreducible nature. Concerning them, Lagrange's point is rather that mathematics should study their mutual relations, and, for this purpose, it should look at them as functions of each other. Still, according to him, this is possible only if a general theory of functions is provided in which functions are considered as such, and identified with abstract quantities, i.e. quantities lacking any specific nature. Algebraic analysis is just this theory. And Lagrange's definition of functions is aimed at fixing what it is about.

Following this definition, quantities enter into expressions. Hence, these expressions are said to include that which, according to our familiar distinction between *designans* and *designatum*, the terms composing them designate.

One could argue that this is simply because of a slip of the pen. But this is implausible, not only because Lagrange would then have been the victim of the same slip in both editions of the *Théorie* and of the *Leçons* as well, but also because this slip would have then been also very common among other contemporary mathematicians. It would have also affected, for example, the definition of function offered in Euler's *Introductio* ([77], vol. I, Sect. 4, p. 4; [83], vol. I, p. 3):

> A function of a variable quantity is an analytical expression composed in any way whatever of this variable quantity and numbers or constant quantities.

Once the possibility of a slip of the pen has been discarded, it only remains to accept that, for both Lagrange and Euler, quantities are not what the terms entering into a "calculational" or "analytical expression" designates, but are these very terms.

A hint for understanding how this is possible can be found in a criticism Lagrange addresses to Newton's conception of the calculus ([134], Sect. 5, p. 4; [140], *Introduction*, p. 3):

> [...] Newton considered mathematical quantities as generated by motion [...]. But [...] introducing motion in a calculation whose object is nothing but algebraic quantities is the same as introducing an extraneous idea [...].

The calculation Lagrange is referring to is the calculus. His point is thus that the calculus should not concern quantities generated by motion, but algebraic quantities. Criticising Newton for introducing motion within pure mathematics was usual. But arguing that the calculus should concern algebraic quantities was new. For Lagrange, the calculus is to be immersed within algebraic analysis. Hence, for him, also algebraic analysis should be about algebraic quantities. But what are algebraic quantities?

In Lagrange's setting, there is no room for identifying them as that which algebra is about, supposing that this is independent of (and prior to) algebraic analysis. In other words, there is no room for taking the functions entering into algebraic analysis to involve quantities that algebra supplies. This would result in a structural duplication (algebra on one side, with its own formulas, and algebraic analysis on the other side, with its functions), of which there is no trace either in the *Théorie* or in the *Leçons*. Rather, Lagrange repeatedly claims or implies that algebra and algebraic analysis do not differ essentially. The following quotation, from his treatise on numerical equations, provides an example ([135], p. vii; [139], p. 15):

> Taken in the most comprehensive sense, algebra is the art of determining unknowns through functions of known quantities or quantities regarded as known; and general solution of equations consists in finding, for all the equations of any degree, the functions of the coefficients of these equations that are able to represent all their roots.

Furthermore, for Lagrange, functions not only include (algebraic) quantities, but are also (algebraic) quantities. Here is what he writes in the second edition of the *Théorie*:

> Through the character 'f' or 'F' placed before a variable, we shall designate in general any function of this variable, that is, any quantity depending on this variable and which varies with it according to a given law.

Another passage where he is quite clear on this is the following ([134], Sect. 2; [140], *Introduction*, pp. 1–2; cf. also [137], p. 4; [138], p. 4):

> The word 'function' has been employed by the first analysts in order to designate in general the powers of a same quantity. Then its meaning has been extended to any quantity however formed by another quantity. Leibniz and the Bernoullis employed it firstly in this general sense, and it is today generally adopted.

Doubtless, Lagrange takes his definition to be consistent with this "generally adopted" sense, which Johann Bernoulli had fixed already in 1718, by stating that a "function of a variable quantity" is a "quantity however composed by this variable quantity and constants" ([15], p. 106).

There is thus no doubt that, for Lagrange, functions are both expressions that contain quantities and quantities. Insofar as this view is openly incompatible with the *designans*/*designatum* distinction, it can be grasped only if this distinction is thrown away. Two new quotations (respectively from [137], p. 4 and [138], p. 4, and, from [136], 235) suggest a way of doing this:

> [...] one should regard algebra as the science of functions, and it is easy to see that, in general, the solution of equations does not consist but in finding the values of unknown quantities as determined functions of known quantities. These functions represent, then, the different operations that have to be performed on the known quantities in order to obtain the values of those which are sought, and they are properly only the last result of the calculation.
>
> Strictly speaking, algebra in general is nothing but the theory of functions. In Arithmetic, one looks for numbers according to given conditions between these numbers and other

numbers; and the numbers that are found meet these conditions without preserving any trace of the operations that were needed in order to form them. In algebra, instead, the sought after quantities have to be functions of given quantities, that is, expressions representing the different operations that have to be performed on these quantities in order to get the values of the sought after quantities. In algebra *stricto sensu*, one only considers primitive functions that result from ordinary algebraic operations; this is the first branch of the theory of functions. In the second branch, one considers derivative functions, and it is this branch that we simply designate with the name 'theory of analytical functions' […].

Lagrange's terminology is fluctuating and imprecise. But it is clear that he considers algebraic analysis to be a general subject including at least two interrelated branches that are not distinguished because of their objects, which are always functions, but rather because of by the way these objects are considered. The former is the theory of algebraic equations; the latter the theory of analytical functions, i.e. Lagrange's own version of the calculus. Arithmetic results when functions are instantiated on numbers. In this case, they can be computed and this produces new numbers whose operational relations with those from which they result is lost. In algebraic analysis, instead, functions are not instantiated on independently given quantities, but are the relevant quantities themselves; they can only be transformed, and, whatever their form might be, they maintain a trace of the operational relations that link them to the quantities of which they are functions. This is just what makes algebraic analysis pure and general. Its subject matter is the system of relations induced by the (indefinite) composition of some (elementary) operations applied to previously indeterminate arguments. Precisely because these arguments are taken as being previously indeterminate, they are subsequently characterised by nothing other than the network of relations they enter into. These relations are immediately displayed, or, as Lagrange improperly says, "represented",[7] by appropriate expressions, which are taken to constitute a *sui generis* sort of quantity: algebraic quantities, or functions. These are endowed with a purely relational identity and lack any intrinsic nature, though being capable of being studied as such, and of being instantiated on numbers and geometric or mechanical magnitudes.

It follows that, according to Lagrange, neither operations nor their arguments precede symbols: at the beginning there are only symbols submitted to appropriate rules; operations and quantities appear next, whenever these symbols are supposed to acquire a mathematical meaning. For example, the symbol '+' is not taken to designate the independently given operation of addition. This operation is rather fixed by the rules of composition and transformations relative to this symbol: it is not because addition is commutative that '$a + b$' can be transformed into '$b + a$',

---

[7] As Lagrange uses it, the verb 'to represent' is not intended to indicate a relation between two distinct entities, one of which is taken to stand for the other under an appropriate respect. Expressions do not "represent" operations because they stand for them or present them afresh. They display at once these operations, their results, and the corresponding relations: 'x^2' displays, for example, the operation of taking the square, the quantity related to x according to this operation, and the relation between it and x.

but the other way around. The universe of Lagrange's general theory of functions is thus a universe of symbols governed by rules of composition and transformation, not a universe of objects, operations, and relations to which these symbols refer.

All this makes clear that, for Lagrange, the notion of a function is mathematically primitive: his definition is intended as a clarification of this notion that is based on no previous mathematical development. All that is required for understanding this definition is taking for granted an appropriate extension of the algebraic formalism originating with Viète and Descartes. But, as such, this formalism is not yet supposed to be a mathematical system; mathematics only begins when the formulas of this formalism are understood as quantities, i.e. just when functions are introduced.

A last point has to be clarified: if things are this way, how can Lagrange maintain that algebraic quantities or functions are quantities in a genuine sense of this term? Partly, this is because they are arguments of operations with the same formal properties as the usual operations on numbers and geometric magnitudes.[8] Furthermore, this is because Lagrange tacitly assigns to them some properties that do not depend on their being constituted by appropriate expressions: he attributes to them a linear order and some metric relations, and also supposes they comply with continuity conditions. This is essential for his reductionist program to succeed. But it also produces a discrepancy between the understanding of functions as expressions and their understanding as quantities. This is one of the reasons why this program ultimately failed.[9]

2.3 Arbitrary Functions and the Arithmetisation of Analysis

A notorious shortcoming of Lagrange's theory is relevant to my purpose. To see it, consider an example ([134], Sect. 96; [140] Sect. I.84).

Let

$$z = ax + by + c \tag{2.1}$$

be a function of two variables x and y, involving the constants a, b and c. Insofar as

$$z'_x = a \quad \text{and} \quad z'_y = b \tag{2.2}$$

this function provides the complete primitive of the following partial differential equation:

$$z - xz'_x - yz'_y - c = 0 \tag{2.3}$$

---

[8]Though partial, this answer is not simple. It reveals a crucial feature that Lagrange's program shares with any foundational reductionist program in mathematics: this program stipulates a new start for mathematics, without being free to forget what mathematics was before its advent. Hence, this start has both to be taken as primitive and to be so shaped as to allow a reformulation of what was there independently of it.

[9]Arguing for this is one of the main purpose of [89].

This is not the only primitive of this equations, however. To get another primitive, suppose that a be a function of x and y, and b a function of a. From (2.1), by taking the derivatives with respect to x and y, one gets, respectively:

$$z'_x = a + a'_x \left[x + yb'_a \right] \qquad \text{and} \qquad z'_y = b + a'_y \left[x + yb'_a \right]$$

It is, then, enough to also suppose that

$$x + yb'_a = 0 \tag{2.4}$$

to get the equalities (2.2), again. Insofar as we have supposed that b is a function of a, from (2.4) it follows that a is a function of $\frac{x}{y}$, and so is $a\frac{x}{y} + b$. Taking this last function to be $\varphi\left(\frac{x}{y}\right)$, from (2.1) one gets

$$z = y\varphi\left(\frac{x}{y}\right) + c \tag{2.5}$$

This is the other primitive of (2.3) we were looking for.[10] In Lagrange terminology ([137], continuation of lect. XX; [138], lect. XX), this is the "general primitive" of (2.3), and it is, indeed, in a quite clear sense, a primitive much more general than (2.1).

The relevant point here is that the function designated by 'φ', as well as the functions that a is supposed to be of x and y, and b of a, are, as Lagrange himself admits ([134], Sect. 95; [140], Sect. I.83), "absolutely arbitrary", in the sense that they are not only susceptible of being displayed by whatsoever expression, but they are even not required to be displayed by any expression at all. All that is required for the argument to proceed is that b be a function of a, a be a function of x and y, and some operational conditions about derivatives functions be met, so as to ensure, for example, that the derivative of by with respect to x is $yb'_a a'_x$. Hence, all that is required for (2.5) to be a solution of (2.3) is that $\varphi\left(\frac{x}{y}\right)$ be a function of $\frac{x}{y}$ and that a function satisfy these operational conditions, which is perfectly independent of its being displayed by any expression.

In order to account for the existence of general primitives of partial differential equations, Lagrange is then forced to deal with arbitrary functions, which are not so

---

[10]Verification is easy. From (2.5) it follows:

$$z'_x = y\varphi'\left(\frac{x}{y}\right)\frac{1}{y} = \varphi'\left(\frac{x}{y}\right) \qquad \text{and} \qquad z'_y = \varphi\left(\frac{x}{y}\right) - y\varphi'\left(\frac{x}{y}\right)\frac{x}{y^2} = \varphi\left(\frac{x}{y}\right) - \varphi'\left(\frac{x}{y}\right)\frac{x}{y}$$

Replacing in (2.3), one gets, then

$$y\varphi\left(\frac{x}{y}\right) + c + x\varphi'\left(\frac{x}{y}\right) = x\varphi'\left(\frac{x}{y}\right) + y\varphi\left(\frac{x}{y}\right) + c$$

.

because they are waiting for a further possible determination through an appropriate expression, but are rather intrinsically indeterminate insofar as they are not expressions, but just quantities that are supposed to depend on other quantities and respect the appropriate operational conditions.

Lagrange cautiously avoids remarking on this. But the question was not ignored at the time. Euler openly tackled it many years earlier [78, 79, 81]. The details of Euler's arguments are not relevant for the present purpose.[11] It is enough to say that in these works, he takes functions to be "quantities somehow determined by some variable" ([81], p. 3). This fits with the definition he provides in the *Institutiones* ([80], p. VI; [84], p. VI):

> Those quantities that depend on others in [...][such a] way that if these are changed, they also undergo a change, are usually said to be functions of these latter [quantities]. This is a quite broad denomination and encompasses in itself all ways in which one quantity can be determined by others. Hence, if 'x' denotes a variable quantity, all quantities that in any way depend on x or are determined by it, are said to be functions of it.

This definition has often been opposed to those offered by Lagrange and by Euler himself in the *Introductio*.[12] It is also mentioned by Hintikka and Sandu in their paper on Frege's notion of function ([129], pp. 296–297) as an early manifestation of the "concept of arbitrary function". Hintikka and Sandu are interested in the question "whether Frege assumed the standard interpretation of higher-order quantifiers or a non-standard one" (*ibid.*, p. 298), i.e. whether, for him, the range of second-order quantifiers is "the entire power set $P(do(M))$ of the relevant domain $do(M)$ of individuals", or "only some designed subset of $P(do(M))$" (*ibid.*, p. 290). For Hintikka and Sandu, "the conception of the standard interpretation [...] is, to all purposes, equivalent with the notion of an arbitrary function or the notion of an arbitrary set" (*ibid.*, p. 298). They argue that "Frege lacked both the idea of arbitrary function and the idea of arbitrary set, and hence in effect opted for a non-standard interpretation" (*ibid.*). The definition of the *Institutiones* is mentioned as evidence that the "idea of an arbitrary function" dates back to long before Frege (*ibid.*, p. 296).

This suggests that in the evolution of the notion of function, two camps opposed each other: on the one side, those that admitted the notion of an arbitrary function, like the Euler of the *Institutiones*, and many others, among whom Hintikka and Sandu mention Dirichlet, Lobachevsky, and Cantor; on the other side, those who rejected or lacked this notion, like the Euler of the *Introductio*, Lagrange—at least for the definition he explicitly provides—and Frege, to whom they also add Weierstrass and Kronecker (*ibid.*, pp. 296–298). At first glance, my claim that the notion of function plays similar roles in Lagrange's and Frege's foundational programs seems to support this account. This is only partially true, however. What follows will explain why.[13]

---

[11]On these arguments and the mathematical discussion they were part of, cf.: [195], pp. 237–300; [117], pp. 1–21; [64]; [31], pp. 21–33; and [151], 256–264.

[12]For example in [207].

[13]Hintikka and Sandu's theses have generated a sharp controversy: cf. [34, 62, 124], for example. This largely depended on their arguing that it is "unfortunate that philosophers habitually

My first remark is that the apparent generality of the definition of the *Institutiones* is limited by the notion of quantity it is based on. In the same treatise, Euler argues that "every quantity, by its nature, can increase and decrease up to infinity" ([80], p. IV; [84], p. V). This echoes the classical, Aristotelian conception of quantity (*Metaphysics*, Δ, 13, 1020*a*, 7–14, and *Categories*, part 6), on which d'Alembert focuses by claiming that a quantity is "that which can be increased or decreased" ([2], p. 653). This is a quite vague conception, however. When the definition of the *Institutiones* is related to it, all that one understands from it is that a function is anything that can increase or decrease insofar as this depends on the increasing or decreasing of something else. Now, this idea is not only quite different from the modern one involved in the standard interpretation of higher-order quantifiers. More importantly, it is also a poor basis for any mathematical argument. To explain his notion of arbitrary function—which, in fact, reduces to arguing that a solution of a partial differential equation can involve something such as a discontinuous function—Euler is forced to rely on the representation of functions through curves. Hence, though perhaps more general than that of the *Introductio*, the definition of the *Institutiones* is more imprecise and less effective: it is inappropriate as a starting point for a general theory of functions, as algebraic analysis was intended to be.

This is why Lagrange preferred grounding his theory on another definition. His attempt failed, largely because of shortcomings like that mentioned above. But this failure did not result in the general admission of the definition of the *Institutiones*, but rather fostered the shaping of a new notion. Cauchy's *Cours d'analyse* [42] was the manifesto of the new course.[14]

As well as Lagrange's *Théorie*, Cauchy's treatise presents itself, according to its title, as a treatise of algebraic analysis. But this last term here takes a quite different meaning than in Lagrange's treatise. For Cauchy, algebraic analysis is a preliminary part of analysis (to be followed by the calculus), and analysis is a particular branch of mathematics. It is then essentially distinct both from algebra and geometry, but it is expected to be as rigorous as the latter, which is possible only insofar as it never relies on "arguments drawn from the generality" of the former (*ibid.*, p. ii; [32], p. 1).[15]

This is already quite far from Lagrange's conceptions. But a more radical difference depends on the fact that Cauchy does not open his treatise by fixing the notion

---

(Footnote 13 continued)

go to Frege", since "Frege was far too myopic to be a fruitful source for concepts, idea and problems" ([129], p. 315). I shall not deal with this allegation, and confine myself to giving an account of Frege's views in their historical context.

[14] An essentially different reaction was promoted by a group of British mathematicians including Woodhause, Babbage, and Peacock. Though their conceptions were highly influential in the history of logic, considering them is not relevant to my present purpose.

[15] A similar view had been endorsed by Ampère, almost twenty years earlier, in a memoir presented to the *Institut des Sciences* in 1803 and appearing in 1806 ([3], p. 496): "That which is termed a fact of analysis has always to be reduced to the metaphysical principles of this science if one wants to have a right idea of it. It is evident, indeed, that one has always to find the reasons for all the results obtained through calculation in the attentive examination of the conditions of any question, since the use of algebraic characters can add nothing to the ideas that they represent.".

of a function, but rather by independently explaining the notion of a quantity. Though his explanation[16] is far from perspicuous, his general strategy is clear enough.

Analysis starts by inheriting the notion of a magnitude from an independent source. This notion is taken as primitive in analysis, since analysis neither requires nor is capable of providing any further clarification of it. Analysis no longer deals with magnitudes, directly. It is rather concerned with their measures (which is also an unanalysed notion). These result from two sources. Each magnitude can be compared to another of the same species which is taken as a unit; but also its increase or decrease can be taken into account. In the former case, its measure is a number; in the latter, it is a quantity. Taken as such, numbers are neither positive nor negative, but only greater or smaller than each other. Quantities, instead, are either positive or negative: they are so insofar as they are respectively measures of the increase or the decrease of a magnitude. But, insofar as the increase and decrease of any magnitude can only be estimated by comparison with an appropriate unit, quantities are associated with numbers, namely they are signed numbers, which are not positive or negative numbers, but numbers preceded either by the sign '+' or by the sign '−'. Hence, any quantity has a numerical value, which is nothing but the number that is got when its sign is omitted.

It is only after having fixed these notions that Cauchy comes to functions. He begins by distinguishing variable from constant quantities: a quantity is variable if it is supposed "to take on successively several values different from each other", while it is constant if it "takes on a fixed and determined value" ([42], p. 4; [32], p. 6). He then introduces functions as follows ([42], p. 19; [32], p. 17):

> When variable quantities are so related to each other that the value of one of them being given, one can infer the values of all the others, one usually conceives these various quantities to be expressed by means of one of them, which therefore is called the 'independent variable'. The other quantities, expressed by means of the independent variable, are those which one terms functions of that variable.

---

[16]Cf. [42], pp. 1–2, and [32], pp. 5–6:

> First of all, we shall indicate what idea it seems appropriate to us to attach to the two words 'number' and 'quantity'. We shall always take the denomination of numbers in the sense in which it is used in arithmetic, by making the numbers to arise from the absolute measure of magnitudes [*grandeurs*], and we shall only apply the denomination of quantities to real positive or negative quantities, i.e. to numbers preceded by the signs '+' or '−'. Furthermore, we shall regard quantities as intended to express an increase or decrease, so that a given magnitude will simply be represented by a number, if one only means to compare it with another magnitude of the same species taken as a unity, and by the same number preceded by the sign '+' or the sign '−', if one considers it as to be used for increasing or decreasing a fixed magnitude of the same species [*comme devant servir à l'accroissemment ou la diminuition d'une grandeur fixe de la même espèce*]. [...] We shall call: the 'numerical value' of a quantity that number which forms its basis; 'equal quantities' those that have the same sign and the same numerical value; and 'opposite quantities' two quantities with the same numerical value affected by opposite signs.

Immediately after this, an analogous definition is offered for functions of several variables. Much later ([42], pp. 246–247; [32], p. 163), Cauchy makes clear that these explanations only concern "real functions", to which "imaginary" ones are opposed: these latter are defined as "expressions" of the form '$\phi(x, y, z, \ldots) + \chi(x, y, z, \ldots)\sqrt{-1}$', where '$\phi(x, y, z, \ldots)$' and '$\chi(x, y, z, \ldots)$' designate real functions of $x, y, z, \ldots$.

At first glance, Cauchy's idea of a real function seems close to that of the Euler of the *Introductio*: the adverb 'usually', occurring in his definition, suggests that, for him, real functions are quantities depending on other quantities though they are not necessarily expressed in terms of these latter quantities. But a crucial difference appears when one focuses on the notion of quantity: for Cauchy, a quantity is what we would today term a real number (though real numbers are only informally defined by him, as measures of magnitudes). Hence, in modern parlance, his real functions are functions of real variables. Imaginary functions, instead, are a symbolic generalisation of real ones: they are just "symbolic expressions": combinations of algebraic signs that do not mean anything by themselves ([42], p. 173; [32], p. 117).

This provides the starting point of the so called arithmetisation of analysis. Put briefly, and using modern terminology, this is a development of mathematical analysis based on the idea that functions have to be defined on real and complex numbers. This program differs from Lagrange's in many respects. Two of them are relevant for my purpose. On the one side, the notion of function is no longer mathematically primitive: before introducing it, a (more or less) appropriate notion of real and complex numbers has to be fixed. On the other side, this notion is now confined within a quite narrow disciplinary context, i.e. a particular branch of mathematics. The failure of Lagrange's program resulted, then, in the removal of the notion of function from the basic foundational role that his program had conferred on it.

But something else is also relevant. According to Cauchy's definition, a real function is identified with a real number, namely a variable one, whose variation depends (in any way whatsoever) on the variation of another real number. Though this conception was later refined, several manuals of real analysis continued to base on it. A late example is Czuber's *Vorlesungen über Differential- und Integralrechnung* [45]. Here is how he (*ibid.*, Sect. 3, p. 15) defines real functions:

> If to every value of the real variable x that belongs to its domain [*Bereich*] a definite number y is correlated, then in general y also is defined as a variable, and is said to be a function of the real variable x.

The basic idea is the same as Cauchy's: a real function is a variable real number. This conception is flawed, at least if it is not offered a clear explanation of what it is for a real number to be variable (an explanation which neither Cauchy nor Czuber were able to offer). But it also contains the crucial idea of conceiving functions extensionally, that is, not for the way they realise a connection between appropriate items, but for their connecting certain items to certain other items. In other words, the idea is that of making the identity of a function rest on what it connects rather than on the way it realises the connection. This idea depends on the dissociation of *relata* from relations—which in Lagrange's idea of algebraic quantity are instead kept together. Furthermore, it depends on the admission that the *relata* come before the relation. This is the idea that, through a gradual and difficult evolution, has finally resulted in the modern extensional set-theoretic notion of function: the notion on which that of an arbitrary function considered by Hintikka and Sandu is based.

Mentioning Czuber in this respect is relevant, since Frege takes his definition into account and openly rejects it, in his [100], to which I shall return at the end of Sect. 2.4.4. As we shall see, Frege's objection does not head him to suggest some refinement of the conception of real functions as real numbers, but rather results in his rejection of the very extensional conception of functions. In this way, Frege radically contrasts the more fundamental ground the program of the arithmetisation of analysis was based on, and certainly does not do it by taking a set-theoretic perspective. He rather comes back, in a sense, to Lagrange's attitude. Emphasising this double contrast of Frege's ideas on functions with the arithmetisation of analysis on one side, and with a set-theoretic perspective on another side, is the aim of the next section, where I shall try to show that the views Frege expounds in [100] are perfectly in agreement with the way he deals with functions in the *Grundgesetze*.

2.4 Functions in Frege's *Grundgesetze*

In a recent paper, Tappenden [192] has called it a "myth" that, so far as it is relevant to Frege, nineteenth century mathematics could be reduced to the arithmetisation of analysis, this being conceived as a process "exemplified by Weierstrass", and essentially consisting in "a series of reductions", such as those of derivatives to limits of reals, reals and limits of reals to sets of rationals, rationals to sets of pairs of integers, and integers to sets (*ibid.*, pp. 99–101). But, for Tappenden, denouncing this myth should not result in endorsing the "countermyth" that Frege was "crucially different from Weierstrass and, by extension, from nineteenth-century mathematics generally", in that he was moved by "philosophical desiderata" rather than "mathematical considerations" (*ibid.*, p. 102). According to Tappenden, Frege's views did differ from Weierstrass's, but "this does not reflect a divide between Frege and mathematicians", since "Weierstrass differed from many mathematicians", especially from Riemann, and "Frege was in the Riemannian tradition" (*ibid.*, pp. 106–107).

Doubtless, Frege cannot be enrolled in the process of successive reductions just mentioned (though he was certainly concerned with the rigorisation of analysis: [58]). There are various reasons for this. Among many others, one is relevant for my purpose: Frege's foundational program neither involves the reduction of natural numbers to sets, nor indulges in the conviction that a prior definition of natural, real and complex numbers is required for the notion of function to be clarified. The contrary is true: for Frege, natural and real numbers have to be defined within a formal system conceived as a system of logic, to be set up before any sort of mathematics, and to be expounded by the appeal to a few (non-mathematical) fundamental notions, including that of function.

As far as the notion of function is concerned, Frege's association with the Riemannian tradition is doubtful, instead. In recounting the differences between Riemann's and Weierstrass's "styles", Tappenden argues that, whereas Weierstrass's mathematics is concerned with "explicit given representations of functions", Riemann's requires proving the existence of functions having certain properties "without producing an explicit expression", and is then "committed" to a "wider conception", according to which functions are not "connected to available expressions" ([192], pp. 107 and 121). For Tappenden, Frege's "treatment of function quantification presupposes the most general notion of function, irrespective of available expressions and definitions" (*ibid.*, p. 114). I disagree. Tappenden provides several pieces of evidence showing that both Frege's scientific milieu and his intellectual sympathy were with Riemann's (*ibid.*, pp. 123–130), but he recognises that there is no evidence supporting the claim that the notion of a function "which Frege takes as basic and unreduced" is just the Riemannian one. Tappenden seems to suggest that the best clue for this is merely given by Frege's exposure to the "mathematics around him" (*ibid.*, p. 132). Still, there is a good reason for doubting that Frege's notion coincides with Riemann's: the latter is a mathematical notion; the former cannot be so intended.

Undoubtedly, Frege was aware of most of the mathematical discussions taking place around him, and it is highly plausible that the crucial role he assigned to functions resulted from his "reflection on the function concept in mathematical analysis" ([58], p. 238; [106], vol. 1, p. 129). But it does not follow from this that Frege just imported his own notion of a function from the contemporary mathematical discussion. He could not have been able to appeal to the notion of a function in the exposition of his logical system, if this notion had not been both perfectly independent of any sort of number, and not in need of any possible mathematical proof of existence, more generally, if it had not been a non-mathematical notion. Hence, this notion could have been neither Weierstrass's, nor Riemann's one.

2.4.1 Elucidating the Notion of a Function

But no more could it have been Lagrange's. The main reason for this is not that Lagrange's notion is based on a conflation of syntactical items and their *designata*. It rather pertains to Frege's very conception of a formal system. His own formal system, the *Begriffsschrift*, is usually presented as a system of second-order logic. But it is, in fact, quite different from a formal system in the modern sense. A crucial difference is that the syntax/semantic distinction, as we conceive it today, is lacking: there is nothing like a purely syntactical level of symbols, formulas and rules, and a subsequent level in which an interpretation is provided. The *Begriffsschrift* is, *ipso facto*, a meaningful system. Hence, to introduce it, more than a simple presentation of its language (merely fixing the syntactical behaviour of its elements) is required.

Fixing the meaning of the relevant symbols and formulas and justifying the relevant rules is also needed.

The exposition of the *Begriffsschrift* that occupies part I of the *Grundgesetze* ([97], pp. 5–69), and opens with the passage I quoted at the beginning of this chapter, is just devoted to this latter task. This is what Frege sets forth in a short "*Einleitung*" (*ibid.*, pp. 1–4) that precedes it.

Calling on the notion of function is part of this task. Hence, Frege could have not required that understanding this notion depended on taking the *Begriffsschrift* for granted. But, insofar as the *Begriffsschrift* is for him a model, or better a source, for any scientific formalism, neither could he have admitted that understanding this notion depended on taking any previous formalism for granted. The more fundamental difference between Frege's and Lagrange's notions rests on this.

Frege explains that, for his enterprise to succeed, some relevant "notions [*Begriffe*]" have to be "made clear [*scharf gefasst*]" ([97], p. 1). This is especially the case for the notion underlying the use that mathematicians make of the words 'set [*Menge*]', or 'system [*System*]', the latter case being that of Dedekind (*ibid.*). Frege takes some explanations offered by Dedekind and Schröder [49, 178] into critical account, and argues that what is actually meant with this use is the "subordination of a concept under a concept or the falling of an object under a concept" ([97], p. 2; [110], p. 2_1). Similar considerations, he adds, hold for the word 'correlation [*Zuordnung*]', which, in the context of a reduction of arithmetic to logic, would better be replaced with 'relation [*Beziehung*]' ([97], p. 3; [110], p. 3_1). It follows, he says, that at the grounds of his own "construction [*Bau*]" there have to be the logical notions of a concept and a relation (*ibid.*). In other words: founding arithmetic on logic means reducing the mathematical notions of a set and a correspondence to the logical ones of a concept and a relation.[17] This is just the aim of the *Begriffsschrift*. But for Frege, a necessary

---

[17]In Chap. 1 of this book, Benis Sinaceur argues that Dedekind's logicism, if any, should not be assimilated to Frege's. The previous remarks should be enough to confirm that this was also Frege's conviction. These remarks fit, moreover, with another that Frege already makes in the Preface of the same *Grundgesetze*, also quoted by Benis Sinaceur, in Sect. 1.5.3 of Chap. 1, above ([97], *Vorwort*, p. VIII; [110]; p. VIII$_1$): "Mr Dedekind too is of the opinion that the theory of numbers is a part of logic; but his essay barely contributes to the confirmation of this opinion since his use of the expressions 'system' 'a thing belongs to a thing' are neither customary in logic nor reducible to something acknowledged as logical". Frege's point is then that the notions of set and set membership are not logical as such, but should rather be reduced to logical ones, which is just what Dedekind does not do. It follows that, for Frege, Dedekind's view that "the unique and therefore absolutely indispensable foundation […][for] the whole science of numbers" is "the ability of the mind to relate things to things, to let a thing correspond to a thing, or to represent a thing by a thing", and that without this ability "no thinking is possible" ([49], p. VIII; [53], p. 14), do not coincide with the idea that "arithmetic belongs to logic", as Stein maintains, by taking this last claim to be the same as the claim that "the principles of arithmetic are essentially involved in all thought" ([185], p. 246). The ability to which Dedekind refers is, indeed, a basic cognitive capacity, which, for Frege, does not pertains to logic at all.

condition for articulating this reduction is making these basic logical notions clear, which should result in conveying a logical content before the reduction begins. This is Frege's main point: insofar as logic is to come first, it cannot result from a further reduction to something which is prior to it; still, for it to begin, a content is to be conveyed. What is needed is not a reduction; still it is something suitable for conveying a content. This is an "exposition" ([97], *Einleitung*, pp. 3–4; [110], pp. pp. 3_1–4_1)[18]:

> Yet even after the concepts are sharply circumscribed, it would be hard, almost impossible, to satisfy the demands necessarily imposed here on the conduct of proof without special auxiliary means. Such an auxiliary means is my *Begriffsschrift*, whose exposition [*Darlegung*] will be my first task. It will not always be possible to give a proper definition of everything, simply because our ambition has to be to go back to what is logically simple, and this as such allows of no proper definition. In such a case, I have to make do with gesturing at what I mean.

What Frege means here by 'exposition [*Darlegung*]' is close to what elsewhere (for example, in: the same *Grundgesetze*, Sect. I.1, footnote, and I.34–35; [95], p. 193; [101], pp. 301–302 and 305–306; [106], vol. 1, p. 232, and vol. 2, p. 63) he means by 'elucidation [*Erläuterung*]'. The crucial role of elucidation in "Frege's project" has been recently emphasised by Weiner ([201], Chap. 6; [202], especially pp. 58–61). This is neither a logical nor a scientific procedure. Still, it is a necessary "propaedeutic" ([101], p. 301; [109], p. 300) for logic, and, then, for any science, including mathematics. Its task is communicating basic contents that, insofar as they are purported to be part of logic, and even provide grounds for it, cannot be communicated by logical means, that is, through indefectible definitions (that for Frege could only be explicit ones). In some cases, these contents are reducible, and elucidation can be plain and unequivocal (provided of course that other contents, also communicated through elucidation, are grasped), and can even result in some sort of explicit (though informal) definitions. That's the case with the notions of a concept and a relation, since Frege takes both concepts and relations to be functions, respectively of one and several arguments, whose values are truth-values ([97], Sect. I.3–4). In some other cases, these contents are irreducible, or ineffable, with the effect that their elucidation is successful only if one can count "on a little goodwill and cooperative

---

[18]The same point is also made in "Über Begriff unf Gegenstand", concerning concepts: [95], p. 193; [104], pp. 42–43.

understanding, even guessing" ([101], p. 301; [109], p. 301).[19] That's the case with the notion of a function.[20]

Functions are opposed to objects, for Frege. Thus, they cannot be expressions. And, for the very same reason, they cannot be quantities, numbers, or sets. Concerning quantities and numbers, this is also a consequence of the requirement that the notion of a function come before mathematics. Concerning sets, things are more entangled, since it is far from certain that Frege considered the notion of a set to be mathematical (though his considerations about the use of the words 'set' and 'system' suggest he did).[21] In any case, the requirements that the notion of a function be logically primitive, and that its elucidation belong then to a propaedeutic for logic are enough for excluding the possibility of understanding functions as sets of pairs.

But this is not all. For Frege, all that which is not an object is a function, to the effect that there is no room for specifying which sorts of entities functions are. Indeed,

---

[19]Cf. also [100], p. 665; [104], p. 115: "The peculiarity of functional signs, which we here called 'unsaturatedness', naturally has something answering to it in the functions themselves. They too may be called 'unsaturated', and in this way we mark them out as fundamentally different from numbers. Of course this is no definition; but likewise none is here possible. I must confine myself to hinting at what I have in mind by means of a metaphorical expression [*bildlichen Ausdruck*], and here I rely on the charitable discernment of the reader." According to several scholars (cf., for instance: [44, 65]), the view that elucidation can convey ineffable content, and that this is an essential task for philosophy is Frege's, and it manifests an important aspect of Frege's influence on Wittgenstein (this view is often said to go back to [110], though Geach does not explicitly mention elucidation and limits himself to arguing that "Frege already held, and his philosophy of logic would oblige him to hold, that there are logical category-distinctions which will clearly show themselves in a well-constrcuted language, but which cannot properly be asserted in language": *ibid.*, p. 55). Usually, these scholars admit that Frege calls on different species of elucidation, and take the elucidation of "what is logically primitive" ([44], p. 182) to be the species in which ineffable content is conveyed, the prototypical example being the elucidation of the concept/object distinction. Despite this, it seems to me that if the notions of function and truth-value are taken for granted, the claim that concepts are first-level functions of one argument whose values are truth-values is fully unproblematic. The prototypical example of elucidation's conveying ineffable content is rather that of the function/object distinction. The case of the elucidation of the notions of a concept and a relation also shows that, if the exposition of the *Begriffsschrift* is assimilated to elucidation, then elucidation is opposed to definition only if this last term is taken in a quite strict technical sense (which is proper to Frege), according to which it only refers to the explicit formal definitions admitted within the *Begriffsschrift*. In a broader sense, definitions, even explicit ones, can enter in an elucidation.

[20]The question whether the exposition that occupies part I of the *Grundsetze* has or not a semantic extent—namely whether one can take it or not to provide "semantic justifications of axioms and rules" ([122], p. 365, where Heck is arguing for the affirmative, in contrast with what is argued by Ricketts [168])—is not fully relevant here. What seems to me relevant is that this semantic extent, if any, is quite different from that which would be involved in any discussion about the interpretation of a formal system, and, overall, that this exposition not only aims at showing that "the rules of the system are truth-preserving and that the axioms are true" ([122], p. 365), but also includes the elucidation of fundamental notions like those of an object, a function, a truth-value, a concept, and a relation (on this claim, cf. also [169], Sect. 6, esp. pp. 191–193). This elucidation is "required if one is to master the notation of [...][Frege's] symbolism and properly understand its significance" ([44], p. 181), namely it is "necessary for explaining how Frege's notation [i.e. his *Begriffsschrift*] is to be used in the expression of thoughts" ([201], pp. 251).

[21]Cf. footnote (17).

provided that they are not objects, to wonder which sorts of entities they are, would be the same as wondering which sort of functions they are... Hence, elucidating the notion of a function cannot consist in telling us what functions are. All that Frege can do towards elucidating this notion is to try to account for the way functions work in already given languages (the natural one, and its codified versions used in mathematics), and expounding how they are intended to work in the *Begriffsschrift*.

In my view, this is connected with a point that the mere assertion that functions are not objects only partially accounts for: according to Frege, appealing to functions is indispensable in order to fix the way his formal language is to run, but functions are not as such actual components of this language. More generally, functions manifest themselves in our referring to objects—either concrete or abstract—and making statements about them, but they are not as such actual inhabitants of some world of *concreta* and *abstracta*. Briefly: Frege's formal language, as well as ordinary ones, display functions, but there are no functions as such. As he writes to A. Marty on August 29th 1882 ([106], vol. 2, p. 164; [108], p. 101): "A concept is unsaturated, in that it requires something that falls under it; hence it cannot subsist [*bestehen*] by itself ".[22] *Mutatis mutandis*, the same holds for functions in general.

This is not at all to deny Frege's antipsychologism and objectivism about functions (and concepts or relations). It is merely to argue that it is not part of these theses that functions actually exist as such in some realm of *abstracta*. What is part of these theses is that functions pertain to an objective account of the way language actually works, rather than the way we subjectively think, with the effect that one must appeal to them in order to account for the logical structure of language and thought. As pointed out by Picardi ([159], p. 53): functions are to be conceived as "objective pattern[s] that we discern in the world", rather than as "separate ingredient[s] of it".

2.4.2 How (First-Level) Functions Work in the **Begriffsschrift**

To clarify all this, let me briefly sketch the role that functions play in the *Begriffsschrift*.

In Sect. I.5 of *Grundgesetze*, Frege establishes that statements (*Satze*) are formed in this system by letting the special sign ' $\vdash$ ' precede appropriate terms. These are either names of a truth-value— i.e. either of the True or of the False—or appropriate formulas. These latter formulas involve Latin letters and are suitable for being transformed into a name of a truth-value through appropriate replacements of these letters. I shall better specify this condition pretty soon. For the time being it is enough to say that, though he does not say it explicitly, Frege implies that a statement in which the sign ' $\vdash$ ' precedes a name of a truth-value asserts that what makes up this name is

---

[22]Cf. also [99], p. 34; [109], p. 282: "It is clear that we cannot put down [*hinstellen*] a concept as independent, like an object; rather it can occur only in a connection. One may say that it can be distinguished within, but it cannot be separated from the context in which it occurs".

a name of the True obtains,[23] whereas a statement in which the sign ' $\vdash$ ' precedes a formula involving Latin letters asserts that it obtains what makes up this formula transforms into a name of the True under any licensed replacement of these letters. The former case is fundamental; the latter reduces to it through the appropriate stipulations on the replacement of Latin letters. Let us then begin with the former.

For Frege, the True and the False are two peculiar objects whose existence is taken for granted. Hence, a name of a truth-value is a name of an object, or a "proper name [*Eigenname*]" as he says ([97], Sect. I.3; [110], p. 7$_1$). But not any proper name is suitable for yielding a statement of the *Begriffsschrift*, if preceded by the sign ' $\vdash$ ', and it is no more so for any name of a truth-value. For a proper name to be suitable for this, it has to belong to the language of the *Begriffsschrift* (or to an appropriate extension of it), and to be appropriately formed within this language. The former requirement is obvious and already sufficient for excluding names like 'the True' or 'the False', which do not belong to this language. The latter is what matters here. It could be so rephrased: a proper name of the language of the *Begriffsschrift* is suitable for yielding a statement of this system (if preceded by the sign ' $\vdash$ ') if it is formed so as to be the name of the value of a concept or relation, i.e. of a function whose values are truth-values. Hence, such a name not only refers to a truth-value, it also refers to it in a certain way, which depends on the nature of the relevant function (and it is just because of this nature that this name is possibly warranted to be a name of the True, as it happens if the corresponding statement is a theorem).

But what does it mean, in the context of the *Begriffsschrift*, that a function has a certain nature? Though Frege is never explicit on this matter, his exposition leaves no doubt: it means that this function is either one of the few primitive ones admitted in this system, or is generated in a certain way by reiteratively composing these primitive functions[24], and, possibly, by relying on some auxiliary explicit definitions.

This is still not clear enough, since, provided that functions are not actual components of the language of the *Begriffsschrift*, the problem of understanding how something which is not such an actual component can be either a primitive item of this language or be composed by reiteratively composing primitive items of it is still open. Part of the answer is that the foregoing condition has to be understood as follows: in the context of the *Begriffsschrift*, a function has a certain nature if the names of its values are either names of values of a certain primitive function, or are generated in a certain way by reiteratively composing these latter names and, possibly, by relying on some other proper names introduced by explicit definition.

But this is not the end of the story, yet. It is still necessary to explain, what does it mean that a proper name is a name of a value of a certain function. For my present purpose, I can restrict the answer to primitive functions (to pass to composed ones, it would be enough to specify which rules of composition are licensed, which is a question that we can leave aside, here).

---

[23]For example, in the same Sect. I.5, Frege argues that ' $2^2 = 4$ ' is a name of the True, and that the statement ' $\vdash 2^2 = 4$ ' asserts that the square of 2 is 4.

[24]Cf. footnote (4) of the Introduction.

These functions are introduced through appropriate informal but explicit definitions.[25] Four of them (two concepts and two relations) are introduced in Sects. I.5–7, and I.12. They are the horizontal, the negation, the identity, and the implication. These are first-level functions: functions whose arguments are objects. For the time being, I primarily restrict the discussion to these functions. I shall explicitly consider higher-level functions in Sect. 2.4.4 (especially pp. 87–89). Up to that point, I shall use the terms 'function', 'concept' and 'relation' to primarily speak of first-level functions. In order to generalise some of the things I shall say about them to functions of any level, some changes would be necessary. But what I shall later say of higher-level functions is intended to show that these changes would not effect what is essential for my purposes.

Consider then, as examples, the four aforementioned functions. They are defined through the following stipulation schemas:

$$
\begin{aligned}
&\mathrm{-\!\!\!-}\,\Delta\ \text{is}\ \begin{cases} \mathsf{T}\ \text{if}\ \Delta\ \text{is}\ \mathsf{T} \\ \mathsf{F}\ \text{if}\ \Delta\ \text{is not}\ \mathsf{T} \end{cases} &&\mathrm{-\!\!\!\top}\,\Delta\ \text{is}\ \begin{cases} \mathsf{F}\ \text{if}\ \Delta\ \text{is}\ \mathsf{T} \\ \mathsf{T}\ \text{if}\ \Delta\ \text{is not}\ \mathsf{T} \end{cases} \\[2ex]
&\Gamma = \Delta\ \text{is}\ \begin{cases} \mathsf{T}\ \text{if}\ \Gamma\ \text{is}\ \Delta \\ \mathsf{F}\ \text{if}\ \Gamma\ \text{is not}\ \Delta \end{cases} &&\begin{matrix}\\ \top \\ \vdash \Gamma \end{matrix}\Delta\ \text{is}\ \begin{cases} \mathsf{F}\ \text{if}\ \Gamma\ \text{is}\ \mathsf{T}\ \text{and}\ \Delta\ \text{is not}\ \mathsf{T} \\ \mathsf{T}\ \text{if}\ \Gamma\ \text{is not}\ \mathsf{T}\ \text{or}\ \Delta\ \text{is}\ \mathsf{T} \end{cases}
\end{aligned} \tag{2.6}
$$

where 'T' and 'F' refer to the True and the False, respectively, and 'Γ' and 'Δ' are schematic letters for objects.

These definitions involve nothing but schematic names of objects, among which '—Δ', '⊤ Δ', 'Γ = Δ', and '⊤ Δ / Γ' are schematic names of values of the relevant functions. It is then clear that these functions are defined by fixing the reference of the names of all their possible values.

Of course, this definition belongs to the language of the exposition of the *Begriffsschrift*. Indeed, though Frege largely uses Greek capital letters, like 'Γ' and 'Δ' in such an exposition, they are not part of the language of *Begriffsschrift* itself, and this is then neither the case of the schematic names involving them. Within the *Begriffsschrift*, Greek capital letters are replaced either by names of particular objects or by Latin letters. These last letters are used to "express generality" ([97], Sect. I.17; [110], p. 31₁). Some Frege scholars (for example [115], p. 67) take them to be free variables and suggest understanding the formulas involving them as abbreviations of universally quantified statements. It seems to me more faithful to Frege's views to understand them as special schematic letters, differing from the Greek capital ones for being used within the *Begriffsschrift*. Insofar as, in this system, any formula

---

[25]These definitions are informal insofar as they belong to the exposition of the *Begriffsschrift*, rather than to the *Begriffsschrift*, itself. Hence, they reduce to stipulations stated in the natural language, as clearly as possible (under the supposition that what is involved in them has been previously elucidated). This is, thus, another example of the fact that, if the exposition of the *Begriffsschrift* is assimilated to elucidation, the latter is not necessarily opposed to definition in the broad sense: cf. see the footnote (19).

occurs within a statement or an explicit definition—which can be taken to be a sort of statement—within the *Begriffsschrift*, Latin letters only enter into statements. This makes it possible to fix their use by stipulating that a statement of the *Begriffsschrift* in which they occur asserts that things are such that the schematic proper names resulting from this same statement by omitting the sign ' $\vdash$ ', and replacing each Latin letter with a Greek capital one, is a schematic name of the True. For example, the statement

$$\vdash\!\!\!\!\begin{array}{c} a\,, \\ a \end{array}$$ (2.7)

asserts that things are such that ' $\begin{array}{c}\rule{1cm}{0.4pt}\ \Gamma \\ \Gamma\end{array}$ ' is a schematic name of the True.[26] This does not entail that the formula ' $\begin{array}{c}\rule{1cm}{0.4pt}\ a\, \\ a\end{array}$ ' is, in turn, a name of the True. This is simply because, taken alone, it is not a well-formed formula of the *Begriffsschrift*, where it can only occur within a statement. This is the same for any formula involving Latin letters.[27]

This should be enough to make clear how functions are supposed to enter into statements in the *Begriffsschrift*. But this is not by far the end of the story, since the exposition of this system —although not this very system—also involves "names of functions [*Functionsnamen*]" ([97], Sect. I.2; [110] p. 6_1), or f-names, as I shall say from now on. On the one hand, this is natural, since it is easy to imagine a situation in which, by speaking about the *Begriffsschrift*, one has to mention some particular functions, as I have just done myself using the terms 'horizontal', 'negation', 'identity', and 'implication'. On the other hand, this is puzzling, since functions are not objects, and it is then difficult to understand how they can have names. The puzzle has two aspects, at least: a notational and a substantial one.

As far as only the former is taken into account, Frege's solution merely depends on the introduction of a special sort of letter, whose purpose is just that of entering into f-names. These are Greek small letters, like 'ξ' and 'ζ'. By replacing 'Γ' and 'Δ' with them in the left hand sides of stipulations (2.6), one gets the following names of the relevant functions:

$$-\xi \quad ; \quad \longrightarrow \xi \quad ; \quad \xi = \zeta \quad ; \quad \begin{array}{c}\rule{1cm}{0.4pt}\ \xi. \\ \zeta\end{array}$$ (2.8)

Like the Greek capital letters 'Γ' and 'Δ', neither these names nor the Greek small letters 'ξ' and 'ζ' belong to the language of the *Begriffsschrift*, but only to

---

[26]In fact, appropriate conventions relative to the "scope of the generality" have to be also made ([97], Sect. I.8 and I.17; [110], pp. 11_1–12_1, 31_1). Here, I cannot enter into this matter, and merely observe that Frege's use of Latin letters is such that generality cannot be expressed in the *Begriffsschrift* only through them: universal quantifiers are also necessary.

[27]Frege emphasises this fact by stipulating that a Latin letter for objects "indicates [*andeute*]" an object rather than refers to it ([97], Sect. I.17; [110], p. 31_1).

that of its exposition. But, unlike Greek capital letters, Greek small ones are not schematic, and are neither variables nor constants. They are merely used to hold places open for being occupied, both in the language of the *Begriffsschrift* and in that of its exposition, by other appropriate letters, so as to get either names of values of the relevant functions, or formulas suitable for entering into statements. Consider implication: the former case obtains if 'ξ' and 'ζ' are replaced by 'Δ ' and 'Γ' or by names of determined objects like '2' and '3', so as to get the schematic names

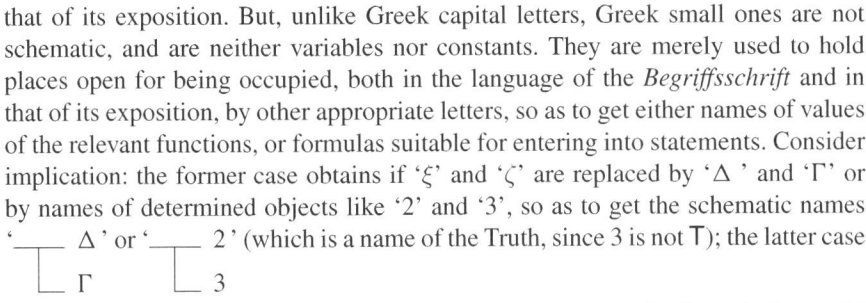

obtains if 'ξ' and 'ζ' are replaced by 'b' and 'a', so as to get the formula '———— b '

suitable for entering into the statement '⊢—— b '. One could then say that f-names are tools to be used in the *Begriffsschrift* for forming proper names and statements, or for analysing them.

This account of the role of functions in the *Begriffsschrift* and in its exposition could be completed in many respects. But from the little I have said, it should be clear that for functions to manifest themselves in the *Begriffsschrift*, there is no need for them to be actual components of its language. Though things are less clear for the language of the exposition of this system, because of the presence of f-names, there is no doubt that, for Frege, these names are tools for forming proper names and statements, or for analysing them. In my understanding, this is just what he means with his well known metaphor about the unsaturated nature of functions.

The point is made, for example, at the very beginning of part I of the *Grundgesetze*, with respect to the example of the numerical function $(2 + 3x^2) x$ ([97], Sect. I.1; [110], pp. 5_1–6_1). Frege claims that the "essence [*Wesen*]" of this function both "reveals itself [...] in the connection [*Zusamengehörigkeit*] it bestows between the numbers whose signs we put for 'x' and the numbers that then result as the reference of the expression" resulting from this replacement, and "lies [...] in the part of the expression that is there besides the 'x'". Then he adds that "the expression of a function is in need of completion, unsaturated" and that 'x' (which, according to him, should be used in mathematics like 'ξ' is used in the *Begriffsschrift*) is there "to hold open places for a numeral", and then to "to make know the particular mode of need for completion that constitutes the peculiar essence" of the function. Despite his using the term 'essence', Frege says nothing here about what he considers functions to be. He only says something about the way the corresponding expressions are intended to work.

2.4.3 (First-level) Functions and Names of Functions

This cannot be all, however, since the substantial aspect of the puzzle about f-names remains still unsettled. Do these names refer to something? And, what does it mean that two f-names are names of the same function or of distinct functions?

These questions concern particular aspects of a more general problem. For Frege, identity only applies to objects. Hence, strictly speaking, no identity condition for functions is conceivable. Does this mean that functions can meet some other sort of sameness conditions, or that there are no such conditions at all[28]?

The matter is connected to the paradox of the concept ⌜horse⌝: in "Über Begriff und Gegenstand" Frege famously holds that "the three words 'the concept ⌜horse⌝' do designate [*bezeichnen*] an object, and, on account of that, they do not designate a concept" ([95], p.195; [104], p. 45). One could think that this merely depends on the awkwardness of natural language, and that this is just what Frege implies by saying that "it is impossible to ignore that there is an unavoidable linguistic hardship [*unvermeidbare sprachliche Härte*] if we claim that the concept ⌜horse⌝ is not a concept" ([95], p. 196; [104], p. 46). Still, this hardship is unavoidable for him, which suggests that he takes the inconvenience of natural language to be a symptom of a deeper problem.

This is confirmed by his raising the problem also in the *Grundgesetze* (in a footnote to Sect. I.4, [97], p. 8; [110], p. 8_1):

> There is a difficulty [...] which can easily obscure the true state of affairs and thereby arouse suspicion concerning the correctness of my conception. If we compare the expression 'the truth-value of Δ's falling under the concept $\Phi(\xi)$' with '$\Phi(\Delta)$' we see that '$\Phi()$' really corresponds to 'the truth-value of ()'s falling under the concept $\Phi(\xi)$', and not to 'the concept $\Phi(\xi)$'. So the latter words do not really designate a concept (in our sense), even though the linguistic form makes it look as if they do. On the inescapable situation [*Zwangslage*] in which language here finds itself, cf. my essay "Über Begriff und Gegenstand".

The relevant language here is that of the exposition of the *Begriffsschrift*. The "inescapable situation" or "unavoidable hardship" in which it finds itself is then a symptom of a problem relative to the basic notions of this system. In this language, 'Φ' works as a schematic letter for functions. Hence '$\Phi(\Delta)$' and 'the truth-value of

---

[28]That identity only applies to objects is a point that Frege makes on many occasions; he often argues as well that a "corresponding relation" applies to concepts or functions. But he does not use a fixed compact vocabulary for this purpose. In his review of Husserl's *Philosophie der Arithmetik* ([98], p. 320; [109], p. 200), he argues that "coincidence [*Zusammenfallen*] in extension is a necessary and sufficient condition for the occurrence between concepts of the relation that corresponds to equality [*Gleichheit*] between objects" (I shall come back later to this claim, at p. 83), then remarks: "it should be noted in this connection that I'm using the word 'equal [*gleich*]' without further addition in the sense of 'not different [*nich verschieden*]', 'coinciding [*zusamenfallend*]', 'identical [*identisch*]'". In "Ausführungen über Sinn und Bedeutung" ([106], vol. 1, pp. 132; [107], p. 122), he also argues that "the word 'the same [*derselbe*]' used to designate a relation between objects cannot properly be used to designate the corresponding relation between concepts". Hence, speaking of sameness conditions for functions is not faithful to Frege's parlance. Still, I use this expression for short, to speak of the conditions under which a certain function is this very function rather than some other one.

Δ's falling under the concept $\Phi\,(\xi)$' are proper names that refer to the same object (either the True or the False). Frege's point is then the following: insofar as the former of these proper names is formed by filling a blank in ' $\Phi()$', the role of 'the concept $\Phi\,(\xi)$' in the latter cannot but be that of contributing to form this name, rather than that of designating a concept. But, then, what does 'the concept $\Phi\,(\xi)$' mean? Or, more generally, what is one speaking about by saying something of the concept $\Phi\,(\xi)$, rather than of some other concept?

To better appreciate the nature of the problem, consider another quandary, only apparently related to it: from the supposition that any function has a value-range, it follows that the concept $\ulcorner$horse$\urcorner$ has an extension; but, if 'the concept $\ulcorner$horse$\urcorner$' refers to an object, the statement 'the concept $\ulcorner$horse$\urcorner$ has an extension' cannot be true. To solve this quandary, it is enough to pass to the language of the *Begriffsschrift*. Let '$Hrs\,(\xi)$' be a name of the concept $\ulcorner$horse$\urcorner$ in an appropriate extension of this language. The statement

$$\text{'}\!\vdash\!\!\underset{}{\overset{\mathfrak{a}}{\rule{2.5em}{0.4pt}}}\!\mathfrak{a}=\grave{\varepsilon}Hrs(\varepsilon)\text{'}$$

is then a rendering of 'the concept $\ulcorner$horse$\urcorner$ has an extension', and it is an immediate consequence of

$$\text{'}\!\vdash\!\!\underset{}{\overset{\mathfrak{f}}{\rule{2em}{0.4pt}}}\!\!\underset{}{\overset{\mathfrak{a}}{\rule{2em}{0.4pt}}}\!\mathfrak{a}=\grave{\varepsilon}\mathfrak{f}(\varepsilon)\text{'}$$

which is a rendering of 'any function has a value-range'.

The problem that Frege tackles in "Über Begriff und Gegenstand" is essentially different, since it is not solvable by passing to the language of the *Begriffsschrift*. It is not about the way some statements have to be appropriately formulated: it is rather about the way functions and f-names have to be understood.

Possibly, a clearer way to state this is the following. Consider the phrase 'the function $\Phi\,(\xi)$', or also the mere f-name '$\Phi\,(\xi)$', by supposing that it is just used for naming a certain function, and replace in them 'ξ' with 'Δ', so as to get 'the function $\Phi\,(\Delta)$' and '$\Phi\,(\Delta)$'. The former expression is misguided. The latter is not, but, clearly, it is no more suitable for naming the relevant function. It follows that both in 'the function $\Phi\,(\xi)$' and in '$\Phi\,(\xi)$'—supposing that this last name is just used for naming a certain function—'ξ' is not used to hold a place open.[29] Hence, in spite of being used to name functions, these expressions are not unsaturated, and are then unsuitable for this purpose.

---

[29]In order to show that the paradox does not depend on the use of expressions like 'the concept _', Wright has stated it as follows ([206], pp. 74–77; for clarity, I adapt his argument to my setting; on this matter, cf. also [68], pp. 212 *seq.*): (i) the expression 'That which is named by '$\Phi\,(\xi)$'' is a singular term; (ii) hence, its reference, if any, is an object; (iii) the reference of 'That which is named by '$\Phi\,(\xi)$'' is that which is named by '$\Phi\,(\xi)$'; (iv) hence, that which is named by '$\Phi\,(\xi)$' is an object. It follows that the problem cannot be solved by merely jettisoning expressions like 'the concept _'.

The solution that Frege offers in "Über Begriff und Gegenstand" matches up with the nature of the problem, since it does not merely consist in suggesting some linguistic tricks. It goes as follows ([95], p. 197; [104], pp. 46–47):

> In logical enquiries one often needs to assert [*auszusagen*] something about a concept, and to shape it in the usual form for it, namely to put the content of the assertion into the grammatical predicate. Consequently, one would expect that the reference of the grammatical subject would be the concept; but, because of its predicative nature, this cannot play this part; it must first be converted into an object, or, speaking more precisely, represented [*vertreten*] by an object, which we designate by the prefix 'the concept', as in 'the concept ⌜man⌝ is not empty'.

The problem with this solution is that it is begging the question, at least partially. As the same point could and should also be made about functions in general, it requires that for each function whose name supplies the grammatical subject of an assertion about itself, there is an object "representing" this same function, to which this name refers, in the context of this assertion. But, for this to provide an effective solution to the problem, one should also require that the truth-conditions of this assertion depend on the relevant function, i.e. that the object representing this function reflects what makes it a certain particular function. And this requires, in turn, that appropriate conditions for singling out this function be provided.

Frege acknowledges that the objects representing functions should be of "a quite special kind" ([95], p. 201; [104]; p. 50). But he is silent not only on their very nature, but also on the way they might reflect the relevant features of the functions they represent, and on the sameness conditions of these functions.

It is quite tempting to take these objects to be the value-ranges of the corresponding functions, and even to argue that this is what Frege himself implies when he claims to have never "identified concept and extension of concept" and adds that he "merely expressed [...][the] view that in the expression 'the number that applies to the concept F is the extension of the concept ⌜like-numbered to the concept F⌝, the words 'extension of the concept' could be replaced by 'concept'" ([95], p. 199; [104]; p. 48). But there are many reasons for resisting this temptation.[30]

Let me advance two of them, both of which depend on taking the relevant problem to be not merely that of providing a reference for 'the function $\Phi(\xi)$' or '$\Phi(\xi)$' in the context of an assertion about a certain function, but rather that of explaining what makes this assertion hold of this very function rather than of some other one. On this understanding, admitting that 'the function $\Phi(\xi)$' or '$\Phi(\xi)$' refer, in the context of this assertion, to the value-range of $\Phi(\xi)$ results in admitting both that, with respect to this context, $\Phi(\xi)$ is to be taken to be the same function as $\Psi(\xi)$ if and only if the value-range of $\Phi(\xi)$ is the same of that of the function $\Psi(\xi)$, and that the truth conditions of this assertion just depend on the value-range of the function $\Phi(\xi)$.

The first reason is that, if this were so, many distinctions and assertions that one would plausibly like to make would collapse and have quite odd truth-conditions. For example, one should conclude that, with respect to the context of an assertion about the function $\_\_\xi$, this last function is to be taken to be the same function as

---

[30] Some of these reasons have been offered in [174, 175]. For a critical discussion of them, cf. [170].

$\xi = (\xi = \xi)$, and that the assertions '$\xi = (\xi = \xi)$ is an elementary function of the *Begriffsschrift*' and 'the function $\xi = (\xi = \xi)$ is called 'horizontal' and enters into any statement of the *Begriffsschrift*' are true insofar as '$\underline{\quad}\xi$ is an elementary function of the *Begriffsschrift*' and 'the function $\underline{\quad}\xi$ is called 'horizontal' and enters into any statement of the *Begriffsschrift*' are true.

These conclusions are not only odd. They also seem to go against Frege's claims. For example, in Sect. I.10 of *Grundgesetze*, he undertakes to offer a "more precise determination of what the value-range of a function is supposed to be" ([97], *Inhaltsverzeichniss*, p. XXVII; [110], p. XVII$_1$). To this purpose, he considers the three functions introduced in the previous sections, namely $\underline{\quad}\xi$, $\underline{\quad}_{\top}\xi$ and $\xi = \zeta$, and remarks that "we can reduce [*zurückführen*] the function $\underline{\quad}\xi$ to the function $\xi = \zeta$ ", since "the function $\xi = (\xi = \xi)$ has the same value as the function $\underline{\quad}\xi$ for every argument" ([97], Sect. I.10; [110]: p. 16$_1$), which seems to imply that, with respect to the context of these assertions, he takes the functions $\underline{\quad}\xi$ and $\xi = (\xi = \xi)$ to be two distinct functions with the same value-range.

The second reason is as follows.[31] Let '$\Phi(\xi)$' and '$\Psi(\xi)$' be two (distinct) f-names. To say that the value-range of $\Phi(\xi)$ is the same as the value-range of $\Psi(\xi)$ means, for Frege, that $\Phi(\Delta)$ is the same object as $\Psi(\Delta)$, whatever the object Δ might be, as Basic Law V prescribes.[32] Hence, admitting that $\Phi(\xi)$ is the same function as $\Psi(\xi)$ if and only if the value-range of $\Phi(\xi)$ is the same as that of $\Psi(\xi)$ results in admitting that $\Phi(\xi)$ is the same function as $\Psi(\xi)$ if and only if $\Phi(\Delta)$ is the same object as $\Psi(\Delta)$, whatever the object Δ might be. But what does it mean that $\Phi(\Delta)$ is the same object as $\Psi(\Delta)$, whatever the object Δ might be? Insofar as Frege has no way to understand the totality of values of a function otherwise than as the value-range of this function, and has no other identity condition for value-ranges of functions than that stated by Basic Law V, according to him this cannot but mean that the proper names '$\Phi(\Delta)$' and '$\Psi(\Delta)$' are identified as the names of the values of two functions $\Phi(\xi)$ and $\Psi(\xi)$ for Δ as argument, and that these functions are associated to appropriate rules, procedures or capabilities which, besides being apt to identify the names of their values, are also apt to warrant that, whatever the object Δ might be, the reference of the proper name '$\Phi(\Delta)$', identified as the name of a value of the function $\Phi(\xi)$, cannot but be the same as the reference of the proper name '$\Psi(\Delta)$' identified as the name of a value of the function $\Psi(\xi)$. It would follow that admitting that 'the function $\Phi(\xi)$' or '$\Phi(\xi)$' refer, in the context of an assertion about the function $\Phi(\xi)$, to the value-range of this function would result in admitting that, with respect to this context, $\Phi(\xi)$ is to be taken to be the same function as $\Psi(\xi)$ if and only if these functions are associated to rules, procedures or capabilities that

---

[31]I develop here a remark of Hintikka and Sandu ([129], p. 299: for Frege, "the extension of a concept can only be apprehended by our logical faculties starting out from the concept".

[32]For simplicity, I only consider here first-level functions with one argument. It is easy to generalise Basic Law V to first-level functions with several arguments. But, if functions of higher-levels are considered, it is not perfectly clear what it would mean, for Frege, that these functions have the same or different value-ranges (on this matter, cf. [174], p. 32), and it would then be hard to allege that, in order to provide sameness conditions for these functions, it would be enough to stipulate that these conditions reduce to the identity conditions of the value-ranges of these functions.

provide such a warrant. But, if it is admitted that functions are associated to such rules, procedures or capabilities, it seems much more natural to maintain that, with respect to the context of an assertion about the function $\Phi(\xi)$, what enforces that this function be taken to be the same as $\Psi(\xi)$ directly pertains to these very rules, procedures or capabilities, without appealing to the value-ranges of these functions.

This looks like a *reductio ad absurdum* of the identification of Frege's objects of a quite special kind with value-ranges of functions. But what about the view that, with respect to the context of an assertion about the function $\Phi(\xi)$, what arranges matters so that this function is to be taken to be the same as $\Psi(\xi)$ directly pertains to the appropriate rules, procedures or capabilities associated to these functions? Answering this question requires taking other elements into account.

The passage of "Über Begriff und Gegenstand" quoted above is not the only one where Frege implies, or even openly claims, that f-names—or, more specifically, concept-words [*Begriffsworten*]—have both sense and reference. He does it, for example, in a letter to Husserl of May, 24th 1891 ([106], vol. 2, pp. 94–98), in "Ausführungen über Sinn und Bedeutung", probably written between 1892 and 1895 ([106], vol. 1, pp. 128–136), and in "Einleitung in die Logik", of August 1906 ([106], vol. 1, p. 208–212; [107], pp. 191–196). In all these cases, he also argues that the reference of a concept-word is the concept itself, and, in the third of these texts, he goes as far as to imply that a function or concept is just the reference of an f-name or a concept-word, respectively.

In this last case, his argument depends on the principle of compositionality, and goes as follows ([106], vol. 1, pp. 209–212; [107], p. 193 and 195). If we say 'Jupiter is larger than Mars', we are saying that the references of 'Jupiter' and 'Mars' stand to one other in a certain relation, and we do this through the words 'is larger than'. Insofar as this relation holds between references of proper names, it "belongs to the realm of references". Hence, one has to admit that also the phrase 'is larger than Mars' is "endowed by reference [*bedeutungsvoll*]". So, if a statement is split up into a proper name and the remainder, then the latter "has for its sense an unsaturated part of a thought, and we call 'concept' its reference". In more generality, there are many proper names that can be analysed into a saturated part, namely, a proper name, and an unsaturated part. If the latter is such that by saturating it with a proper name having a reference, one gets another such proper name, then "we call 'function' the reference of this unsaturated part".[33]

This being said, Frege cannot but remark that claims like these bring us back to the paradox tackled in "Über Begriff unf Gegenstand". In "Einleitung in die Logik", he confines himself to arguing that "language forces upon us" the "mistake [*Fehler*]" or "inaccuracy [*Ungenauigkeit*]" these claims involve, with the result that we cannot avoid them but by bearing the difficulty in mind and insisting that concepts are unsaturated or "predicative in character" ([106], vol. 1, pp. 209–210; [107], p. 193). In "Ausführungen über Sinn und Bedeutung", he says, or at least implies, something

---

[33]We find a similar claim already in "Über Begriff und Gegenstand" ([95], p. 198; [104], pp. 47–48): "We must say in brief, taking [...] 'predicate' in the linguistic sense: a concept is the reference of a predicate".

more. He specifies ([106], vol. 1, p. 128–132; [107], pp. 118–121) that "a concept-word refers to a concept, if the word is used as it is appropriate for logic". Then he adds, as a clarification, that "in any statement, we can substitute *salva veritate* one concept-word for another if they have the same extension, so that it is also the case that in relation to inference and to the laws of logic, concepts differ only insofar as their extensions are different". To reinforce these claims, Frege observes that the unsaturatedness of functions also comes out in the case of concepts entering into the subject of a statement [*Subjektsbegriffen*], such as in 'all equilateral triangles are equiangular', which he takes to be the same as 'if anything is an equilateral triangle, than it is an equiangular triangle'. To consider a simpler example, this means that a statement like 'the morning star is a planet' should be rephrased, in good logic, as 'the object that is the morning star is a planet', with the result that its subject involves the concept-word '...is the morning star', whose reference is the concept ⌜Morning Star⌝. Finally, Frege goes on to argue that the identity of the extensions of concepts results in a second-level relation holding between the concepts themselves and corresponding to the identity of objects.[34]

What seems to me important here is that Frege relativises his claims to the case where concept-words, and plausibly f-names in general, are "used as it is appropriate for logic" and inferences and laws of logic are concerned, which means, I suggest, that these names occur (as unsaturated components) within some proper names or statements used for affirming and inferring truths about objects, as always happens in the *Begriffsschrift*.[35] Hence, his point seems to be that, when language is used in order to affirm and infer truths about objects and f-names occur as unsaturated parts of proper names and sentences, the former names have references and refer to functions, and functions differ only if their value-ranges differ, so that a second-level relation analogous to the identity between objects applies to functions when they have the same value-range.

These claims should not be taken as evidence for identifying Frege's objects of a quite special kind with the value-ranges of functions, and even less as evidence for arguing that, for Frege, the identity of value-ranges provides the sameness of the corresponding functions. It seems quite clear, indeed, that these claims only apply insofar f-names are used as it is appropriate for logic, i.e only insofar as functions are involved in affirming and inferring truths about objects, namely about their values. This leaves open the problem of understanding what makes it that an assertion about a function is about this function rather than some other function, or, more generally,

---

[34]Frege even arrives at suggesting a special sign for this relation (to be used, of course, in the language of the exposition of the *Begriffsschrift*). Let $\Phi\,(\xi)$ and $\Psi\,(\xi)$ two concepts with the same extension. Frege suggests writing '$\Phi\,(\alpha) \stackrel{\alpha}{\smile} \Psi\,(\alpha)$' arguing that this expresses the same thing as '$\underset{\alpha}{\smile} \Phi\,(\alpha) = \Psi\,(\alpha)$'.

[35]That logic is concerned with truths about objects is, in my view, the distinctive mark of Frege's extensionalist conception of logic (which he emphasises in "Ausführungen über Sinn und Bedeutung" by repeatedly observing that his remarks favour the "logician of extension against that of intension" ([106], vol. 1, p. 128 and 133–134; [107], p. 118 and 122–123). But this conception does not entail at all an extensionalist conception of functions.

what makes it that a certain function is this very function rather than some other function.

Frege argues that a concept-word has a reference and this is just what he calls 'concept' also in the 1903 paper on "Über die Grundlagen der Geometrie". But in this case, he adds that "this is not a definition, since the decomposition [of a proper name or statement] into a saturated and an unsaturated part must be considered as a logically primitive phenomenon that must simply be recognised but not reduced to something simpler". This is a hint for a better understanding of Frege's view. In the language of the exposition of the *Begriffsschrift*—the only one in which Frege grants to himself the licence to speak about functions—one can describe what functions and f-names do in the language of the *Begriffsschrift*, or in any other language used for affirming truths about objects. But one cannot say what functions are, since, though being at work in these latter languages, functions are not, as such, actual components of them. If the account of what functions and f-names do is intended to be fine-grained enough for identifying the contribution of single functions, then unavoidably we fall into inaccuracy. However, this should not be so bad as to blur what is essential, namely that functions manifest themselves in the way we refer to objects through proper names and use statements to affirm truths about objects. Saying that the reference of f-names used in these languages (as unsaturated expressions) are functions is then nothing more than saying that f-names contribute to form (molecular) proper names and statements, or can be recognised through an analysis of (molecular) proper names and statements, and that functions "establish connections"[36] between the objects whose names are recognised as (saturated) parts of the relevant (molecular) proper names and statements and those that these latter proper names refer to and these statements are about. This suggests that the sense of an f-name depends on the way the references of the (molecular) proper names involving this f-name (as an unsaturated part of it) are to be determined on the basis of the references of the proper names which are recognised as (saturated) parts of the former proper names, i.e. on the way functions establish connections between objects. The value-ranges of functions merely depend, instead, on which objects are connected to which others. And, insofar as the same objects can be connected in different ways, two f-names can have different senses though referring to functions with the same value-range.

In this picture, the sense of an f-name essentially differs from the function this name refers to, since the former depends on the way the latter does what it does: in other words, functions act, and senses differ if the ways they act differ. And, both the sense and the reference of an f-name differ from the value-range of the corresponding function, since value-ranges neither act, nor differ if the ways the functions act differ.[37] Still, it seems obvious that the same function cannot connect the same objects in two different ways. Hence, though functions differ (i.e. produce different outcomes) only insofar as their value-ranges differ, when their names are used as it is appropriate for logic, when these same names are used in the context

---

[36]Cf. the quote from Sect. I.1 of *Grundgesetze* at the end of Sect. 2.4.2 below.

[37]I'm indebted to F. Schmitz for this account of the distinction between sense and reference of an f-name and the value-range of the corresponding function.

of an assertion about particular functions (necessarily made in the language of the exposition of the *Begriffsschrift*), these functions differ insofar as the senses of these names (when used as it is appropriate for logic) differ.

Now for Frege, in the language of the *Begriffsschrift*, and in any other language appropriate for expressing and inferring truths about objects, *f*-names cannot be used appropriately unless it is determinate which objects the relevant functions connect to which other objects. This is a requirement that Frege often advances. For example in the "Ausführungen über Sinn und Bedeutung": "it must be determinate [*bestimmet*] for every object whether it falls under a concept or not; a concept-word which does not meet this requirement on its reference is not endowed with a reference [*bedeutungslos*]" ([106], vol. 1, p. 133; [107], p. 122). I do not see any other way to understand this requirement than by taking it as demanding that an appropriate use of *f*-names requires the capability of deciding which object the corresponding function connects to any given object. But, at least in the context of a codified language suitable for being used in science (like the *Begriffsschrift* whatsoever extended), this capability cannot be conceived as a mere subjective ability, but rather depends on the availability of appropriate rules or procedures. And, if this is so, it is natural to admit that the sense of an *f*-name (when used as it is appropriate for logic) just depends on these rules or procedures, so that, in the context of an assertion about functions (which cannot but be an assertion about what functions and *f*-names do in the language *Begriffsschrift*), also the sameness of functions depend on these same rules or procedures.

This brings us back to the view I have above contrasted to the identification of Frege's objects of a quite special kind with value-ranges of functions. According to this view, these objects should somehow reflect the distinctive features of these rules or procedures (i.e. the differences among them), even when they connect the same objects to the same other objects and result then in the same value-ranges.[38]

---

[38]In a letter to Husserl of October 30th–November 1st, 1906 ([106], vol. 2, pp. 101–105), Frege argues both that the thought expressed by a statement is what it has in common with any other equipollent [*äequipollent*] statement (*ibid.*, p. 102) and that 'if *A* then *B*' and 'it is not the case that *A* without *B*' are equipollent (*ibid.*, pp. 103–104). Insofar as the thought expressed by a statement is its sense, this means that these last statements have the same sense, so that also the *f*-names 'if ξ then ζ' and 'it is not the case that ξ without ζ' should have the same sense. This might appear to conflict with the view that two statements have different senses if one can "understand" both of them at the same time "while coherently taking different [epistemic] attitudes towards them", that Evans has ascribed to Frege ([85], pp. 18–19). To solve this conflict, C. Penco has suggested distinguishing the semantic from the epistemic sense of a statement, arguing that the latter "could be represented by the different procedures through which each formula is given a truth condition" ([158], pp. 104–105). This suggests that 'if ξ then ζ' and 'it is not the case that ξ without ζ' have the same semantic sense but different epistemic senses, since these *f*-names are related to different procedures. Though Penco's notion of the epistemic sense of a statement fits with my understanding of Frege's notion of the sense of an *f*-name, it seems to me relevant to observe that, in the language of the *Begriffsschrift*, conjunction is expressed through implication and negation ([97], Sect. I.12), so that 'ξ and ζ' is, by convention, a shortcut for 'it is not the case that if ξ then non ζ'. One could then argue that the previous statements have the same

2.4.4 Compositionality of Functions, Higher-Level Functions, and the Notion of an Arbitrary Function

This picture seems to fit perfectly with the compositional approach to functions that is at work in the *Grundgesetze*. This approach is evident from the way the exposition of the *Begriffsschrift* proceeds. Here I cannot but limit myself to consider another example that manifests this approach quite clearly and should be enough to complete what I have said on this matter so far.

I have already mentioned the Sect. I.10 of *Grundgesetze*, where Frege tries to determine as precisely as he can what is the value-range of a function, since, he says ([97], Sect. I.10; [110], p. 16_1), "we have admittedly by no means yet completely fixed the reference of a name such as '$\dot{\varepsilon}\Phi(\varepsilon)$' ". This lack of determination, he argues, can be overcome if, "for each function, it is determined, when it is introduced, what values it takes on for value-ranges as arguments". This claim makes it already clear that, for Frege, any function that receives a name in his system is to be introduced in a way that makes it possible to determine its values. But this is not all. What is also relevant, to show Frege's attitude towards functions, is that he considers appropriate to tackle the problem by considering the three first-level functions considered up to that point, namely $\xi = \zeta$, $\underline{\quad}\xi$ and $\underline{\top}\,\xi$. The argument Frege develops concerning these functions and the reasons for he takes this argument appropriate as a response to the problem are far from crystal-clear. R. Heck has submitted both Frege's argument and the response he draws from it to a very subtle analysis ([121]; [123], Chap. 4). I cannot enter this matter here. What is relevant is that, as Heck observes, the argument "works only because Frege's formal language has certain expressive resources, and does not have others—because, that is, for each of the functions introduced before Sect. 10, the question what values it takes on for value-ranges as arguments can be reduced, in one way or another, to the corresponding question about identity" ([121], p. 277; [123], pp. 98–99), where 'identity' refers, of course, to the function $\xi = \zeta$. It is, then, the specific nature of the logical formalism that has been chosen, and, in particular, the nature of its primitive first-level functions, that, in Frege's mind, allows him to begin to respond to a general question about functions and their value-ranges. And the initial response is, moreover, capable of generalisation, just because of the way other functions are formed out from the primitive ones or are explicitly introduced thanks to appropriate stipulations. Since, in concluding his argument, after having remarked that what he has established through it is enough for determining "the *value-ranges* as far as is possible here", he remarks ([97], Sect. I.10; [110]: p. 18_1):

> Only when the further issue arises of introducing a function that is not completely reducible to the functions already known will we be able to stipulate what values it should have for

---

(Footnote 38 continued)
sense (without specification), since they correspond to the same procedure, so that one could also say (in the language of the exposition of the *Begriffsschrift*) that the f-names 'it is not the case that ξ without ζ' and 'if ξ then ζ' refer, once appropriately rendered, to the same function.

value-ranges as arguments; and this can then be viewed as a determination of the value-ranges as well as of that function.

This claim and the way Frege reasons in Sect. I.10, before concluding this way clearly manifests a compositional approach to functions. But they still do not provide enough evidence for concluding that the universe of Frege's functions includes only functions that are introduced or purported to be introduced in such a way that they result in being *ipso facto* associated with a rule or procedure to be used for determining the references of the names of their values.

After all, all what has been said up to now only applies to functions that have a name in the language of the exposition of the *Begriffsschrift* and whose values have a name in the *Begriffsschrift* itself. The fact that these functions are associated to such a rule or procedure is an obvious consequence of the fact that these functions are elementary functions introduced through stipulations like (2.6), or functions formed out from elementary functions so introduced. Hence—one could argue—wondering about the sameness conditions of these functions is essentially different from wondering about the sameness conditions of all functions whatsoever.

But is this distinction appropriate for the case of Frege? For us, the notion of any function whatsoever is not to be reduced to that of a function having any name whatsoever in an appropriate language, or whose values have any whatsoever name in an(other) appropriate language. But it seems to me that this cannot be also the case with Frege. Insofar as functions are not actual components of some world of *concreta* and *abstracta*, but merely manifest themselves in the way we refer to objects, they can only be distinguished by looking at the way their names contribute to the formation of proper names and statements, or can be recognised as (unsaturated) components of proper names and statements.

To this, one can retort that in the language of the *Begriffsschrift*, functions are also supposed to provide arguments of other functions of a higher-level. To see the problem, take the way Frege introduces the first-order universal quantifier in Sect. I.8 of *Grundgesetze* ([97], Sect. I.8; [110]: p. 12_1):

'$\underset{\mathfrak{a}}{\smile}\ \Phi(\mathfrak{a})$' refers to the True if the value of the function $\Phi(\xi)$ is the True for every argument, and otherwise to the False.

This stipulation introduces a second-level concept: the concept $\underset{\mathfrak{a}}{\smile}\ \varphi(\mathfrak{a})$. The empty place in the name of this concept is marked by 'φ', which works in names of second-level functions as 'ξ' works in names of first-level ones. Now, it seems that, as a stipulation introducing first-level functions implicitly relies on the totality of objects (which provides the range of the relevant schematic letters for objects), a stipulation introducing a second-level function implicitly relies on the totality of first-level functions with the appropriate number of arguments (which would provide the range of the relevant schematic letters for functions, like 'Φ' in the foregoing stipulation). If this were so, the question would be obvious: does Frege hold that this totality includes only functions having names, or whose values have names—these names being either appropriately introduced in the relevant languages, or composed on the basis of names appropriately introduced—or does he hold that this totality is larger? This question is similar to that which Hintikka and Sandu answer in their paper mentioned in Sect. 2.3.[39] There is then no need to consider second-order quantifiers to advance such a question. The consideration of first-order quantifiers, or, more generally, of second-level functions, is enough.

But, is this question appropriate? The following quotation drawn again from "Über Begriff und Gegenstand" makes me doubt that it is ([95], p. 201; [104]: pp. 50–51):

> [...] the assertion that is made about a concept does not suit an object. Second-level concepts, under which concepts fall, are essentially different from first-level concepts, under which objects fall. The relation of an object to a first-level concept under which it falls is different from the relation, certainly analogous, of a first-level to a second-level concept. To do justice at once to that distinction and to the analogy, we might perhaps say that an object falls *under* a first-level concept, a concept falls *within* a second-level concept. The distinction of concept and object thus still holds, with all its sharpness.

At first glance, this is a quite vague distinction. But one could perhaps clarify it by suggesting that what Frege means here is that, in the language of the *Begriffsschrift*, names of second-level functions occur within proper names or statements, as unsaturated components of them, insofar as these latter names result, or are taken to result, from saturating the former names with appropriate names of first-level functions. If any possible saturation of a name of a second-level function with a name of a fist level one results in a name of a truth-value, the corresponding second-level function is a concept or a relation, and the relevant first-level functions fall *within* it if these names refer to the True. Consider the previous example. The f-name ' $\underset{\frown}{}\mathfrak{a}\underset{\frown}{}\, \varphi(\mathfrak{a})$ ' of a second-level function occurs in the proper name ' $\underset{\frown}{}\mathfrak{a}\underset{\frown}{}\, \mathfrak{a} = \mathfrak{a}$ ' insofar as the latter results from saturating the former with the name ' $\xi = \xi$ ' of a first-level function.[40] Now, insofar as ' $\underset{\frown}{}\mathfrak{a}\underset{\frown}{}\, \mathfrak{a} = \mathfrak{a}$ ' is a name of a truth-value, and this is also the case for any other proper name resulting from saturating ' $\underset{\frown}{}\mathfrak{a}\underset{\frown}{}\, \varphi(\mathfrak{a})$ ' with a name of a first-level function, $\underset{\frown}{}\mathfrak{a}\underset{\frown}{}\, \varphi(\mathfrak{a})$ is a concept. Furthermore, insofar as ' $\underset{\frown}{}\mathfrak{a}\underset{\frown}{}\, \mathfrak{a} = \mathfrak{a}$ ' refers to the True, the function ' $\xi = \xi$ ' falls *within* this concept.

More generally, if a proper name results, or is taken to result, from saturating the name of a second-level function with appropriate names of first-level functions, then the first-level functions named by these latter f-names are said to be arguments of the second-level function named by the former f-name. The difference from the case of first-level functions is clear: for a proper name resulting from saturating a name of a first-level function with names of objects to belong to the language of the

---

[39]The question has also been considered by Dummett, who has argued ([74], pp. 219–220) that "there is meagre evidence" for attributing to Frege the conception that his function-variables "range over the entire classical totality of [appropriate] functions", and that "his formulation make it more likely that he thought of his function-variable as ranging over only those functions that could be referred to by functional expressions in his symbolism".

[40]To understand what I mean by speaking of proper names or statements which are taken to result (rather than merely resulting) from saturating names of second-level functions with appropriate names of first-level functions, consider the example of a proper name like ' $\Phi(\Delta)$ ' or ' $\Psi(\Delta, \Gamma)$ '. These can be either taken to result from saturating the names ' $\Phi(\xi)$ ' and ' $\Psi(\xi, \zeta)$ ' of first-level functions with the proper names ' Δ ' and ' Γ ', or taken to result from saturating the names ' $\varphi(\Delta)$ ' and ' $\psi(\Delta, \Gamma)$ ' of second-level functions with these same names of first-level functions.

Begriffsschrift, these names of objects have in turn to belong, as such, to this same language; a proper name belonging to the language of the *Begriffsschrift* cannot result, instead, from saturating a name of a second-level function with f-names belonging, as such, to this same language, for the simple reason that this language does not include f-names (otherwise than as unsaturated components of proper names or statements).

This syntactical difference is structurally relevant but does not undermine what is essential for my present purpose: Frege's treatment of functions both of first and of higher-level, in the exposition of the *Begriffsschrift*, focuses on the way proper names (namely names of values of functions) and statements are formed by saturating f-names with other appropriate names. Hence, though one could and should say that a second-level function connects first-level functions to objects, rather than objects to objects, the work that first and second-level functions carry out within the language of the *Begriffsschrift* is essentially the same, and essentially depends on their having a name, or on the fact that their values have a name.

Mutatis mutandis, all that has been said for second-level functions also applies to higher-level ones. An example is given by the second-order universal quantifier, which Frege introduces in all its generality in the *Grundgesetze*, I.24 as the third-level function $\underline{\quad\mathfrak{f}\quad}\mu_\beta\,(\mathfrak{f}\,(\beta))$ defined by stipulating that '$\Omega_\beta\,(\Phi\,(\beta))$' refers to the True whatever the first-level function $\Phi\,(\xi)$ might be.[41] Hence, the possibility of functions which are not endowed with a name, or are not at least associated to appropriate rules to be used for forming the names of their values and determining their reference, merely lies outside the horizon of Frege's use of the notion of function in the exposition of the *Begriffsschrift*. Any argument to be used for arguing that he admits functions like these would be not only merely speculative, but also intrinsically vague, since Frege give us no hint for understanding what he could mean by claiming that certain functions exist, if this were not merely intended as a metaphoric way for saying that their names or the names of their values are at work in some appropriate language.

In Sect. I.2 of *Grundgesetze*, Frege describes a process through which the mathematical notion of a function was gradually extended ([97], Sect. I.2; [110]: p. 6_1): firstly, functions were taken to be formed only by the fundamental arithmetic operations; then operations involving a passage to a limit were admitted; finally, "the word 'function' was so generally understood that in some cases the connection between the argument and the value of a function could no longer be expressed through the signs of analysis, but only through words",[42] and complex numbers were admitted both as arguments and values of functions. Frege add then that, "in both direction [...][he has] gone still further", for having introduced new signs or used old ones for a new purpose, and for having admitted other objects than numbers as arguments

---

[41] The letter 'μ' is here used here to hold a place empty for a second-level function with one argument, the index 'β' is used to make it clear that the arguments whose places are indicated by 'β', both in '$\mathfrak{f}\,(\beta)$' and in '$\Phi\,(\beta)$ ', are bound, and 'Ω' and 'Φ' serve as schematic letters for functions of the second and first-level, respectively.

[42] This is enough evidence for concluding that, contrary to what Hintikka and Sandu seem to imply ([129], p. 311) Frege admitted the possibility of non-differentiable functions in real analysis (on this matter, cf. also [124], p. 41, and [36], pp. 90, 99–100).

and values of functions. But he neither says nor implies that functions and/or their values could lack names or not be associated to rules or procedures to be used for forming the names of their values and determining their reference.

In another passage, close to this one, drawn from "Function und Begriff" ([94], p. 12; [104] p. 28), Frege mentions Dirichlet's function as an example of a mathematical function merely described through ordinary language, by describing it as a function "whose value is 1 for rational and 0 for irrational arguments". According to Burgess, this is enough for inferring that it is not part of Frege's notion of function "that a function must be definable in a *symbolic* language". Hence, he continues, even if Frege had been convinced that "there is a finitary *symbolic* language [...] in which for every function there is an expression", he could have not based this conviction on purely conceptual grounds, being rather forced to appeal, at least partially, to inductive evidence relative to the possibility of defining, in some appropriate language, suitable functions to be used as witnesses for the existence theorems of contemporary mathematical analysis ([35], p. 106). This would have been a mistake, but, for Burgess, it is plausible to ascribe such a mistake to Frege, provided that it was only some years later that "it became more or less established orthodoxy in the mathematical community that functions are not restricted to be definable", and he "was largely unaware of the bearing of Cantor's cardinality theorems" entailing that "there are more Fregean concepts than Fregean objects": *ibid.*, p. 107 and 101–102).

This mention of Cantor's cardinality theorems suggests that, for Burgess, Frege's notion of a function could have been undermined by considerations about sets' cardinality, which seems to have been possible only if this notion had been extensional in nature. But all that I have said up to now should provide evidence for concluding that this is not so.

Burgess's discussion is largely based on a passage—also quoted by Hintikka and Sandu as a major piece of evidence for their main thesis ([129], p. 312–313)— drawn from "Was ist eine Funktion?", the 1904 paper that I mentioned at the end of Sect. 2.3 ([100], pp. 662–663; [104], pp. 112–113), where Frege argues against Czuber's definition of function. In this paper, he remarks that the idea of function as a law of correlation expressed by an equation "has been found too narrow", but suggests that the difficulty "could be easily avoided by introducing new signs into the symbolic language of arithmetic". This passage is open to many interpretations not necessarily fitting with Hintikka and Sandu's thesis.[43] But it seems to me that another passage drawn from this same paper is much more explicit.[44] In my view, it makes manifest in a nutshell the main feature of Frege's notion of function, by making it clear that it is not extensional at all. This is just the passage where Frege critically discusses Czuber's definition of real functions. Here is what he says ([100], p. 661–662; [104], pp. 111–12):

---

[43] According to [62], pp. 142–145, the context of this passage suggests that Frege is here merely arguing for the possibility of extending the class of analytically representable functions so as "to include all functions of a particular class": a quite common view among mathematicians of his time. Heck and Stanley ([124], p. 419–421) have considered, instead, that Frege's only point, here, is that functions are unsaturated, which seems to me a quite implausible interpretation.

[44] This passage is also partially quoted in ([129], p. 312).

It would be simpler and clearer to state the matter as follows. 'With every number of an *x*-domain is correlated a number. I call the totality of these numbers the *y*-domain'. Here we certainly have a *y*-domain, but we have no *y* of which we could say that it is a function of the real variable *x*. Now, the delimitation of the domain appears irrelevant to the question of the nature of the function [*Wesen der Funktion*]. Why could we not at once take the domain to be the totality of real numbers, or the totality of complex numbers, including real numbers? The heart of the matter really lies in a quite different place, viz. hidden in the word 'correlated'. Now, how do I acknowledge whether the number 5 is correlated with the number 4? The question is unanswerable unless it is somehow completed. […] Correlation […] takes place according to a law [*Gesetz*], and different laws of this sort can be thought of. Hence, the expression '*y* is a function of *x*' has no sense, unless it is completed by the statement [*Angabe*] of the law according to which the correlation takes place. This is a mistake in the definition. And is not the law, which this definition treats as not being given, the main thing? […] Distinctions between laws of correlation will go along with distinctions between functions; and these cannot any longer be regarded as quantitative. If we just think of algebraic functions, the logarithmic function, elliptic functions, we conceive ourselves immediately that these are qualitative differences […].

2.5 Concluding Remarks

In a sense, my account of Frege's notion of function fits with Hintikka and Sandu's conclusions.[45] But it hinges on different concerns. In my view, the relevant question is not whether Frege endorsed the standard or a non-stardard interpretation of second-order logic. What is relevant is rather the way functions are supposed to work both in his formal system and in his exposition of it.[46] For Frege, functions are neither defined on sets, nor conceived as pairs of sets. So, it is out of order to wonder whether he takes the range of second-order quantifiers to coincide or not with the whole power set of the range of the first order-quantifiers. The notion of set (however conceived) is not a resource Frege considers himself to be licensed to appeal to in the exposition of his formal system, which is not, by the way, in need of any semantic interpretation, since it is *ipso facto* presented as an already interpreted system ([188], p. 4).

In replying to Hintikka and Sandu's paper, Heck and Stanley have argued that "Frege would not have accepted any of the familiar arguments in favour of a non-

---

[45]More recently, Sandu has reiterated his and Hintikka's major theses, and added that, for Frege, intensions have "logical primacy" over extensions ([173], pp. 241–243). To support this, Sandu argues that Ramsey's efforts for reforming logicism [165] were manly motivated by an understanding of Frege's and Russell's conceptions about functions, which is close to that outlined in his joint paper with Hintikka. In opposition to this, Ramsey would have aimed to conciliate logicism with "the extensional attitude of the mathematics of his days (Cantorian set theory)", and this resulted in his grasping of "the concept of arbitrary function in extension" ([173], pp. 238 and 250). This has convinced Demopoulos to reconsider the objections to Hintikka and Sandu's theses advanced in a joint paper with Bell [62], and conclude that "Frege's functions" should be distinguished from "arbitrary correspondences" i.e. arbitrary functions in set-theoretic sense ([61], especially p. 6).

[46]I agree then with Demopoulos, according to whom, "the interest of Hintikka and Sandu's paper has less to do with standard versus nonstandard interpretations of second-order logic than with Frege's concept of a function" ([61], p. 6).

standard interpretation", with the result that, if he "did interpret his higher-order quantifiers non-standardly, then a study of his reasons for doing so would presumably provide a entirely new set of motivations for rejecting the standard interpretation" ([124], p. 417–418). This argument depends on the admission that Frege could have had positive reasons for rejecting the standard interpretation, and could have then conceived it. But this is just what he could not have done. He was faced neither with the choice between the standard and any nonstandard interpretation of second-order logic, nor with the choice between accepting or rejecting our extensional set-theoretic notion of arbitrary function. His way of conceiving functions was simply such as to make this idea unavailable to him.[47]

Rather than projecting Frege's conception onto the modern (set-theoretic) setting, we should instead try to understand the intrinsic motivations of his own approach, by placing it in the appropriate historical and philosophical framework. This framework is provided by his reaction to the program of the arithmetisation of analysis, namely to the requirement that numbers and magnitudes should be defined within a formal logical system whose exposition is to depend on the appeal to a quite general, and then non-mathematical notion of function, the elucidation of which should moreover result in the elucidation of the notions of concept and a concept's extension, and, then, in the clarification of the very nature of logic.

In Sect. 2.3, we saw that the program of the arithmetisation of analysis came in turn from a reaction to Lagrange's foundational program, which also ascribed a basic role to the notion of function. A comparison between Lagrange's and Frege's views on functions is then quite natural.

Though both of them take the notion of function to be primitive, they take it to be so in two quite different ways. For Lagrange, functions are objects, namely expressions of an appropriate formalism which is taken for granted. The essential purpose for focusing on them is that of providing a purely relational construal of the notion of quantity, whose aim is to free this notion from any specific essence and make it perfectly formal, and then general. For Frege, functions are opposed to objects, so that they cannot be expressions. Furthermore, they are supposed to act in any language appropriate for expressing truths about objects, and any language is supposed to be meaningful, so that the elucidation of the notion of function is conceived as a prerequisite for the exposition of any appropriate formalism. The essential purpose for focusing on this notion is that of making clear the way in which we refer to objects and express truths about them, so as to provide a formal construal of logic, just conceived as the general framework in which truths about objects can be expressed.

Despite these crucial differences, Lagrange's and Frege's conceiving the notion of function as primitive and taking it as a basis for developing the respective foundational programs also results in important analogies and makes both these programs

---

[47]Despite their focusing on the question of whether Frege adopted the standard or a nonstandard interpretation of second-order logic, Hintikka and Sandu also suggest something like this when they claim ([129], p. 313) that "there is no niche in [...][Frege's] world for [...][our] notion of an arbitrary function", and that in Frege's logic "there is no room for the idea of a arbitrary function-in-extension".

essentially different from both the arithmetisation of analysis and from set-theoretic reduction. What opposes the two former programs to the latter ones is, so to say, an intensional approach: the idea that an appropriate foundation of mathematics necessarily depends on the clarification of the way the relevant items mathematics is about are related to each other. Furthermore, both for Lagrange and Frege, this clarification depends on the identification of a formal expression for the relevant relations within an appropriate formalism. In both cases, mathematics is then conceived as a system of appropriate expressions. For Lagrange, these expressions are functions; for Frege, they merely make the role and nature of functions manifest. Still, in both cases the idea of a function detached from any appropriate expression is merely inconceivable. If my account is correct, this is not the effect of Lagrange's and Frege's intellectual myopia.[48] It rather depends on the intrinsic nature of their respective programs.

---

[48]cf. see the footnote (13).

Chapter 3
Frege, Russell, Ramsey and the Notion of an Arbitrary Function

Gabriel Sandu

3.1 The Background

In *Frege's Philosophy of Language*, Dummett claims that Frege's notion of a function coincides with the notion of an arbitrary correspondence ([68], pp. 223 and 177):

> [...] Frege had not the slightest qualm about the legitimacy or intelligibility of higher-order quantification: he used it from the first, in *Begriffsschrift*, freely and without apology, and did not even see first-order logic as constituting a fragment having any special significance.

> [...] it is true enough, in a sense, that, once we know what objects there are, then we also know what functions there are, at least, so long as we are prepared, as Frege was, to admit all "arbitrary" functions defined over all objects.

Against this background, I claimed with Hintikka, in [129], that Frege's notions of a function and a class cannot be that of an arbitrary correspondence or arbitrary collection of objects and that Frege favoured, instead, some variety of non-standard interpretation, for which the domain of the function variables is something less than the characteristic functions of all subsets of the domain over which the individual variables range. When we wrote our paper, we were unaware of Dummett's argument in *Frege's Philosophy of Mathematics* which shows that the author changed his mind *vis à vis* his earlier position emerging from the above quote. Here is what he write there ([74], pp. 219–220):

> [...] Frege fails to pay due attention to the fact that the introduction of the [class] abstraction operator brings with it, not only new singular terms, but an extension of the domain. [...] [I]t may be seen as making an inconsistent demand on the size of the domain D, namely that, where D comprises n objects, we should have $n^n \leq n$, which holds only when $n = 1$, whereas we must have $n \geq 2$, since the two truth-values are distinct: for there must be n^n extensionally non-equivalent functions of one argument and hence n^n distinct value-ranges. But this assumes that the function-variables range over the entire classical totality of functions from D into D, and there is meagre evidence for attributing such a conception to Frege. His formulations make it more likely that he thought of his function-variables as

© Springer International Publishing Switzerland 2015
H. Benis-Sinaceur et al., *Functions and Generality of Logic*,
Logic, Epistemology, and the Unity of Science 37,
DOI 10.1007/978-3-319-17109-8_3

ranging over only those functions that could be referred to by functional expressions of his symbolism (and thus over a denumerable totality of functions), and of the domain D of objects as comprising value-ranges only of such functions.

The last two sentences, which show Dummett attributing to Frege a non-standard interpretation of his function-variables, are alike in spirit to some of the considerations we put forward in [129]. For example this one ([129], p. 292):

> Sometimes, a non-standard interpretation is guided by the idea that only such properties, relations, and functions can be assumed to exist as can be defined or otherwise captured by a suitable expression of one's language. In the case of theories with infinite models, this leads inevitably to a non-standard interpretation, for there can be only a countable number of such definitions or characterisations available for this purpose. Hence they cannot capture all the subsets of $do(M)$, for there is an uncountable number of them.

In a rejoinder to our paper, Bell and Demopoulos [62] took side with Dummett's standard interpretation of Frege's function variables in [68], and argued that Frege's concept of a function coincides with the set-theoretic notion of an arbitrary correspondence. The main idea behind that paper is summarised in [61], p. 5:

> Our thought was that whatever covert role the neglect of Cantor's theorem might have played in the inconsistency of [...][*Grundgesetze*], it is unlikely that Frege sought to ignore the theorem by assuming that the totality of functions, like the totality of expressions, is countably infinite. But we sided with Dummett in [...] [74] and supposed that Frege might very well have been misled into assuming that what holds for certain countable interpretations of the function variables holds in general; hence we agreed with Dummett's evaluation of the sense in which Frege missed the significance of the possibility of different interpretations for his program.

For me and Hintikka the definability of functions in one's suitable symbolism is just one possible manifestation of the basic idea underlying the non-standard interpretation: the connection between an argument and the corresponding value of a function is determined by a formal law, norm or property. This idea stands in contrast to the conception underlying the standard interpretation according to which the correlation between values and arguments is purely arbitrary and not determined by such a law. For this reason, the main argument of our paper was intended to focus on the distinction between, on one side, the idea of arbitrary variation between values and arguments, and the idea of a correlation as determined by a formal law, on the other. We argued that Frege could not have had a standard interpretation of his function-variables given that the notion of a law was important for him when characterising functions. Part of our argument was Frege's discussion of the inadequacies of the definition of a function proposed by Czuber. We contrasted Czuber's notion of correlation which involves no assertion as to the law of correlation, and which can be set up in the most various ways, with Frege's conception of correlation which focuses on the idea of a law ([100], p. 662; [104], p. 112):

> Correlation, then, takes place according to a law, and different laws of this sort can be thought of. In that case, the expression y is a function of x' has no sense, unless it is completed by the law of correlation.

To the question of how such a law is specified, Frege answers (*ibid.*):

> Our general way of expressing such a law of correlation is an equation in which the letter
> '*y*' stands on the left side whereas on the right there appears a mathematical expression
> consisting of numerals, mathematical signs, and the letter '*x*', e.g. '$y = x^2 + 3x$'.

Frege also remarks that with the introduction of the notion of a law, "variability has
dropped out of sight, and instead generality comes into view, for that is what the word
'law' indicates" (*ibid.*). One of Frege's conclusions is that the notion of a function
has nothing to do with variation, that '*x*' does not denote an "indefinite" or "variable"
number, but serves to express generality.

In another rejoinder to our paper, Heck and Stanley [124] claimed that we placed
too much emphasis on Frege's remarks. They admit that Frege manifests a tendency
to explain the notion of a function in terms of the nature of functional expressions,
but that this should not obscure the fact that functions, for Frege, are the kind of
unsaturated entities which only need to have arguments and values.

In [173] I considered the notion of an arbitrary correlation in the context of
Ramsey's criticism of *Principia*'s notion of classes and his moving away from a
predicative notion of a function towards the notion of a function-in-extension, which
is an arbitrary correlation between arguments and propositions. The idea was to
bring another, indirect evidence to my earlier claim with Hintikka to the effect that
Frege could not have defended the idea of arbitrary correlation, for that would have
placed him in the same camp with Ramsey, against Russell. In fact, I thought that
Russell's notion of a propositional function and Frege's notion of a concept stand in
deep contrast to Ramsey's notion of a function in extension in his "Foundations of
Mathematics" [165]. Some of the arguments in my paper determined Demopoulos
to reconsider, in [61], his earlier position with Bell which had attributed to Frege a
standard interpretation. The present paper contains some reflections on these matters.
The main focus will be on the notion of an arbitrary correlation, but let me start by
saying few things on the connection between this notion and Dedekind theorem.

3.2 The Standard versus Non-standard Distinction and Dedekind Theorem

In [129] I and Hintikka claimed that it is the standard interpretation which is the
most important for foundations of mathematics, for it is the only one which allows
one to formulate descriptively complete categorical axiomatisations of mathematical
theories such as number theory and the theory of real numbers (*ibid.*, p. 295). The
only concrete example we gave was Dedekind's characterisation of real numbers
by means of the cut principle, which says that every bounded set of reals has a
least upper. This characterisation is a categorical one only if the sets involved are
arbitrary and not restricted, as in Frege's system, to courses of values of concepts
expressible in the language of arithmetic. We concluded that "there is a deep sense
in which Frege's system is not adequate for interpreting results in contemporary set

theory and mathematical theorising, for instance in real analysis" (*ibid.*, p. 314). It is this connection between Frege's non-standard interpretation and the failure of his system to formulate categoricity results that irritated some of our critics, including Demopoulos ([61], pp. 4–5):

> But although Dummett shares Hintikka and Sandu's conclusion that Frege tended toward a non-standard interpretation, his analysis does not support Hintikka and Sandu's evaluation of Frege's foundational contributions. If we follow Dummett, Frege missed the fact that the consistency of [...][*Grundgesetze*], relative to a non-standard interpretation, does not necessarily extend to its consistency when the logic is given a full interpretation. This is certainly an oversight, but it is not the oversight that is appealed to in those of Hintikka and Sandu's criticisms of Frege that so offended some of their critics, as for example, whether, without having isolated the notion of a standard interpretation, Frege could have even conceptualised results like Dedekind's categoricity theorem.

In [61], Demopoulos refers to [120], who showed that Frege proved an analogy of Dedekind's theorem using an axiomatisation of arithmetic that is only a slight variant of the Peano-Dedekind axiomatisation. He also refers to [62] for an argument which questions the systematic dependence of categoricity results on the standard interpretation. The argument shows that categoricity proofs can also be given in a suitably rich first-order theory such as Zermelo-Fraenkel set theory, and these proofs have pretty much the same form as categoricity proofs in second-order logic. His conclusion is this ([61], p. 5):

> Hintikka and Sandu's claim that Frege could not even have formulated (let alone appreciated) these results because of their dependence on the standard interpretation is therefore incorrect both historically and methodologically. It is incorrect historically because Frege successfully proved a categoricity theorem like Dedekind's. And it is incorrect systematically because essentially the same argument establishes the categoricity of second-order arithmetic in any of the usual systems of set theory. And surely it is implausible that only someone familiar with the categoricity of the Peano-Dedekind Axioms as a theorem of second-order logic has really grasped the theorem or its proof. At most, Frege might be charged with having missed a subtlety concerning the distinction between formal and semi-formal systems; but this is hardly surprising for the period in which he wrote.

3.3 The Isomorphism Theorem

In [129] we did not explicitly claim that Frege was unable to formulate, let alone to appreciate, results like Dedekind's theorem. But it is true that our paper suggested it. In the light of [120], that was certainly an oversight. Few things should be said, however. As pointed out by Heck, the statement of this theorem, that is, that two structures which satisfy the Dedekind-Peano axioms are isomorphic, does not appear in *Grundgesetze*. Heck shows how it can be extracted from the proof of the famous theorem 263, which states the conditions under which the number of objects falling under a concept G is *Endlos* (cf. the introduction to the present volume, p. ix, above). The conditions state that there exists a relation Q which is functional, and thus determines a sequence, no object follows after itself in this sequence, each G stands in the relation Q to some object in the series, and the G's are the members of the Q-sequence beginning with some object.

In the proof of this theorem, Frege builds up by induction a binary relation which maps the natural numbers into the members of the Q-sequence, and vice versa. That is, the members of this relation are the pairs $(0, x_0)$, $(1, x_1)$, ..., where $x_0, x_1, \ldots$ are the G's in the order determined by Q. This relation is functional and it preserves both the orderings of the natural numbers and the Q-ordering. For this reason, Heck proposes to call theorem 263 (or rather the theorem 254 which proves the general result that all simple and endless series are isomorphic) 'the Isomorphism Theorem'. It can be proved in second-order logic augmented by the ordered pair axiom. Heck shows the ordered pairs to be dispensable and also suggests that Frege knew his use of ordered pairs to be dispensable so that finally this is a "theorem of second-order arithmetics and logic simpliciter" ([120], p. 322). And when the conditions of the Isomorphism Theorem are rewritten so that one can easily derive from them the more familiar Dedekind-Peano axioms, then the proof of theorem 263 shows that "any two structures satisfying *Frege*'s axioms for arithmetic are isomorphic" (*ibid.*, pp. 324–325).[1]

But although we are told that a modern reader should take the Isomorphism Theorem to show that any two structures satisfying certain conditions are isomorphic, this theorem is not put to much use in *Grundgesetze*. I take Frege's proof and Heck's reconstruction of it to give us a derivation in second-order logic. In the remaining of this section I will look at a more recent argument about categoricity proofs in second-order logic and set theory that seems to support Demopoulos' conclusion.

According to Väänänen ([198], p. 378):

> [...] the situation is entirely *similar* in second-order logic and in set theory. [...] All the usual mathematical structures can be characterised up to isomorphism in set theory by appeal to their second-order characterisation but letting the second-order variables range over sets that are subsets of the structure to be characterised. The only difference to the approach of second-order logic is that in set theory these structures are indeed explicitly defined while in second-order logic they are merely described. In this respect second-order logic is closer to the standard mathematical practice of not paying attention to what the "objects" e.g. complex numbers really are, as long as they obey the right rules.

In the perspective of second-order logic, in mathematics one studies statements of the form

$$\mathbb{M} \vDash \varphi \tag{3.1}$$

where $\mathbb{M}$ is a mathematical structure and φ is a mathematical statement written in second-order logic. Väänänen remarks, that if $\mathbb{N}$ is the structure $\mathbb{N} = (N, +, \times, <)$ and $\varphi_{\mathbb{N}}$ is a second-order axiomatisation of arithmetic, so that we have

$$\forall \mathbb{M}(\mathbb{M} \vDash \varphi_{\mathbb{N}} \Leftrightarrow \mathbb{M} \cong \mathbb{N})$$

---

[1] What Heck calls 'Frege's axioms for arithmetic' are just the four conditions stated above for the number of objects falling under G to be *Endlos*, where Q is instantiated by the successor relation and G by the concept of a natural number: cf. [119], Sect. 6.

the statement (3.1) can be expressed as a second-order logical truth

$$\vDash \varphi_N \rightarrow \varphi \tag{3.2}$$

The problem knowingly is that the second-order logical truth is not recursively axiomatisable. But, he continues, there are two stronger versions of (3.2), one in set theory and the other one in second-order logic:

$$ZFC \vdash \forall \mathbb{M}(\mathbb{M} \vDash \varphi_N \rightarrow \mathbb{M} \vDash \varphi) \tag{3.3}$$

and

$$CA \vdash \varphi_N \rightarrow \varphi \tag{3.4}$$

where CA is a second-order axiomatisation of second-order logic including a comprehension axiom and the axiom of choice. He thinks, then, that it is reasonable to give this later statement as a justification of (3.2). And he immediately adds this (*ibid.*, p. 375):

> I have called (3.3) and (3.4) *stronger* forms of (3.2) because I take it for granted that ZFC and CA are *true* axioms. It is not the main topic of this paper to investigate how much ZFC and CA can be weakened in this or that special instance of (3.3) and (3.4), as such considerations do not differentiate second-order logic and set theory from each other in any essential way.

Väänänen (*ibid.*, p. 376) notices, in fact, an apparent difference between (3.4) and (3.1): (3.1) is about the material truth of the statement φ in a (standard) model $\mathbb{N}$, whereas (3.4) seems to assert something that holds in all the models of CA, standard and non-standard. But he immediately points out this is only an appearance, as it can be seen by considering two versions of the sentence φ_N for the structure $\mathbb{N} = (N, +, \times)$: φ_N^1 in the vocabulary $\{+_1, \times_1\}$, and φ_N^2 in the vocabulary $\{+_2, \times_2\}$. If 'CA' now denotes the axiomatisation of second-order logic in a vocabulary that includes both $\{+_1, \times_1\}$ and $\{+_2, \times_2\}$, then we have

$$CA \vdash \varphi_N^1 \wedge \varphi_N^2 \rightarrow Isom_{1,2}$$

where '$Isom_{1,2}$' denotes the statement of second-order logic stating that there is a bijection f such that

$$\forall x \forall y [f(x +_1 y) = f(x +_2 y)]$$
$$\forall x \forall y [f(x \times_1 y) = f(x \times_2 y)]$$

The conclusion is as follows (*ibid.*):

[…] in this subtle sense, (3.4) really asserts the truth of φ in one and only one model, namely the standard model. […] Naturally, *CA* itself has non-standard models but they should not be the concern in connection with (3.4) because we are not studying *CA* but the structure […][$\mathbb{N}$]. In fact the whole concept of a model of *CA* is out of place here as *CA* is used as a medium of evidence for (3.2). We can convince ourselves of the correctness of the evidence by simply looking at the proof given in *CA* very carefully. There is no infinitistic element in this.

The situation is similar in set theory. In this perspective, in mathematics one studies statements of the form

$$\Phi(a) \tag{3.5}$$

where "$\Phi(x)$ is a first-order formula with variables ranging over the universe of sets, and a is a set" (*ibid.*, p. 377). Now we are told that (*ibid.*):

If we compare (3.1) and (3.5), we observe that the former is restricted to one presumably rather limited structure […][$\mathbb{M}$] while (3.5) refers to the entire universe. This is one often quoted difference between second order-logic and set theory. Second-order logic takes one structure at a time and asserts second-order properties about that structure, while set theory tries to govern the whole universe at a time.

Two qualifications are added. The first is this, "while it is true that (3.5) refers to the entire universe, typical mathematical propositions are really statements about some V_α such that $a \in V_\alpha$ (*ibid.*). The second qualification concerns the justification of (3.5), which raises the same worries as the justification of (3.1) in second-order logic. The justification is given by the stronger statement

$$ZFC \vdash \Phi(a) \tag{3.6}$$

where a is assumed to be a definable set. Väänänen concludes that there is no fundamental difference between set theory and second-order logic.

But, then, Väänänen wonders, "which is the right way to do mathematics: second-order logic or set theory?" (*ibid.*, p. 379). And here is, finally, his answer (*ibid.*):

Let us leave aside the question whether the higher ordinals that exist in set theory are really needed. The point is that set theory is just a "taller" version of second-order logic, and if one does not need (or like) the tallness, then one can replace set theory by second - (or higher-) order logic. However, this does not yield more categoricity, for both second-order logic and set theory are equally "internally categorical". If we look at second-order logic and set theory from the outside we enter meta-mathematics. Then we can build formalisations of the semantics of either second-order logic or set theory and prove their categoricity in "full" models as well as their non-categoricity in "Henkin" models.

This ends my exposition of Väänänen's arguments. The point I want to underline, against their author, which is the same as the point I raised earlier in connection with Demopoulos's and Heck's discussion of Frege's proof of Dedekinf theorem, is that (3.4), like its set-theoretical counterpart (3.6), is just a formal derivation. None of them stands by itself in the context of justification. One cannot "take for granted that

ZFC and *CA* are true axioms" and assert in the same time that "the whole concept of a model of *CA* is out of place here as *CA* is used as a medium of evidence for (3.2)". The point is not so much that of looking at second-order logic or set theory from the inside or outside, but rather that of a derivation having a content or not. If it does not, then it cannot serve as a "medium of evidence", and this for the simple reason that it does it not refer to the concepts it purports to refer.

3.4 Ramsey's Notion of a Predicative Function in "Foundations of Mathematics"

In [173], I suggested to look at the question of the notion of arbitrary correspondence from a different angle: Ramsey's criticisms of the notion of propositional function in Whitehead and Russell's *Principia* [203]. My punching line was that Frege's functions (concepts) are as predicative as *Principia*'s propositional functions and thereby Ramsey's criticisms of the logic of *Principia* and his conclusion that this logic is inadequate for the logicism programme (the reduction of mathematics to logic) apply *mutatis mutandis* to Frege's logic.

Ramsey criticises *Principia*'s notion of a propositional function arguing for the need of its extension, in the context of the logicist reduction of mathematics to logic. Ramsey anticipates Carnap's distinction between two kinds of logicist reductions, respectively depending on:

1. the definition of all concepts of a mathematical theory in terms of logical notions.
2. (1) plus the derivation of the axioms of the resulting theory from purely logical axioms.

Carnap [41] points out that Russell operated a reduction of type (1). In his anticipation of this distinction, Ramsey observes that a reduction of type (1) would show the generality of mathematics, while a reduction of type (2) would illustrate the necessity of mathematics. The sense of necessity Ramsey is concerned with in his remarks is that according to which tautologies, in Wittgenstein' sense (namely sentences true in every universe of discourse), are necessary. Ramsey observes that in order to perform a reduction of type (2), one would have to give up the notion of propositional function to be found in *Principia*.

Let me comment first on Ramsey's notion of a predicative function in [165]. I will rely heavily on Trueman's reconstruction of it in [194].

We start with a class of propositions built from a stock of atomic propositions of the form 'John is tall', through (possibly an infinite application of) truth-functional connectives. Ramsey ([165], p. 35) defines a propositional function of individuals, as "a symbol of the form '$f(\hat{x}, \hat{y}, \hat{z}, \ldots)$'" such that every replacement in it of '$\hat{x}$','$\hat{y}$','$\hat{z}$', …with names of individuals yields a proposition of the initial class. Ramsey takes, moreover, a propositional function '$f(\hat{x}, \hat{y}, \hat{z}, \ldots)$' to be identical with '$g(\hat{x}, \hat{y}, \hat{z}, \ldots)$' if the substitution of the same set of names in one and the other yields the same proposition, that is, if '$f(\hat{x}, \hat{y}, \hat{z}, \ldots)$' and '$g(\hat{x}, \hat{y}, \hat{z}, \ldots)$' have the same truth-table. The definition extends to higher-order propositional functions.

To specify the subclass of propositional functions that Ramsey calls 'predicative functions' we need first to specify atomic predicative function of individuals. They are "the result of replacing by variables any of the names of individuals in an atomic proposition expressed by using names alone" (*ibid.*, p. 38). Thus '$\hat{x}$ is tall' is an atomic predicative function.

This notion is then extended to cover truth-functions of propositional functions and propositions. Ramsey's definition is as follows (*ibid.*):

> Suppose we have functions $\phi_1(\hat{x}, \hat{y})$, $\phi_2(\hat{x}, \hat{y})$, etc., then saying that a function $\psi(\hat{x}, \hat{y})$ is a certain truth-function [...] of the functions $\phi_1(\hat{x}, \hat{y})$, $\phi_2(\hat{x}, \hat{y})$, etc. and the proposition p, q, etc., we mean that any value of $\psi(\hat{x}, \hat{y})$, say $\psi(a, b)$, is that truth-function of the corresponding values of $\phi_1(\hat{x}, \hat{y})$, $\phi_2(\hat{x}, \hat{y})$, etc., i.e. $\phi_1(a, b)$, $\phi_2(a, b)$, etc. and the propositions p, q, etc.

Hence, '$F(\hat{x_0}, \hat{x_1}, \ldots, \hat{x_n})$' is a truth-function of some propositional functions and propositions if and only if any of its values for some appropriate arguments is the corresponding truth-function of the values of these propositional functions for these arguments and of these propositions. To take an example '$\left[G(\hat{x_1}, \hat{x_2}) \vee p\right] \wedge \left[H(\hat{x_1}, \hat{x_2}) \vee q\right]$' is a certain truth-function of '$G(\hat{x_1}, \hat{x_2})$', '$H(\hat{x_1}, \hat{x_2})$', p, and q, since for whatever names 'a', 'b', 'c', 'd', '$[G(a, b) \vee p] \wedge [H(c, d) \vee q]$' is this same truth-function of '$G(a, b)$', '$H(c, d)$', p, q.

Finally Ramsey defines predicative functions of individuals as follows (*ibid.*, p. 39):

> A *predicative function* of individuals is one which is any truth-function of arguments which, whether finite or infinite in number, are all either atomic functions of individuals or propositions.

Hence, a predicative function of individuals is "a (perhaps infinite) truth-function of atomic predicati[...][ve] functions of individuals and propositions", and, vice versa such a truth-function is a predicative function of individuals ([194], p. 294).

What needs to be emphasized is that, according to this definition, the predicativity of a propositional function '$F(\hat{x})$' consists, as generally agreed, in the fact that the proposition '$F(a)$', which '$F(\hat{x})$' assigns to 'a', says or predicates the same thing of a as the proposition '$F(b)$', which '$F(\hat{x})$' assigns to 'b', does of b.

I regard Ramsey's definition of a predicative function as a manifestation of the phenomenon we discussed in connection with Frege: the specification of a function by a formal law. In this case the "glue" which keeps the arguments and values together is a propositional function. A good example is our earlier propositional function '$\hat{x}$ is tall', which maps 'Socrates' to 'Socrates is tall' and 'Plato' to 'Plato is tall', i.e.,

'F(Socrates)' is 'Socrates is tall'
'F(Plato)' is 'Plato is tall'.

Let me finally mention that Ramsey uses his notion of propositional function to give an account of quantification in the *Tractatus*. The proposition '$\forall x F(x)$' is conceived of as the conjunction of all the values of '$F(\hat{x})$', and the proposition '$\exists x F(x)$' as the disjunction of all these propositions. Similarly, the higher-order '$\forall \varphi f (\varphi (\hat{x}))$' is the conjunction of all the values of '$f (\varphi (\hat{x}))$' ([165], p. 40). In addition, the use of quantifiers is governed by what is known as the exclusive interpretation of quantifiers, e.g. '$\exists x R(x, a)$' is conceived of as a disjunction of all the values of '$R(\hat{x}, a)$' except for '$R(a, a)$', and similarly for '$\forall x R(x, a)$', which is conceived of as the conjunction of these values.

3.5 Ramsey's Reduction of Type (2)

Reduction (2) is achieved in two steps, following Whitehead and Russell. In the first step mathematics is reduced to the theory of classes, e.g., each natural number n is defined as the class of all n-membered classes of individuals. For instance, 1 is defined as the class of all singletons, 2 as the class of all doubletons, etc. In the second step, the theory of classes is reduced to logic. It is here that propositional functions are needed, as class-terms are partially eliminated in favour of propositional functions. As a result of this process, every class is presented as the extension of a propositional function. But as Trueman observes Ramsey realised that if the only admissible propositional functions are predicative functions, then there can be no reduction of mathematics to logic. As logical truths are tautologies, then the failure of this reduction would also be a failure to show that mathematical truths are tautologies in Wittgenstein's sense.

Trueman spells out nicely what is at stake here ([194], p. 296):

> If 1 is defined as the class of singletons of individuals and 2 as the class of doubletons of individuals, then the mathematical truth that $1 \neq 2$ requires that there be a singleton or a doubleton: otherwise, 1 and 2 would both be empty and hence identical. If, in turn, the existence of a singleton or doubleton of individuals is to be reduced to logic then, assuming [...][that all logical truths are tautologies, and vice versa], it must be a tautology that some propositional function is true of exactly one or exactly two individuals. But, if every propositional function is predicati[...][ve] then this is not a tautology, and this is because predicati[...][ve] functions do not logically discriminate between individuals, meaning that it is not contradictory for every individual to satisfy exactly the same predicati[...][ve] functions as every other individual

These considerations illustrate the kind of challenge Ramsey faced. Assume there are only two individuals, a and b. If it is a tautology that some propositional function is true of only these two individuals, then it must be a contradiction that, say, every atomic predicative propositional function '$F(\hat{x})$' is satisfied by both 'a' and 'b'. But the fact that every atomic predicative propositional function '$F(\hat{x})$' may be satisfied by both 'a' and 'b' is something that follows from the logical independence of atomic propositions: atomic propositional functions do not discriminate between individuals.

The argument extends then to the general case. As every predicative function is a truth-function of atomic predicative functions and propositions, it is not a contradiction that every individual which satisfies one function, satisfies also another. So as Trueman points out it cannot be a tautology that some predicative function is true of exactly one individual, or exactly two individuals, etc.

3.5.1 Logical Necessity versus Analytical Necessity

It has been emphasised (*ibid.*) that the argument establishing that predicative functions do not logically discriminate between individuals at no point appeals to the Tractarian assumption that all necessity is logical necessity. We could, for instance, introduce a different notion of necessity, call it 'analytic necessity', which is necessity in virtue of meaning. This will not rule out the possibility that there are two individuals who satisfy the same predicative functions, provided the atomic propositions would remain logically independent in the above sense of logical necessity.

Such a notion of necessity has been considered, among others, by the Finnish logician Erik Stenius [186]. According to Stenius, a statement is analytic if it is true in virtue of the semantic conventions for certain of its symbols. Alternatively, a statement is analytic if, according to the semantic conventions for some of its expressions, no state of affairs is a truth restriction for it (that is, no state of affairs makes it false).

The statement 'If a is red, then a is not green' symbolised by

$$`R(a) \rightarrow \neg G(a)`$$
(3.7)

can be shown to be analytic in Stenius's sense. Its truth-table is

$R(a)$	$G(a)$	$\neg G(a)$	$R(a) \rightarrow \neg G(a)$
T	T	F	F
T	F	T	T
F	T	F	T
F	F	T	T

This truth-table seems to possess what Stenius calls, a truth-restriction, that is, a state of affairs which renders the proposition '$R(a) \rightarrow \neg G(a)$' false. But for Stenius this truth restriction is not a state of affairs because the colours green and red are logically incompatible. Therefore the first line must be erased and the truth restriction vanishes.

We witness here a violation of the logical independence of the atomic propositions which is of a different kind than the one we have considered so far: the truth of '$R(a)$' is incompatible with that of '$G(a)$', that is, one and the same individual cannot simultaneously be the argument of both propositional functions '$R(\hat{x})$' and '$G(\hat{x})$'. Given that (3.7) is analytic for each a, then so is

'No red objects are green'

symbolised by

$$`\forall x\, [R(x) \to \neg G\,(x)]\,`$$

Thus the failure of logical independence in this new sense leads to the new notion of analytical necessity. But allowing for this kind of non-logical, analytical necessity is perfectly compatible with the logical independence of atomic propositions from the previous section.

Ramsey wanted to show that mathematical statements are logically necessary in the Tractarian sense. For this he needed to give up the kind of logical independence of atomic propositions we considered in the previous section and find a notion of propositional function which would discriminate between individuals and would not be grounded in the notion of analytical necessity illustrated in this section. He did that by introducing the notion of a propositional function in extension. Before discussing it, let me point out that Frege went a different way: although he wanted to show that mathematical (arithmetical) statements are reducible to logic, he did not conceive of logical statements as necessary in the Tractarian sense. For Frege logical statements are the most general statements about a universe of discourse. No wonder that the modern discussion around Frege's logicism ended up in debating whether the ultimate logical principles to which arithmetics is reduced are analytic or not.

3.5.2 Ramsey's Propositional Functions in Extension

Ramsey needed, then, a new notion of a propositional function which would allow him to distinguish between individuals. In order to do this, he needed to extend the notion of a function to cover also non-predicative propositional functions, where predicativity is understood as above. Here is how he expresses himself ([165], p. 52):

> The only practicable way to do it as radically and drastically as possible; to drop altogether the notion that $\varphi\,(a)$ says about a what $\varphi\,(b)$ says about b, to treat propositional functions like mathematical functions, that is, to extensionalise them completely. Indeed it is clear that, mathematical functions being derived from propositional, we shall get an adequate extensional account of the former only by taking a completely extensional view of the latter.

Ramsey is well aware that he cannot give an explicit definition of a function in extension and for this reason he contents himself to explain this notion rather than define it. His explanation is given in terms of the notion of correlation, that is, a relation in extension between propositions and individuals, which associates to each individual a unique proposition. In specifying the nature of this correlation, he remarks that it may be "practicable or impracticable" (*ibid.*). I take this to be just another way for Ramsey to say that the correlation is not determined by a formal law, but is an arbitrary association between individuals and propositions.

Ramsey uses propositional functions in extension to define identity:

$$x = y \;\; =_{df} \;\; \forall \varphi_e\, [\varphi_e\,(x) \equiv \varphi_e\,(y)]$$

where φ_e takes propositional functions in extension as values. We notice that when a and b are the same individual, then '$a = b$' is the conjunction of '$p \equiv p$', '$q \equiv q$', ..., which is a tautology. On the other side, when a is distinct from b, then there is a propositional function '$\varphi_e(\hat{x})$' such that '$\varphi_e(a)$' is 'p' while '$\varphi_e(b)$' is '$\neg p$'. In this case '$a = b$' is a conjunction of propositions which includes '$p \equiv \neg p$', that is, a contradiction.

After having defined identity, Ramsey can introduce set-theoretical notions. In order to introduce singletons, he considers the propositional function '$\hat{x} = a$', where a is an arbitrary individual. When identity is defined as above, then '$a = a$' is a tautology, and for any other b, '$b = a$' is a contradiction. Hence it is a tautology that some propositional function is true of exactly one individual. By a similar reasoning one can introduce doubletons. The propositional function '$\hat{x} = a \vee \hat{x} = b$' is true of exactly two individuals. (I am indebted to Trueman for this argument.)

It is perhaps worth comparing Ramsey's definition of identity to the modern model-theoretic definition of identity in second-order logic with the standard interpretation:

$$x = y \Leftrightarrow \forall X\,[X(x) \Leftrightarrow X(y)]$$

Here X is a second-order variable ranging over sets. By the standard interpretation we mean that every model for the second-order language is such that the range of the second-order variables is the full power set of the set which is the range of the first-order variables. In this setting, instead of showing that '$a = a$' is a tautology, and for any individual b distinct from a, '$b = a$' is a contradiction, we can show that in every model in which a and b are the same individual, then '$\forall X\,[X(x) \Leftrightarrow X(y)]$' is (trivially) true. And in every model in which a is distinct from b, '$\forall X\,[X(x) \Leftrightarrow X(y)]$' is false. Indeed, the first claim is true: it follows from the principle of extensionality of sets. As for the second claim, the set $\{a\}$ falsifies the formula '$\forall X\,[X(x) \Leftrightarrow X(y)]$'. Given the standard interpretation, this set exists. Notice that the only principle we need to rely on is the extensionality of sets.

We can achieve the same result by using functions. In this case the definition of identity would be

$$x = y \Leftrightarrow \forall f\,[f(x) \Leftrightarrow f(y)]$$

where we may take f to be a function from individuals to truth-values. Then '$a = a$' is a tautology and '$b = a$' is false in every model in which a and b are distinct individuals: take a function f which maps a to T and b to F. Here we need the standard interpretation of function variables and the notion of function in extension. Then, we can go on and reconstruct singletons and doubletons as Ramsey did.

In the remaining of mys paper let me consider two objections against the notion of propositional functions in extension discussed in [194]. One of them is due to Sullivan, the other to Wittgenstein.

3.5.3 Sullivan's Objection to the Notion of Propositional Function in Extension: Containment

According to Sullivan [187], the main difference between propositional functions and propositional functions in extension lies in the fact that the former are contained in their values in a way in which the latter are not. In other words, a propositional function in extension needs all its values to be individuated, whereas one single value suffices for the individuation of a (predicative) propositional function. It is not difficult, intuitively, to see why this is so. Take any argument and consider the proposition which is the value of the propositional function for that argument. By deleting the argument, you can recover the propositional function. To take an example, if '$F(\hat{x})$' is a propositional function and you know that 'F (John)' is 'John is tall', then you also know that 'F (Peter)' is 'Peter is tall', etc. On the other side, if you know that 'φ_e (John)' is 'Paris is beautiful' then you cannot infer anything about 'φ_e (Peter)', when φ_e is a function in extension.

Trueman ([194], Sect. 4) gives an example of a propositional function which shows Sullivan's argument to be invalid. Here it is take the function

$$`P(\hat{x}) \vee \exists y \left[T \text{ (Plato)} \wedge \neg P (\hat{y})\right]`\qquad (3.8)$$

where '$T(\hat{x})$' is

$$`P(\hat{x}) \vee \neg P(\hat{x})`$$

Consider first the value of this function for an argument 'a' other than 'Plato', i.e.

$$`P (a) \vee \exists y \left[T \text{ (Plato)} \wedge \neg P (\hat{y})\right]`\qquad (3.9)$$

By the convention governing the use of quantifiers, '$\exists y \left[T \text{ (Plato)} \wedge \neg P (\hat{y})\right]$' is a disjunction of the values of 'T (Plato) $\wedge \neg P (\hat{y})$' for every argument other than 'Plato'. But given that 'T (Plato)' is a tautology, and the conjunction of a proposition with a tautology is that proposition itself, this conjunction is equivalent to the conjunction of the values of '$\neg P (\hat{y})$', for every argument other than 'Plato', one conjunct of which will be '$\neg P (a)$'. Hence (3.9) will be a disjunction including both '$P (a)$' and '$\neg P (a)$' as disjuncts, and will, then, be a tautology. On the other side, the value of (3.8) for 'Plato' is the the disjunction of 'P (Plato)' and the values of '$\neg P (\hat{y})$' for every argument other than 'Plato'. It will, then, be the disjunction of an atomic proposition with the negation of another atomic proposition, and will not be a tautology.

It follows that (3.8) maps every name other than 'Plato' to a tautology, and 'Plato' to a non-tautology. Trueman concludes that we need all the values of this propositional functions in order to establish its identity, and thereby Sullivan's claim should be restricted to atomic predicative propositional functions: only in this case the propositional function may be recovered by whatever value of the function one considers.

As nice as this example is, one should not overestimate its importance, though (I also take this to be Trueman's position). Its particularity is due to the convention governing the use of quantifiers that we discussed above. Even if the property of containment held only for atomic predicative propositional functions, it would still explain why these functions are more accessible than their extensional relatives and how our conceptual system can somehow integrate and manipulate potentially infinite correlations of arguments and values. For in the absence of properties like containment or other mechanisms which perform a similar function, the question still remains: How are we to understand the notion of mapping? Moreover, is there any way for us to grasp potentially infinite correlations?

3.5.4 Substitution

I take the notion of containment to provide an answer to the second question. One possible answer to the first question that Trueman considers is to understand mapping in terms of substitution. This is an expected move: after all, we needed substitution when we explained the notion of atomic predicative propositional function in the first place. We took such a function to be the result of replacing by variables any of the names of individuals in an atomic proposition. An example may help. For '$F(\hat{x})$' standing for the atomic propositional function '$\hat{x}$ is wise', when we substitute '$\hat{x}$' with 'Socrates' we thereby generate 'Socrates is wise' in which 'Socrates' occurs as a name of Socrates. The sense in which non-atomic predicative functions "map" names to propositions is explained analogously. It is quite clear that substitution, as a mechanical operation on expressions in an underlying language, explains the property of containment and thus also answers the second question considered above.

The operation of substitution cannot obviously ground the notion of mapping that underlies propositional functions in extension. One has to try something else. Returning to our last example, we notice that the operation of substitution generates a table:

$F(\hat{x})$	
'Socrates'	'Socrates is wise'
'Plato'	'Plato is wise'

Following the same idea, we could also introduce atomic predicative propositional functions in extension by tables, e.g.:

$F_e(\hat{x})$	
'Socrates'	'Queen Anne is dead'
'Plato'	'Einstein is a great man'

As expected, this suggestion is not shared by those who oppose arbitrary correlations. For the whole matter of dispute is the nature of the relation between the name on the left side, and the corresponding proposition on the right side. Trueman endorses an argument by Wittgenstein ([204], part II, Chap. 16) who points out that the name 'Socrates' appears here only to direct us to a line of the table. We could have marked instead the lines of this table with any signs we liked: numerals, letters or squares of colour, etc. The fact that we chose to mark each line of this table with strings which look like the names of Socrates and Plato should not mislead us into thinking that they are those names. Consequently, if '$F_e\left(\hat{x}\right)$' is a predicative function defined as in the table above, then the first and second occurrences of the string 'Socrates' in '$F(\text{Socrates}) \wedge F_e(\text{Socrates})$' have different significances, the first is an occurrence of the name of Socrates and the latter is not.

We are back to square one. Wittgenstein's criticism is nothing else but a milder expression of the requirement that we have seen at work in the case of predicative propositional functions. According to that requirement, the proposition '$F(a)$' that the propositional function '$F(\hat{x})$' assigns to 'a', must say or predicate the same thing of a as the proposition '$F(b)$', which '$F(\hat{x})$' assigns to b, predicates of b. The present version is milder because it only asks for the value that the function assigned to 'a' to say something about a. Still it is obvious that in both cases we witness the refusal to accept the idea that what is important for individuating a function is an arbitrary correlation of values and arguments.

3.5.5 Arbitrary Functions

This is, then, Wittgenstein's criticism of Ramsey's notion of a propositional function in extension: the argument of such a function is there "only to direct us to a line of the table". In other words, when a function in extension is introduced, one abstracts from the nature of the connection between arguments and values and makes sure only that, to each argument, there is a line in the table.

Wittgenstein's criticism sounds surprisingly similar to Frege's criticism of the extensional notion of a set and of the individuation of sets through their members. In [129], p. 301, we pointed out two different reasons for Frege to reject the individuation of classes through their members. The first one concerns the definition of the empty class. Here is what Frege writes in an undated letter to Peano ([106], vol. II, p. 177; [108], p. 109):

> Of course, one must not then regard a class as made out by the objects (individual, entities), that belong to it; for removing the objects one would then also be removing the class constituted by them. Instead, one must regard the class as made out by the characteristic marks, i.e., the properties which an object must have if it is to belong to it. It can then happen that these properties contradict one another, or that there occurs no object that combines them in itself. The class is then empty but without being logically objectionable for that reason.

The second reason concerns the individuation of infinite classes. According to Frege, from the finiteness of the human intellect it follows that an infinite class cannot be given solely by its members. The only way it can be given is by deriving it from a concept, that is, by takings it as "yielded by thought". Here is what Frege writes in "Booles rechnende Logik und die Begriffsschrift" ([106], vol. I, p. 38; [107], p. 34; notice that this passage illustrates the first reason, too). This is also made clear in the following passage:

> But it is surely a highly arbitrary procedure to form concepts merely by assembling individuals, and one devoid of significance for actual thinking unless the objects are held together by having characteristics in common. It is precisely these which constitute the essence of the concept. Indeed one can form concepts under which no object form, where it might perhaps require lengthy investigation to discover that this was so. Moreover, a concept, such as that of number, can apply to infinitely many individuals. Such a concept would never be attained by logical addition. Nor finally may we presuppose that the individuals are given *in toto*, since some, such as e.g. the numbers, are only yielded by thought.

As we observed in [129], p. 305, "what made possible the conception of an arbitrary set was the gradual disentanglement of the notion of set from intensional ingredients such as concepts, properties, etc., and the definition of sethood in an alternative way". In a parallel development, the modern notion of arbitrary function emerged through the gradual disentanglement of the notion of correlation from Fregean concepts, equations and other formal rules, or from requirements like predicativity. Ramsey's notion of a propositional function in extension is one step in this process of emancipation. Modern logic has developed Ramsey's idea and taken functional dependencies as arbitrary correlations between values and arguments. Here is one example which illustrates this trend taken from [116].

The idea, made possible by the development of model-theoretical semantics, is not to define arbitrary functional correlations, but to introduce a new logical constant in the object language and then give its meaning through a semantical clause. More specifically, the syntax of first-order logic is extended with atomic formulas of the form

$$=\!\left(\vec{x}, y\right)$$

intended to express arbitrary functional dependence: the (values of the) variables $\vec{x}$ totally (functionally) determine (the value of) y. Such an atom is interpreted in a model by a set X of (partial) assignments in the universe of the model. The semantical clause that we need is:

1. X makes the formula '$=\!\left(\vec{x}, y\right)$' true if and only if for any two distinct assignments s and s' in X, whenever s and s' agree on the values of the variables in $\vec{x}$, they also agree on the values of y.

The right-hand side of this double implication defines a functional correlation in purely extensional terms, without appealing to any particular relation between an argument and its value. Here is an example ([197], p. 11), which also illustrates the kind of extensional correlation Wittgenstein objected to:

	x_0	x_1	x_2
s_0	1.5	4	0.51
s_1	2.1	4	0.55
s_2	2.1	4	0.53
s_3	5.1	4	0.54
s_4	8.9	4	0.53
s_5	21	4	0.54
s_6	100	4	0.54

The set X consisting of the six assignments $s_0, \ldots, s_6$ makes both $=(x_0, x_1)$ and $=(x_0, x_1)$ true.

Grädel and Väänänen define also independence. To this purpose, the syntax of first-order logic is extended with atomic formulas of the form

$$x \perp y$$

with the intended interpretation: the (values of the) variable y is (are) independent of the (values of the) variable x. Such a formula is interpreted by a set X of assignments, as in the previous case, but now the interpretative semantical clause is:

2. X makes '$x \perp y$' true if and only if for any two assignments s and s' in X there is a third assignment s'' such that s'' agrees with s on the value of x and it agrees with s' on the value of y.

This definition tells us that the value $s(x)$ of x alone does not determine the value $s(y)$ of y, for there may be another assignment s' in X which assigns to y a distinct value, i.e. $s'(y) \neq s(y)$. But then according to the proposed definition, there is a third assignment s'' such that $s''(y) = s'(y)$ and $s''(x) = s(x)$. That is, just when we thought that on the basis of $s(x)$ we can conclude that the value of y is $s(y)$, we discover s'' which gives the same value for x but a different value for y. In other words, borrowing Wittgenstein's jargon, there is an argument which "points to two lines" in the table. In our example, X makes both '$x_1 \perp x_2$' and '$x_0 \perp x_2$' true.

3.6 Conclusion

In [62], Bell and Demopoulos accept Dummett's initial view to the effect that Frege's interpretation of the function variables is the standard one and that Frege's concept of a function coincides with the set-theoretic notion of an arbitrary correspondence, in which case the domain of the function variables is in one-one correspondence with the power-set of the domain of the individual variables. In [61], reconsiders this matter (*ibid.*, pp. 5–6):

> More recently, reflection occasioned by reading [173] has convinced me that the equation
> of Frege's concept of a function with the notion of an arbitrary correspondence should be
> reconsidered, and that it might be fruitful to reconsider it from the perspective of Ramsey's
> interpretation of *Principia*'s's propositional functions.

Demopoulos's conclusion is that Frege's assimilation of concepts to functions which map into truth values is as predicative as Russell. The correspondence is not arbitrary, but is constrained by the principle that if a function maps two objects to the True, they must fall under a common concept. But he also observes that Fregean functions and concepts lack the explicit association with propositions that is characteristic of Russellian propositional functions. *Principia*'s propositional functions "map to the truth values only by 'passing through' a proposition" whereas "Frege's concepts map directly to the truth values" (*ibid.*, p. 16). Despite his acknowledgement that Fregean concepts are constrained in the way mentioned above, Demopoulos is reluctant to explicitly admit that Frege's notions of a function is not extensionalist in nature. He prefers to close his paper in a rather ambiguous way, as follows (*ibid.*, pp. 16–17):

> Fregean functions and concepts [...] lack the explicit association with propositions that is
> characteristic of propositional functions; an extensionalist interpretation of a Fregean concept
> as an arbitrary mapping of objects to truth values is arguably still a Fregean concept. However
> its utility for Frege's theory of classes is unclear. According to a theory like Frege's, concepts
> provide the principle which gives classes their 'unity', and they also serve the epistemological
> function of providing the principle under which a collection of objects can be regarded as
> a separate object of thought. A class that is generated by an arbitrary pairing of individuals
> with truth values might be one that is 'determined by a concept', but the concept which
> determines it seems no more epistemically accessible than the collection itself. Even if it can
> be convincingly argued that such concepts sustain the unity of the classes they determine,
> it can hardly be maintained that they are capable of playing the epistemological role which
> the predicative interpretation can claim for its functions and concepts.

The overall conception that dominates the present paper as well as the ideas developed in [129] and [173], is that Russell's notion of a propositional function and Ramsey's notion of predicative function are one more manifestation, albeit a special one, of the same phenomenon which governs Fregean concepts: their determination by a norm (rule, equation, concept). If that were not the case, then they would not be able to perform, the epistemological function that Demopoulos attributes to them. Now, in the last quote Demopoulos speculates with the idea that a class that is generated by an arbitrary function might still be generated by a concept which is epistemically inaccessible. I take the point of this remark and of those that follow it to be that of emphasizing that there is still a considerable gap between Fregean concepts (functions) on one side, and Russell's and Ramsey's predicative functions, on the other.

A detailed comparison between predicative functions and Fregean concepts is outside the purpose of this paper. The point I tried to defend here and elsewhere is that both Frege's and Russell's conceptions of a function stand in clear contrast to Ramsey's notion of a propositional function in extension and to the extensionalist notion of a function illustrated by clause **1** in Sect. 3.5.5, above. There is no doubt that Frege could not have such a conception, for he tells us ([97], Sect. I.10; [110], p. 161):

We have only a way always to recognise a value-range as the same if it is designated by a name such as $\grave{\epsilon}\Phi(\epsilon)$, whereby it is already recognisable as a value-range. However, we cannot decide yet whether an object that is not given to us as a value-range is a value-range or which function it may belong to; nor can we decide in general whether a given value-range has a given property if we do not know that this property is connected with a property of the corresponding function.

As this passage illustrates, For Frege, "an object that is not given to us as a value-range", i.e. that is not introduced as the extension of a law or concept, does not tell us what function that value-range corresponds to. True, Frege was possibly thinking here of any object whatsoever, and not necessarily of one that is easily identifiable as a value-range of some indeterminate function; his point seems to be that taking the value-range of a function $\Phi(\xi)$ to be the same as the value-range of a function $\Psi(\xi)$ if and only if the values of these functions are the same for any argument does not allow us to decide whether a certain table, the Mount Blanc, or Julius Caesar are value-ranges. But, it is a matter of fact that his claim is general, and it also applies, then, to objects that are easily identifiable as value-ranges, namely to classes. In this case, the point becomes that, when a class is given to us as such and not as a value-range of a determinate function, there is no vantage point from which we can say what function it is the value-range of. Ramsey's notion of propositional function in extension and the notion of functional dependence illustrated by clause **1** in Sect. 3.5.5, above may be seen as the perfect target of Frege's critical remark: the set of assignments, or, as we may call it, the value-range X in that clause, may be the extension, as we all know, of many functional laws.

What Frege and Russell ignored and Ramsey realized, is that one can and needs to talk about a function even when one is not able to individuate it through the law that generates is, like for instance when one talks about the properties *all* functions have. In that case one abstracts from the nature of the formal law that generates the corresponding extension. The framework outlined in the previous section allows one to do just that. According to clause **1**, a functional dependence is, indeed, just a set satisfying appropriate conditions (namely Armstrong axioms in data base theory, as showed by Väänänen in [197], Sects. 8.1 and 8.2). This provides an extensionalist notion of a function, akin in spirit to Ramsey's notion of a function in extension, which anticipates its treatment in contemporary mathematics: a notion that stands opposite both to Frege's conception, according to which a function is constrained by a law, and to Russell's idea of a propositional function.

Bibliography

1. AA. VV, *Séances des écoles normales, recueillies par des sténographes et revues par les professeurs*, nouvelle édition. (Impr. du Cercle Social, Paris, 1800–1801). -13 vols.: 1–10, Leçons + 1–3, Débats
2. J.B. le R. d' Alembert, Quantity, in *Encyclopédie, ou dictionnaire raisonné des sciences, des arts et des métiers*, (Briasson, David l'aîné, le Breton, Durand, Paris, 1751–1780), 35 vols. Vol. 13, 1765, pp. 653–655
3. A.M. Ampère, Recherches sur l'application des formules générales du calcul des variations aux problèmes de la mécanique. Mémoires présentés à l'Institut des sciences, lettres et arts, par divers savans, et lus dans ses assemblées. Sciences mathématiques et Physiques, I:493–523, January 1806. Presented on 26th Floréal an XI (May, 16th, 1803)
4. J. Avigad, Methodology and metaphysics in the development of Dedekind's theory of ideals, in *The Architecture of Modern Mathematics*, ed. by J. Ferreirós, J. Gray (Oxford University Press, Oxford, 2006), pp. 159–186
5. G. Baker, 'Function' in Frege *Begriffsschrift*: dissolving the problem. Br. J. Hist. Philos. **9**, 525–544 (2001)
6. M. Beaney (ed.), *The Frege Reader* (Blackwell Publishing Ltd., Oxford, 1997)
7. J.P. Belna, *La notion de nombre chez Dedekind, Cantor, Frege* (Vrin, Paris, 1996)
8. P. Benacerraf, What numbers could not be. Philos. Rev. **74**, 47–73 (1965). Also in [10], pp. 172–294
9. P. Benacerraf, H. Putnam (eds.), *Philosophy of Mathematics* (Prentice-Hall Inc., Englewood Cliffs, 1964)
10. P. Benacerraf, H. Putnam (eds.), *Philosophy of Mathematics* (Cambridge University Press, Cambridge, 1983)
11. H. Benis-Sinaceur, *Jean Cavaillès. Philosophie Mathématique* (PUF, Paris, 1994)
12. H. Benis-Sinaceur, *Cavaillès* (Les Belles Lettres, Paris, 2013)
13. J. van Benthem, Logical constants across varying types. Notre Dame J. Form. Log. **30**, 315–342 (1986)
14. J. van Benthem, Invariance and definability: two faces of logical constants, in *Reflections on the Foundations of Mathematics. Essays in Honor of Sol Feferman*, ASL Lecture Notes in Logic. ed. by W. Sieg, R. Sommer, C. Talcott, pp. 426–446
15. J. Bernoulli, Remarques sur ce qu'on a donné jusqu'ici de solutions des problèmes sur les isopérimètres. Histoire de l'Académie Royale des Sciences [de Paris] Avec les Mémoires de Mathématiques et Physique pour la même année, pp. 100–138 of the Mémoires, 1718; publ. 1722
16. B. Bolzano, *Wissenschaftslehre, Versuch einer ausführlichen und grösstentheils neuen Darstellung der Logik mit steter Rücksicht auf deren bisherige Bearbeiter* (J. E. v. Seidel, Sulzbach, 1837)

© Springer International Publishing Switzerland 2015

H. Benis-Sinaceur et al., *Functions and Generality of Logic*,

Logic, Epistemology, and the Unity of Science 37,

DOI 10.1007/978-3-319-17109-8

17. D. Bonnay, Logicality and invariance. Bull. Symb. Log. **14**, 29–68 (2008)
18. G. Boolos, On second-order logic. J. Philos. **72**, 509–527 (1975). Also in [26], pp. 37–53
19. G. Boolos, To be is to be a value of a variable (or to some values of some variables). J. Philos. **81**, 430–449 (1984). Also in [26], pp. 54–72
20. G. Boolos, Nominalist platonism. Philos. Rev. **94**, 327–344 (1985). Also in [26], pp. 73–87
21. G. Boolos, Saving Frege from contradiction. Proc. Aristot. Soc. **87**, 137–151 (1986–1987). Also in [26], pp. 171–182. We refer to this last edition
22. G. Boolos, The consistency of Frege's Foundations of arithmetic, in *On Being and Saying: Essays in Honor of Richard Cartwright*, ed. by J. Thomson (MIT Press, Cambridge, 1987), pp. 3–20. Also [26], pp. 183–201
23. G. Boolos, The standard of equality of numbers, in *Meaning and Method: Essays in Honor of Hilary Putnam*, ed. by G. Boolos (Cambridge University Press, Cambridge, 1990), pp. 261–277. Also in [56], pp. 234–254, and [26], pp. 202–219. We refer to this last edition
24. G. Boolos, The advantages of honest toil over theft, in *Mathematics and Mind*, ed. by A. George (Oxford University Press, Oxford, 1994), pp. 27–44. Also in [26], pp. 255–274. We refer to this last edition
25. G. Boolos, 1879? in *Reading Putnam*, ed. by P. Clark, B. Hale (Basil Blackwell, Oxford, 1994). Also in [26], pp. 237–254. We refer to this last edition
26. G. Boolos, Frege's theorem and the Peano postulates. Bull. Symb. Log. **1**, 317–326 (1995). Also in [26], pp. 291–300. We refer to this last edition
27. G. Boolos, Is Hume's principle analytic? in *Logic, Language and Thought: Essays in Honour of Michael Dummett*, ed. by R.J. Heck Jr. (Clarendon Press, Oxford, 1997), pp. 245–262. Also in [26], pp. 301–314. We refer to this last edition
28. G. Boolos, *Logic, Logic and Logic* (Harvard University Press, Cambridge, 1998). With introductions and afterword by J.P. Burgess; ed. by R. Jeffrey
29. G. Boolos, Gottlob Frege and the foundations of arithmetic. In [26], pp. 143–154
30. G. Boolos, R.G. Heck Jr., Die Grundlagen der Arithmetik, §§82−3, in *Philosophy of Mathematics Today* (Clarendon Press, Oxford, 1998), pp. 407–428. Also in [26], pp. 315–338. We refer to this later edition
31. U. Bottazzini, *The Higher Calculus. A History of Real and Complex Analysis from Euler to Weierstrass* (Springer, New York, 1986)
32. R.E. Bradley, C.E. Sandifer, *Cauchy's Cours d'analyse. An Annotated Translation* (Springer, Dordrecht, 2009)
33. J.P. Burgess, Review of [205]. Philos. Rev. **93**, 638–640 (1984)
34. J.P. Burgess, Hintikka and Sandu versus Frege in re arbitrary functions. Philos. Math. 3rd Ser. **1**, 50–65 (1993)
35. J.P. Burgess, Frege on arbitrary functions, in *Frege's Philosophy of Mathematics*, ed. by W. Demopoulos (Harvard University Press, Cambridge, 1995), pp. 89–107. This is a revised version of [32]
36. T.W. Bynum, The evolution of Frege's logicism, in *Studien zu Frege I. Logic und Philosophy der Mathematik*, ed. by M. Schirn (Stuttgart and Bad Cannstatt, Frommann-Holzboog, 1976), pp. 276–286
37. M. Cadet, M. Panza, The Logical System of Frege's Grundgesetze: A Rational Reconstruction. Manuscrito. **38**, 5–94 (2015)
38. G. Cantor, *Grundlagen einer allgemeinen Mannigfaltigkeitslehre in mathematisch-philosophischer Versuch in der Lehre des Unendlichen* (Teubner, Leipzig, 1883)
39. G. Cantor, *Gesammelte Abhandlungen mathematischen und philosophischen Inhalts* (Springer, Berlin, 1932). Herausgegeben von E. Zermelo
40. R. Carnap, Die alte und die neue Logik. Erkenntnis **1**, 12–26 (1930)
41. R. Carnap, Die logizistische Grundlagen der Mathematik. Erkenntnis **2**, 91–105 (1931). An English Translation is included in [9], pp. 31–41 and [10], pp. 41–52
42. A.L. Cauchy, *Cours d'analyse de l'École royale polytechnique [...], 1re partie. Analyse algébrique* (Debure frères, Paris, 1821)
43. J. Cavaillés, *Sur la logique et le théorie de la science* (PUF, Paris, 1947)

44. J. Conant, Elucidation and nonsense in Frege and early Wittgenstein, in *The New Wittgentein*, ed. by A. Crary, R. Read (Routledge, London, 2000), pp. 174–217

45. E. Czuber, *Vorlesungen über Differential- und Integralrechnung*, vol. I (Teubner, Leipzig, 1898)

46. U. Dathe, Gottlob Frege und Johannes Thomae. Zum Verhältnis zweier Jenaer Mathematiker, in *Frege in Jena. Beiträge zur Spurensicherung*, ed. by G. Gabriel, W. Kienzler (Verlag Königshausen & Neumann GmbH, Würzburg, 1997), pp. 87–103; bad 2 of Kritisches Jahrbuch der Philosophie

47. R. Dedekind, *Stetigkeit und irrationale Zahlen* (F. Vieweg und Sohn, Braunschweig, 1872)

48. R. Dedekind, Sur la théorie de nombres entiers algébriques. Bulletin des sciences mathématiques et astronomiques, 1st Series **11**, 278–288 (1876), and 2nd Series, **1**, 1–121 (1877). Parts of this paper are also edited in [52], vol. III, pp. 262–296, in [53], pp. 239–256

49. R. Dedekind, *Was sind und was sollen die Zahlen?* (F. Vieweg und Sohn, Braunschweig, 1888)

50. R. Dedekind, *Was sind und was sollen die Zahlen?*, 2nd edn. (F. Vieweg und Sohn, Braunschweig, 1893)

51. R. Dedekind, Über die Begründung der Idealtheorie. Nachrichten von der Königlichen Geselshaft der Wissenschaften zu Göttingen. Math.-Phys. Klasse 106–113 (1995). Also in [52], vol. II, pp. 50–58. We refer to this last edition

52. R. Dedekind, Über Zerlegungen von Zahlen durch ihre grössten gemeinsamen Teiler, in *Festschrift der Technischen Hochschule zu Braunschweig bei Gelegenheit der 69. Versammlung Deutscher Naturforscher und Ärtze* (1897), pp. 1–40. Also in [52], vol. II, pp. 104–147

53. R. Dedekind, *Essays on the Theory of Numbers* (The Open Court P. C., Chicago, 1901). Translated by W.W. Beman

54. R. Dedekind, *Was sind und was sollen die Zahlen?*, 3rd edn. (F. Vieweg und Sohn, Braunschweig, 1911)

55. R. Dedekind, *Gesammelte mathematische Werke* (Vieweg, Braunschweig, 1930–1932). Herausgegeben von E. Noether, R. Fricke, and Ö. Ore. 3 vols

56. R. Dedekind, *La création des nombres* (Vrin, Paris, 2008). French translation of [46], [44], and pieces relative to both writings, with Introductory Notes and footnotes by H. Benis-Sinaceur

57. R. Dedekind, H. Weber, Theorie der algebraischen Funktionen einer Veränderlichen. J. für die reine und angewandte Mathematik **92**, 181–290 (1982). Also in [52], vol. 8, pp. 238–350

58. W. Demopoulos, Frege and the rigorization of analysis. J. Philos. Log. **23**, 225–245 (1994)

59. W. Demopoulos (ed.), *Frege's Philosophy of Mathematics* (Harvard University Press, Cambridge, 1995)

60. W. Demopoulos, On the origin and status of our conception of number. Notre Dame J. Form. Log. **41**, 210–226 (2000)

61. W. Demopoulos, On logicist conceptions of functions and classes, in *Logic, Mathematics, Philosophy: Vintage Enthusiasms. Essays in Honour of J.L. Bell*, ed. by D. De Vidi, M. Hallett, P. Clark (Springer, Dordrecht, 2011), pp. 3–18

62. W. Demopoulos, J.L. Bell, Frege's theory of concepts and objects and the interpretation of second-order logic. Philos. Math. Ser. 3 **1**, 139–156 (1993)

63. W. Demopoulos, P. Clark, The logicism of Frege, Dedekind and Russell, in *Oxford Handbook of Philosophy of Mathematics and Logic*, ed. by S. Shapiro (Oxford University Press, Oxford, 2005), pp. 129–165

64. J. Dhombres, Un texte d'Euler sur les fonctions continues et le fonctions discontinues, véritable programme d'organisation de l'analyse au 18e siècle. Cahiers du Séminaire d'Histoire des Mathématiques **9**, 23–97 (1988). Including a French translation of [78]

65. C. Diamond, Inheriting from Frege: the work of reception, as Wittgenstein did it, in *The Cambridge Companion to Frege*, ed. by M. Potter, T. Ricketts (Cambridge University Press, Cambridge, 2010), pp. 550–601

66. P.G. Lejeune Dirichlet, *Vorlesugen über Zahlentheorie*, 3rd edn. (F. Vieweg und Sohn, Braunschweig, 1879). Herausgegeben un mit Zusätzen versehen von R. Dedekind

67. P. Dugac, *Richard Dedekind et les fondements des mathématiques* (Vrin, Paris, 1976)
68. M. Dummett, *Frege: Philosophy of Language* (Duckworth, London, 1973)
69. M. Dummett, Frege as a realist. Inquiry **19**, 455–468 (1976). Also in [70], pp. 79–96. We refer to this last edition
70. M. Dummett, Objectivity and reality in Lotze and Frege. Inquiry **25**, 95–114 (1982). Also in [70], pp. 97–125. We refer to this last edition
71. M. Dummett, Frege's myth of the third realm. Untersuchungen zur Logik und zur Methodologie **3**, 24–38 (1986). Also in [70], pp. 249–262
72. M. Dummett, Thought and perception: the views of two philosophical innovators, in *The Analytic Tradition: Meaning, Thought and Knowledge*, ed. by D. Bell, N. Cooper (Blackwell, Oxford, 1990), pp. 83–103. Also in [70], pp. 263–288. We refer to this last edition
73. M. Dummett, *Frege and Other Philosophers* (Clarendon Press, Oxford, 1991)
74. M. Dummett, *Frege. Philosophy of Mathematics* (Duckworth, London, 1991)
75. M. Dummett, P. Neumann, S. Adeleke, On a question of Frege's about right-ordered groups. Bull. Lond. Math. Soc. **18**, 513–521 (1987). A summarized version with final comments by M. Dummett is included in [70]
76. H. Edwards, Dedekind's invention of ideals. Bull. Lond. Math. Soc. **15**, 8–17 (1983)
77. L. Euler, *Introduction in analysisn infinitorum* (M.-M. Bousquet & Soc., Lausanne, 1748). 2 vols Re-edited in [79], ser. 1, vols. VIII and IX
78. L. Euler, De vibratione chordarum exercitatio. Nova Acta Eruditorum, pp. 512–527 (1749). Re-edited in [79], ser. 2, vol. 10, pp. 50–62
79. L. Euler, Remarques sur les mémoires précédens de M. Bernoulli. Histoire de l'Académie Royale des Sciences [de Berlin] **9**, 196–222 (1753); publ. 1755. Re-edited in [79], ser. 2, vol. 10, pp. 233–254
80. L. Euler, Institutiones calculi differentialis cum eius usu in analysi finitorum ac doctrina serierum. Impensis Acad. imp. sci. Petropolitanæ, ex off. Michaelis, Berolini (1755). Re-edited in [79], ser. 1, vol. X. We refer to this last edition
81. L. Euler, De usu functionum discontinarum in analysi. Novi Commentarii academiae scientiarum Imperialis Petropolitanæ **11**, 3–27 abstract at pp. 3–7 of Summarium Dissertationum (1765). Re-edited in [79], ser. 1, vol. 23, pp. 74–91
82. L. Euler, *Leonhardi Euleri Opera omnia* (Soc. Sci. Nat. Helveticæ, Leipzig, 1911–...), 76 volumes published to date
83. L. Euler, *Introduction to Analysis of the Infinite* (Springer, New York, 1988–1990). Transalated by J.D. Blanton, 2 vols
84. L. Euler, *Foundation of Differential Calculus* (Springer, New York, 2000). Transalated by J.D. Blanton
85. G. Evans, *The Varieties of Reference* (Clarendon Press, Oxford, 1982)
86. S. Feferman, Logic, logics, and logicism. Notre Dame J. Form. Log. **40**, 31–54 (1999)
87. S. Feferman, Set theretical invariance criteria for logicity. Notre Dame J. Form. Log. **51**, 3–20 (2010)
88. S. Feferman, Which Quantifiers are Logical? A combined semantical and inferential criterion. Forthcoming. Available online at http://math.stanford.edu/~feferman/papers.html
89. G. Ferraro, M. Panza, Lagrange's theory of analytical functions and his ideal of purity of method. Arch. Hist. Exact Sci. **66**, 95–197 (2012)
90. J. Ferreirós, On the relations between Georg Cantor and Richard Dedekind. Hist. Math. **20**, 343–363 (1993)
91. J. Ferreirós, *Labyrinth of Thought. A History of Set Theory and Its Role in Modern Mathematics*, 2nd edn. with added postscript, 2007 (Birkhäuser, Basel, 1999)
92. G. Frege, *Begriffsschrift, eine der Arithmetischen nachgebildete Formelsprache des reinen Denkens* (Nebert, Halle, 1879). English translation in [122], pp. 1–82
93. G. Frege, *Die Grundlagen der Arithmetik* (W. Köbner, Breslau, 1884)
94. G. Frege, *Function und Begriff. Vortrag, gehalten in der Sitzung vom 9. Januar 1891 der Jenaischen Gesellschaft für Medicin und Naturwissenschaft* (H. Pohle, Jena, 1891)

95. G. Frege, Über Begriff und Gegenstand. Vierteljahresschrift für wissenschaftliche Philosophie **16**, 192–205 (1892)

96. G. Frege, Über Sinn und Bedeutung. Zeitschrift für Philosophie und philosophische Kritik, NF **100**, 25–50 (1892)

97. G. Frege, *Grundgesetze der Arithmetik*. (H. Pohle, Jena, 1893–1903), 2 volumes

98. G. Frege, Rezension von Dr. E. G. Husserl: Philosophie der Arithmetik [...]. Zeitschrift für Philosophie und philosophische Kritik [...] **103**, 313–332 (1894)

99. G. Frege, Über die Grundlagen der Geometrie. Jahresbericht der Deutschen Mathematiker-Vereinigung **12**, 319–324, 375 (1903)

100. G. Frege, Was ist eine Funktion?, in *Festschrift Ludwig Boltzmann gewidmet zum sechzigsten Geburtstage, 20. Februar 1904*, ed. by S. Meyer (J. A. Barth, Leipzig, 1904), pp. 656–666

101. G. Frege, Über die Grundlagen der Geometrie. Jahresbericht der Deutschen Mathematiker-Vereinigung **15**, 293–309, 377–403, 423–430 (1906)

102. G. Frege, Der Gedanke. Beiträge zur Philosophie de deutschen Idealismus **I**(58–77) (1918–1919)

103. G. Frege, *The Foundations of Arithmetic* (Blackwell, Oxford, 1953). Translated by J.L. Austin

104. G. Frege, *Translations from the Philosophical Writings of Gottlob Frege*, ed. by P. Geach, M. Black (Basil Blackwell, Oxford, 1960)

105. G. Frege, *Kleine Schriften*, ed. by I. Angelelli (Olms, Hildesheim, 1967)

106. G. Frege, *Nachgelassene Schriften und Wissenschaftlicher Briefwechsel* (Meiner, Hamburg, 1969–1976) (2 vols; 2nd edn. of, vol. 1, 1983). Vol. 1, *Nachgelassene Schriften*, ed. by H. Hermes, F. Kambartel, F. Kaulbach; vol. 2, *Wissenschaftlicher Briefwechsel*, ed. by G. Gabriel, H. Hermes, F. Kambartel, C. Thiel, A. Veraar

107. G. Frege, *Posthumous Writings* (Basil Blackwell, Oxford, 1979). English Translation of vol. 1 of [103]

108. G. Frege, *Philosophical and Mathematical Correspondence* (Basil Blackwell, Oxford, 1980). English Translation of vol. 2 of [103]

109. G. Frege, *Collected Papers on Mathematics, Logic and Philosophy*, ed. by B. McGuinness (Blackwell, Oxford, 1984). English edition of [102]

110. G. Frege, *Basic Laws of Arithmetic* (Oxford University Press, Oxford, 2013). Translated and ed. by P.A. Ebert and M. Rossberg, with C. Wright

111. P. Geach, Saying and showing in Frege and Wittgenstein. Acta Philosophica Fennica **28**, 54–70 (1976)

112. G. Gentzen, Untersuchungen über das logische Schlissen. Mathematische Zeitschrift **39**, 176–210 and 405–431 (1935). English translation in G. Gentzen, *Collected Papers*, ed. by M.E. Szanbo (North-Holland, Amsterdam 1969), pp. 68–131

113. K. Gödel, What is Cantor's continuum problem? In [9], pp. 258–273. Also in [10], pp. 470–485, and in K. Gödel, *Collected Works*, ed. by S. Feferman (Oxford University Press, Oxford, 1986–1995) (3 vols), vol. II, pp. 254–270. We refer to to this last edition

114. K. Gödel, Russell's mathematical logic, in *The Philosophy of Bertrand Russell*, ed. by P.A. Schlipp (Northwestern University Press, Evanston (Ill.), 1944), pp. 125–153. Also in [9], pp. 447–469, [10], pp. 447–469, and in K. Gödel, *Collected Works*, ed. by S. Feferman (Oxford University Press, Oxford, 1986–1995) (3 vols), vol. II, pp. 119–141. We refer to to this last edition

115. W. Goldfarb, Frege's conception of logic, in *The Cambridge Companion to Frege*, ed. by M. Potter, T. Ricketts (Cambridge University Press, Cambridge, 2010), pp. 63–85

116. E. Grädel, J. Väänänen, Dependence and independence. Studia Logica **101**, 399–410 (2013)

117. I. Grattan-Guinness, *The Development of Foundations of Mathematical Analysis from Euler to Riemann* (MIT Press, Cambridge, 1970)

118. B. Hale, C. Wright, *The Reason's Proper Study* (Clarendon Press, Oxford, 2001)

119. R. Heck, The development of arithmetic in Frege's Grundgesetze der arithmetic. J. Symb. Log. **58**, 579–601 (1993). Also in [56], pp. 295–333

120. R. Heck, Definition by induction in Frege's Grundgesetze der Arithmetik. In [56], pp. 295–233 (1995)

121. R. Heck, Grundgesetze der Arithmetik I §10. Philosophia Mathematica, Ser. 3 **7**, 258–292 (1995)
122. R. Heck, Frege and semantic, in *The Cambridge Companion to Frege*, ed. by M. Potter, T. Ricketts (Cambridge University Press, Cambridge, 2010), pp. 342–378
123. R. Heck, *Reading Frege's Grundgesetze* (Clarendon Press, Oxford, 2012)
124. R. Heck, J. Stanley, Reply to Hintikka and Sandu: Frege and second-order logic. J. Philos. **90**, 416–424 (1993)
125. J. Van Heijenoort (ed.), *From Frege to Gödel: A Source Book in Mathematical Logic, 1879–1931* (Harvard University Press, Cambridge, 1967)
126. G. Hellman, Structuralism, in *Oxford Handbook of Philosophy of Mathematics and Logic*, ed. by S. Shapiro (Oxford University Press, Oxford, 2005), pp. 536–562
127. D. Hilbert, *Die Grundlagen der Geometrie. Teubner, Leipzig, 1899*, 10th edn. (Teubner, Stuttgart, 1968)
128. D. Hilbert, Über die grundlagen der Logik und der Arithmetik, in *Verhandlungen des dritten internationalen Mathematiker-Kongresses: in Heidelberg vom 8. bis 13. August 1904* (Teubner, Leipzig, 1905), pp. 174–285. English translation in [122], pp. 129–138
129. J. Hintikka, G. Sandu, The skeleton in Frege's cupboard: the standard versus nonstandard distinction. J. Philos. **89**, 290–315 (1992)
130. E. Husserl, *Formale und transzendentale Logik, Versuch einer Kritik der logischen Vernunft.* (Niemeyer, Halle, 1929). Also edited as, vol XVII of *Husserliana*, Martinus Nijhoff, Den Haag, 1974
131. E. Husserl, *Formal and Transcendental Logic* (Martinus Nijhoff, The Hague, 1969). English translation of [127], by D. Cairns
132. H. Hodes, Logicism and the ontological commitments of arithmetic. J. Philos. **81**, 123–149 (1984)
133. P. Kitcher, Frege, Dedekind, and the philosophy of mathematics, in *Frege Synthetized*, ed. by L. Haaparanta, J. Hintikka (D. Reidel, Dordrecht, 1986), pp. 299–343
134. J.-L. Lagrange, *Théorie des fonctions analytiques [...]* (Impr. de la République, Paris, 1797). Prairial an V: May-June 1797
135. J.-L. Lagrange, *De la résolution des équations numériques de tous les degrés* (Duprat, Paris, 1798). an VI: 1798
136. J.-L. Lagrange, Discours sur l'objet de la théorie des fonctions analytiques. Journal de l'École Polytechnique, 2(6th cahier):232–235, Thermidor an VII: July-August 1799. Re-edited in [138], vol. VII, pp. 325–328
137. J.-L. Lagrange, Sur le calcul des fonctions. Impr. du Cercle Social, 1801. Vol. X Leçons of [1]; reprinted with slight revisions as the 12th cahier of the Journal de l'École Polytecnique, Thermidor, an XII: July-August 1804. We refer to this last edition
138. J.-L. Lagrange, *Leçons sur le calcul des fonctions.* nouvelle édition revue, corrigée et augmentée par l'auteur (Courcier, Paris, 1806). Re-edited in [138], vol. X
139. J.-L. Lagrange, *Traité de la résolution des équations numériques de tous les degrés. Avec des Notes sur plusieurs points de la Théorie des équations algébriques* (Courcier, Paris, 1808). Reprinted in [138], vol. VIII
140. J.-L. Lagrange, *Théorie des fonctions analytiques [...]* (Courcier, Paris, 1813). Re-edited in [138], vol. IX
141. J.-L. Lagrange, *Euvres de Lagrange* (Gauthier-Villars, Paris, 1867–1892). 14 vols.; ed. by M.J.-A. Serret [et G. Darboux]
142. E. Landau, Richard Dedekind-Gedächtnisrede. Nachrichten von der königlichen Gesellschaft der Wissenschaften zu Göttingen. Geschäftliche Mitteilungen 50–70 (1917)
143. R. Lipschitz, *Lehrbuch der Analysis. Erster Band: Grundlagen der Analysis* (Von Max Cohen & Sohn, Bonn, 1877)
144. P. Martin-Löf, On the meanings of the logical constants and the justifications of the logical laws. Nordic J. Philos. Log. **1**, 11–60 (1996)
145. D.C. McCarty, The mysteries of Richard Dedekind, in *From Dedekind to Gödel*, ed. by J. Hintikka (Kluwer Academic Press, Dordrecht, 1995), pp. 53–96

146. V. McGee, Logical operations. J. Philos. Log. **25**, 567–580 (1996)
147. C. McLarty, 'Mathematical Platonism' versus gathering the dead: what Socrates teaches Glaucon. Philosophia Mathematica, 3rd Series **13**, 115–134 (2005)
148. C. McLarty, What structuralism achieves, in *The Philosophy of Mathematical Practice*, ed. by P. Mancosu (Oxford University Press, Oxford, 2008), pp. 354–369
149. E. Nœther, Abstrakter Aufbau der Idealtheorie in algebraischen Zhal- und Funktionenkörper. Mathematische Annalen **96**, 26–61 (1927)
150. E. Nœther, Hyperkomplexe Grössen und Darstellungstheorie. Mathematische Zeitschrift **30**, 641–692 (1929)
151. M. Panza, *La forma della quantità. Analisi algebrica e analisi superiore: il problema dell'unità della matematica nel secolo dell'illuminismo, Cahiers d'historie et de philosophie des sciences*, vols. 38 and 39 (Paris, 1992). 2 vols
152. M. Panza, Euler's *Introductio in analysin infinitorum* and the program of algebraic analysis: quantities, functions and numerical partitions, in *Euler Reconsidered, Tercentenary essays*, ed. by R. Backer (Kendrick Press, Heber City Utah, 2007), pp. 119–166
153. C. Parsons, Frege's theory of number, in *Philosophy in America*, ed. by M. Black (Allen & Unwin, London, 1964), pp. 180–203. Also in C. Parsons, *Mathematics in Philosophy* (Cornell University Press, Ithaca, 1983[1], 2005[2]), pp. 150–175. We refer to this last edition
154. C. Parsons, Some remarks on Frege's conception of extension, in *Studien zu Frege I, Logic und Philosophy der Mathematik*, ed. by M. Schirn (Stuttgart and Bad Cannstatt, Frommann-Holzboog, 1976), pp. 265–278
155. C. Parsons, What is the iterative conception of set? in *Logic, Foundation of Mathematics, and Computability Theory*, ed. by R.E. Butts, J. Hintikka (Reidel, Dordrecht, 1977), pp. 335–367. Also in [10], pp. 503–529. We refer to this last edition
156. C. Parsons, The structuralist view of mathematical objects. Synthese **84**, 303–346 (1990)
157. V. Peckhaus, Formalistische Taschenspielertriks? Frege un Hankel, in *Frege in Jena. Beiträge zur Spurensicherung*, ed. by G. Gabriel, W. Kienzler (Verlag Königshausen & Neumann GmbH, Würzburg, 1997), pp. 111–122; bad 2 of Kritisches Jahrbuch der Philosophie
158. C. Penco, Frege: two theses, two senses. Hist. Philos. Log. **24**, 87–109 (2003)
159. E. Picardi, Frege and Davidson on predication, in *Knowledge, Language and Interpretation. On the Philosophy of Donald Davidson*, ed. by C. Amoretti, N. Vassallo (Ontos Verlag, Frankfurt, 2008), pp. 49–79
160. H. Poincaré, Sur la nature du raisonnement arithmétique. Revue de Métaphysique et de Morale **2**, 371–384 (1894)
161. H. Poincaré, Les mathématiques et la logique Revue de Métaphysique et de Morale **13**, 815–835 (1905) and **14**, 17–34 and 294–317 (1906)
162. D. Prawitz, *Natural Deduction. A Proof-theoretical Study* (Almqvist & Wiksell, Stockholm, 1965) New edn. (Dover Publications, New York, 2006)
163. H. Putnam, Mathematics without foundations. J. Philos. **64**(1), 5–22 (1967). Also in H. Putnam, *Mathematics, Matter, and Method: Philosophical Papers I* (Cambridge University Press, Cambridge, 1975), pp. 43–59
164. H. Putnam, The thesis that mathematics is logic, in *Bertrand Russell, Philosopher of the Century*, ed. by R. Schoenman (Allen & Unwin, London, 1967), pp. 273–303. Also in H. Putnalm, *Mathematics, Matter, and Method: Philosophical Papers I* (Cambridge University Press, Cambridge, 1975), pp. 12–42
165. F.P. Ramsey, The foundations of mathematic. Proc. Lond. Math. Soc. Ser. 2 **25**(Part 5), 338–384 (1925). Re-edited in F.P. Ramsey, *The Foundations of Mathematics* (Routledge and Kegan Paul, London, 1931), pp. 1–61. We refere to this edition
166. E.H. Reck, Dedekind's structuralism: an interpretation and partial defense. Synthese **137**, 369–419 (2003)
167. M. Resnik, Second-order logic still wild. J. Philos. **85**, 75–87 (1988)
168. T. Ricketts, Objectivity and objecthood: Frege's metaphysics of judgement, in *Frege Synthesized*, ed. by L. Haaparanta, J. Hintikka (Reidel, Dordrecht, 1986), pp. 65–95

169. T. Ricketts, Concepts, objects and the context principle, in *The Cambridge Companion to Frege*, ed. by M. Potter, T. Ricketts (Cambridge University Press, Cambridge, 2010), pp. 149–219
170. M. Ruffino, Extensions as representative objects in Frege's logic. Erkenntnis **52**, 232–252 (2000)
171. B. Russell, *The Principles of Mathematics* (George Allen & Unwin Ltd., London, 1903)
172. B. Russell, *Introduction to Mathematical Philosophy* (Allen & Unwin, London, 1919). 2nd edn. (1920). We refer to this last edition
173. G. Sandu, Ramsey and the notion of arbitrary function, in *Critical Reassessment*, ed. by M.J. Frápolli, F.P. Ramsey (Continuum, London, 2005), pp. 237–256
174. M. Schirn, Frege's objects of a quite special kind. Erkenntnis **32**, 27–60 (1990)
175. M. Schirn, On Frege's introduction of cardinal numbers as logical objects, in *Frege: Importance and Legacy. Perspectives in Analytical Philosophy*, ed. by M. Schirn (De Gruyter, Berlin, 1996), pp. 114–173
176. D. Schlimm, Richard Dedekind: Axiomatic Foundations of Mathematics. M.A. thesis, Carnegie Mellon University, Pittsburgh (2000)
177. E. Schröder, *Lehrbuch der Arithmetik und Algebra für Lehrer und Studirende* (Teubner, Leipzig, 1873)
178. E. Schröder, *Vorlesungen über die Algebra der Logik*, vol. I (Teubner, Leipzig, 1890)
179. S. Shapiro, *Foundations Without Foundationalism: A Case for Second-order Logic* (Clarendon Press, Oxford, 1991)
180. S. Shapiro, Do not claim too much: second-order logic and first-order logic. Philosophia Mathematica, 3rd Ser. **7**, 42–64 (1999)
181. S. Shapiro, Frege meets Dedekind: a neologicist treatment of real analysis. Notre Dame J. Formal Log. **41**(4), 335–364 (2000)
182. G. Sher, *The Bounds of Logic. A Generalized Viewpoint* (MIT Press, Cambridge, 1991)
183. W. Sieg, D. Schlimm, Dedekind's analysis of number: systems and axioms. Synthese **147**, 121–170 (2005)
184. M.A. Sinaceur, L'infini et les nombres. Commentaires de R. Dedekind à "Zahlen". La correspondance avec Keferstein. Revue d'Histoire des Siences **27**, 251–278 (1974). Including a transcription (with French translation) of Dedekind's correspondence with H. Keferstein
185. H. Stein, Logos, logic, and logistiké: Some philosophical remarks on nineteenth-century transformation of mathematics, in *History and Philosophy of Modern Mathematics History and Philosophy of Modern Mathematics*, ed. by W. Aspray, P. Kitcher (University of Minnesota Press, Minneapolis, 1988), pp. 238–259
186. E. Stenius, *Critical Essays* (North-Holland Publishing Corporation, Amsterdam, 1972)
187. P.M. Sullivan, Wittgenstein on "the foundations of mathematics", June 1927. Theoria **61**, 105–142 (1995)
188. G. Sundholm, Virtues and vices of interpreted classical formalisms: some impertinent questions for Pavel Materna on the occasion of his 70th birthday. in *Between Words and Worlds*, ed. by T. Childers, J. Palomaki (Filosofia [The Institute of Philosophy, Academy of Sciences of the Czech Republic], Prague, 2000), pp. 3–12
189. W.W. Tait, Truth and proof: the platonism of mathematics. Synthese **69**, 341–370 (1986)
190. W.W. Tait, Some recent essays in the history of the philosophy of mathematics: a critical review. Synthese **96**, 293–331 (1993)
191. W.W. Tait, Frege versus Cantor and Dedekind: on the concept of number, in *Early Analytic Philosophy*, ed. by W.W. Tait (Open Court, Chicago, 1997), pp. 213–248
192. J. Tappenden, The Riemannian background to Frege's philosophy, in *The Architecture of Modern Mathematics. Essays in History and Philosophy*, ed. by J. Ferreirós, J.J. Gray (Oxford University Press, Oxford, 2006), pp. 97–132
193. A. Tarski, What are logical notions. Hist. Philos. Log. **7**, 143–154 (1986). Posthumous paper ed. by J. Corcoran
194. R. Trueman, Propositional functions in extension. Theoria **77**, 292–311 (2011)

195. C. Truesdell, *The Rational Mechanics od Flexible or Elastic Bodies, 1638–1788* (1995). Vol. IX of [79]

196. W.V.O. Quine, *Philosoohy of Logic* (Harvard University Press, Cambridge, 1970). 2nd edn. (1986)

197. J. Väänänen, *Dependence Logic. A New Approach to Independence Friendly Logic* (Cambridge University Press, Cambridge, 2007)

198. J. Väänänen, Second order logic, set theory and foundations of mathematics, in *Epistemology versus Ontology: Essays on the Philosophy and Foundations of Mathematics in Honour of Per Martin-Löf*, ed. by P. Dybjer, S. Lindström, E. Palmgren, G. Sundholm (Springer, Dordrecht, 2012), pp. 371–380

199. H. Wang, *Reflections of Kurt Gödel* (MIT Press, Cambridge, 1987)

200. H. Wang, *A Logical Journey, From Gödel to Philosophy* (MIT Press, Cambridge, 1996)

201. J. Weiner, *Frege in Perspective* (Cornell University Press, Ithaca, 1990)

202. J. Weiner, Understanding Frege's project, in *The Cambridge Companion to Frege*, ed. by M. Potter, T. Ricketts (Cambridge University Press, Cambridge, 2010), pp. 32–62

203. A.N. Whitehead, B. Russell, *Principia Mathematica* (Cambridge University Press, Cambridge, 1910–1913). 2 vols. 2nd edn. (1925)

204. L. Wittgenstein, *Philosophische Grammatik* (Blackwell, Oxford, 1969). Herausegegeben von R. Rhees

205. C. Wright, *Frege's Conception of Numbers as Objects* (Aberdeen University Press, Aberdeen, 1983)

206. C. Wright, Why Frege does not deserve his grain of salt [...], in *Grazer Philophische Studien 55, New Essays of the Philosophy of Michael Dummett*, ed. by J. Brandl, P. Sullivan (Rodophi, Amsterdam, 1998), pp. 239–263. Also in [115], pp. 72–90. I refer to this later edition

207. A. Youschkevitch, The concept of function up to the middle of the 19th century. Arch. Hist. Exact Sci. **16**, 37–85 (1976)

208. E. Zermelo, Untersuchungen über die grundlagen der mengenlehre i. Mathematische Annalen **65**, 261–281 (1908). English translation in [122], pp. 199–215

209. E. Zermelo, Sur les ensembles finis et le principe de l'induction complète. Acta Math. **32**, 185–193 (1909)

International Political Economy Series

Series Editor

Timothy M. Shaw, Visiting Professor, University of Massachusetts Boston, USA, and Emeritus Professor, University of London, UK

The global political economy is in flux as a series of cumulative crises impacts its organization and governance. The IPE series has tracked its development in both analysis and structure over the last three decades. It has always had a concentration on the global South. Now the South increasingly challenges the North as the centre of development, also reflected in a growing number of submissions and publications on indebted Eurozone economies in Southern Europe.

An indispensable resource for scholars and researchers, the series examines a variety of capitalisms and connections by focusing on emerging economies, companies and sectors, debates and policies. It informs diverse policy communities as the established trans-Atlantic North declines and 'the rest', especially the BRICS, rise.

Titles include:

Andrei Belyi and Kim Talus
STATES AND MARKETS IN HYDROCARBON SECTORS

Dries Lesage and Thijs Van de Graaf
RISING POWERS AND MULTILATERAL INSTITUTIONS

Leslie Elliott Armijo and Saori N. Katada (*editors*)
THE FINANCIAL STATECRAFT OF EMERGING POWERS
Shield and Sword in Asia and Latin America

Md Mizanur Rahman, Tan Tai Yong, Ahsan Ullah (*editors*)
MIGRANT REMITTANCES IN SOUTH ASIA
Social, Economic and Political Implications

Bartholomew Paudyn
CREDIT RATINGS AND SOVEREIGN DEBT
The Political Economy of Creditworthiness through Risk and Uncertainty

Lourdes Casanova and Julian Kassum
THE POLITICAL ECONOMY OF AN EMERGING GLOBAL POWER
In Search of the Brazil Dream

Toni Haastrup, and Yong-Soo Eun (*editors*)
REGIONALISING GLOBAL CRISES
The Financial Crisis and New Frontiers in Regional Governance

Kobena T. Hanson, Cristina D'Alessandro and Francis Owusu (*editors*)
MANAGING AFRICA'S NATURAL RESOURCES
Capacities for Development

Daniel Daianu, Carlo D'Adda, Giorgio Basevi and Rajeesh Kumar (*editors*)
THE EUROZONE CRISIS AND THE FUTURE OF EUROPE
The Political Economy of Further Integration and Governance

Karen E. Young
THE POLITICAL ECONOMY OF ENERGY, FINANCE AND SECURITY IN THE UNITED ARAB
EMIRATES
Between the Majilis and the Market

Monique Taylor
THE CHINESE STATE, OIL AND ENERGY SECURITY

Benedicte Bull, Fulvio Castellacci and Yuri Kasahara
BUSINESS GROUPS AND TRANSNATIONAL CAPITALISM IN CENTRAL AMERICA
Economic and Political Strategies

Leila Simona Talani
THE ARAB SPRING IN THE GLOBAL POLITICAL ECONOMY

Andreas Nölke (*editor*)
MULTINATIONAL CORPORATIONS FROM EMERGING MARKETS
State Capitalism 3.0

Roshen Hendrickson
PROMOTING U.S. INVESTMENT IN SUB-SAHARAN AFRICA

Bhumitra Chakma
SOUTH ASIA IN TRANSITION
Democracy, Political Economy and Security

Greig Charnock, Thomas Purcell and Ramon Ribera-Fumaz
THE LIMITS TO CAPITAL IN SPAIN
Crisis and Revolt in the European South

International Political Economy Series
Series Standing Order ISBN 978–0–333–71708–0 hardcover
Series Standing Order ISBN 978–0–333–71110–1 paperback

You can receive future titles in this series as they are published by placing a standing order. Please contact your bookseller or, in case of difficulty, write to us at the address below with your name and address, the title of the series and one of the ISBNs quoted above.

Customer Services Department, Macmillan Distribution Ltd, Houndmills, Basingstoke, Hampshire RG21 6XS, England

Governing Climate Induced Migration and Displacement

IGO Expansion and Global Policy Implications

Andrea C. Simonelli
Founder, Adaptation Strategies International (ASI), USA

First published 2016 by
PALGRAVE MACMILLAN

Palgrave Macmillan in the UK is an imprint of Macmillan Publishers Limited,
registered in England, company number 785998, of Houndmills,
Basingstoke, Hampshire RG21 6XS.

Palgrave Macmillan in the US is a division of St Martin's Press LLC,
175 Fifth Avenue, New York, NY 10010.

Palgrave Macmillan is the global academic imprint of the above companies
and has companies and representatives throughout the world.

Palgrave® and Macmillan® are registered trademarks in the United States,
the United Kingdom, Europe and other countries.

ISBN: 978–1–137–53865–9

This book is printed on paper suitable for recycling and made from fully
managed and sustained forest sources. Logging, pulping and manufacturing
processes are expected to conform to the environmental regulations of the
country of origin.

A catalogue record for this book is available from the British Library.

A catalog record for this book is available from the Library of Congress.

Contents

List of Tables

Acknowledgments

I would like to extend a very sincere thanks to Professor Timothy Shaw, Christina Brian, Judith Allan, and the Palgrave Macmillan team for making this process smooth and pleasant. I could not have asked for a better experience. In this same vein, Dr. Timothy Cadman also deserves special recognition for getting me involved with Palgrave through his IPE series book, *Climate Change and Global Policy Regimes: Towards Institutional Legitimacy* in which an earlier abridged synopsis of this work is a chapter. This book could not have been possible without my time at Oxford's Refugee Studies Centre (2009); their Summer School on Forced Migration was invaluable for properly conceptualizing climate displacement within the frames of forced migration and refugee studies. Similarly, I would not have been able to complete this book without my time spent at the United Nations University Institute for Environment and Human Security (2010) and their hard-working staff who provided much of the research on the UNFCCC WIM – especially my colleague and friend Koko Warner. Additionally, certain references to the Maldives were possible because of Ilan Dr. Kelman and my field work on his project funded by the Norwegian Research Council. Finally, I'd like to thank my family for their continued support (Mary, Ric, and Nick Simonelli) and my wonderful friend and ex-intern, Amy, for sending me a HUGE care package of healthy snacks while I was trapped in the snow reworking the manuscript.

1
Introduction

Climate change is a topic most often broached by environmental scientists and its effects discussed in terms of animal populations and atmospheric events. The quintessential image accompanying this discussion is the sad-looking polar bear on a lonely iceberg. However, its direct effect on human life is yet to garner such attention. Many do not yet associate the consequences for wildlife with similar consequences for humanity. A changing climate will affect how people are able to use their environment as the locations of arable land and water supplies will shift. In some places, sea level rise and desertification will forcibly displace current human populations. How the world seeks to deal with this shift is yet to be seen. Climate change is also publically discussed in terms of sterile statistics. What tends to be missing is how climate change relates to humanity as a whole. What does a 2 degree Celsius rise in temperature mean in the life of the average person? Can that person conceive of what X tons of carbon in the atmosphere looks like? Without a direct relationship to its effect on humans, these estimates cannot be fully understood. They are vague descriptors at best and useless at worst. Gigatons of invisible gasses cannot be adequately internalized by the minds of most people; it is too abstract. In addition, a rise in temperature effects the whole globe, but with a wide variance across regions, longitudes, and zones of habitation. Thus, how can climate science be connected to the changes seen in individuals' daily lives? This is a difficult challenge and even more so in countries where climatic effects are less visible. The Intergovernmental Panel on Climate Change (IPCC) provides a source intended to parse out these effects in the Working Group II Assessment Reports "Impacts, Adaptation and Vulnerability". Each report contains a "Summary for Policy Makers", which is an annotated version with more accessible language and summarized results designed for those who are not scientists in the formal sense. Its language describes the risks and changes to the natural environment, but with minimal emphasis on how climate affects humanity. This means that any reader needs to be able to extrapolate in order to further connect how the likelihood of climate trends will affect specific human sectors. The report suggests generalities over regions and time which need to be specified further in order to completely connect the earth's physical and biological changes to human activity. Science can only estimate the future in general terms.

The Summary proposes some examples of major projected proposed impacts by sector. Table 1.1 presents an annotated version which focuses on climate trends that the IPCC identify and their likelihoods in both the Assessment Reports 4 and 5 (AR4 and AR5) from 2007 and 2013 respectively.

The trends explicated here are long-term changes to typical weather events based on two different time frames: early in this century and on the cusp of the next century. If the latest two Assessment Reports are considered, these trends are either stable or more certain over time. Additionally, the most recent report, AR5, shows that climate science models more strongly predict changes than do previous reports. For example, there are two trends that are described in more specific terms: drought and tropical cyclone activity. Drought had not been adequately projected in terms of changes in soil moisture for early in this century, but is deemed "likely" for late in the century in AR4. However, AR5 adds a generalized descriptor. For cyclones, we see the same low confidence early in the century, but it changes from "likely" to "more likely than not" later and even a specific location where this will be a consideration. This is helpful, in that it can more specifically designate where changes will occur, but again, a purely scientific explanation is still woefully vague. However, Table 1.2 is much more descriptive than Table 1.1. Below is a list of weather-related trends, again, but paired with effects on human health and (separately) industry, settlement, and society from AR4. These descriptions still need to be fitted to individual regions, countries, and localities, but begin to better define the impact of climate trends on human life and livelihoods.

Here, the effects on human health and industry, settlement, and society provide a much broader basis for understanding the impacts of climate trends. These effects vary, but relate to large-scale economic disruptions, personal livelihood issues, infrastructure, vulnerabilities, and potential for migration. Not every locale will be affected by all of these trends, but identifying how an intensification of tropical storms (for example) will

Table 1.1 IPCC proposed major climate change impacts

Direction of Climate Trends from Assessment reports 4 and 5

Direction of Trend	Likelihood of further changes: Early 21st Century (AR5)	Likelihood of further changes: Late 21st Century (*AR4* and AR5)
Warmer and/or fewer cold days and nights over most land areas	Likely	*Virtually certain*, Virtually certain
Warmer and/or more frequent hot days and nights over most land areas	Likely	*Virtually certain*, Virtually certain
Warm spells/heat waves. Frequency increases over most land areas	Not formally assessed	*Very Likely*, Very Likely
Heavy precipitation events. Frequency increases over most areas	Likely over many land areas	*Very Likely*, Very Likely
Areas affected by drought increases	Low confidence	*Likely*, Likely (on a regional to global scale)
Intense tropical cyclone activity increases	Low confidence	*Likely*, More Likely than Not (in the Western North Pacific and North Atlantic)
Increased incidence of extreme high sea level (excludes tsunamis)	Likely	*Likely*, Very Likely

Note: For changes in the early 21st century the dates include 2016–2035 and for the late 21st century the dates include the years 2081–2100. Additionally, *Virtually Certain* refers to a likelihood of outcome greater than 99% probability and *Very Likely* refers to a likelihood of outcome 90 to 99% probability, and *Likely* refers to a likelihood of outcome 66 to 90% probability. Finally, there is low confidence related to areas affected by drought increases, because there is low confidence in projected changes in soil moisture specifically.

affect human habitation is a starting point for an assessment of how to govern and plan for such changes. It is not that climate science is uncertain, but that there is a need to combine the "hard" and "soft" sciences to further develop responses to climate effects. While those scientists who live in a world of computer models and atmospheric statistics can demonstrate how likely a region is to face certain trends, social scientists are needed to determine how vulnerable a location is to large-scale disruption, how resilient is the society/ecosystem to this disruption, and what kinds of adaptation will be needed. Social science researchers

Table 1.2 IPCC climate effects on humans

Direction of Trend	Human Health	Industry, Settlement, and Society
Over most land areas, warmer and fewer cold days and nights, warmer and more frequent hot days and nights	Reduced human mortality from decreased cold exposure	Reduced energy demand for heating; increased demand for cooling; declining air quality in cities; reduced disruption to transport due to snow, ice; effects on winter tourism
Warm spells/heat waves. Frequency increases over most land areas	Increased risk of heat-related mortality, especially in the elderly, chronically sick, very young, very socially isolated	Reduction in quality of life for people in warm areas without appropriate housing; impacts on the elderly, very young, and the poor
Heavy precipitation events. Frequency increases over most areas	Increased risk of deaths, injuries, and infectious respiratory and skin diseases	Disruption of settlements, commerce, transport, and societies due to flooding; pressures on urban and rural infrastructures; loss of property
Area affected by drought increases	Increased risk of food and water shortage; increased risk of malnutrition; increased risk of water- and food-borne diseases	Water shortages for settlements, industry, and societies; reduced hydropower generation potentials; potential for population migration
Intense tropical cyclone activity increases	Increased risk of deaths, injuries, water- and food-borne diseases; post-traumatic stress disorders	Disruption by flood and harsh winds; withdrawal of risk coverage in vulnerable areas by private insurers, potential for population migrations, loss of property
Increased incidence of extreme high sea level (excludes tsunamis)	Increased risk of death and injuries by drowning in floods and migration-related health effects	Costs of coastal protection versus costs of land-use relocation; potential for movement of populations and infrastructure (also see tropical cyclones above)

interested in the societal and political effects of climate change have to use a literature base that can parallel the types of risks that will slowly occur. Though one cannot study how an increase of temperature or storm surge occurrence will affect people, one can study the effects of high temperatures and storm surge from past events. This link will allow for a connection between scientific data, measures, and models to those who will inevitably experience them. The risks to humanity have begun to be described in terms of coastlines, buildings, and lost tourist revenue (Arifin, 1997; BBC News, 2009; Wright, 2009; Morton, 2009; Reuters, 2009). While these examples are mostly economic, increases in extreme weather events affect human settlements, health, and personal security, among other things. Thus, how climate change will influence humanity is still yet to be a lived reality for most. Scientific projections and prob-abilities only provide an ambiguous framework under which to begin to plan, prepare, mitigate, and adapt.

Migration as a form of adaptation to climate change needs to be addressed, because the nations with the highest carbon emissions are not doing enough to curb their global impact. Therefore, there is an increasing need to develop a governance structure to tackle the spon-taneous and planned climate induced migration and displacement already occurring. A 2009 report by the World Wildlife Fund (WWF) Australia suggested that only three out of 20 industries are moving fast enough to deliver the transformation to the greener economy needed by 2014 to stay under a 2°C rise in temperature (Clarke, 2009). As of 2015, the goal of a minimum 2°C temperature rise is still elusive. If the global temperature rises beyond 2°C, certain nations currently facing growing climate-related pressures will have no recourse other than to migrate; this will be a sentence of extinction for some. As the pressures of a new Kyoto commitment period loom for the COP 21 in Paris, it is clear that in order to slow the need for migration, the deal has to make significant gains in the mitigation sector.

Meanwhile, the global governance of climate change induced dis-placement is currently at the stage of ad hoc development. Legal and conceptual categorization of this phenomenon has been difficult and slow moving. The mainstay of most research on the topic of climate induced migration and displacement has come from the field of inter-national human rights and refugee law. In this vein is how/which cur-rent international legal norms and protections can assist those who will need to migrate or are already being displaced. Due the fact that there is no formal legal standard or even set of policies to guide action on this phenomenon, legal analysis mostly entails international soft law instruments. Legal and non-legal scholars alike use inconsistent language to describe what is happening; many authors have begun to define those affected by climate change in terms of refugeehood; "climate refugee", "climate change refugee", "environmental refugee", "disaster refugee", and "ecological refugee" are most often cited. Legally speaking, the word "refugee" defines a very specific identification which carries with it certain rights and obligations; a concrete meaning and privilege. These rights do not apply equally to all persons fleeing their homes simply because the term "refugee" has been presupposed onto their condition. This grouping is also referred to as "climate change migrants", "climate migrants", "environmental migrants", and "climate displacees". These inconsistencies occur because there has been no common academic or policy-based consensus as to where this group fits into the current discourse on climate change or migration. While a case can be made for many of these labels, their varying use has been problematic for governance. To adequately place those being displaced under the most appropriate governance structure, what is needed is a concrete definition which can be applied through policy. If they are refugees, there is a place for them under the United Nations High Commission for Refugees (UNHCR). If they are migrants, they belong under the treaties of the International Organization for Migration (IOM). However, if they are not currently "refugees", should they be? Does "migration" adequately describe their predicament and its drivers? Or are they "displacees", those who are pushed out of their original environments? Being driven out of one's homeland by the actions of others can also be considered a humanitarian problem. If so, they can also find a home under the United Nations Office for the Coordination of Humanitarian Affairs (UN OCHA). Or rather, should their plight be governed somewhere else? With this phenomena being an unintended consequence of climate change, should the United Nations Framework Convention on Climate Change (UNFCCC) be involved? Thus, conceptualization of this phenomenon is crucial for adequate governance.

UNHCR, IOM, and UN OCHA currently handle many types of human migration, from assisting refugees to economic migrants to those affected by natural disasters. These structures have expanded their reach over time as drivers for migration continue to be identified; adding another group of migrants could be seen as a natural progression. Nevertheless, a major impediment to the addition of climate change induced migrants or displacees into current governance systems is determining who is responsible for them. Responsibility has been an essential component when dealing with other types of migrants. Specific protections and statuses are based on either a nation's responsibility to its own people or the world's responsibility to those whose governments fail to assist them. Responsibility refers to those who caused the impetus to migrate and thus should pay for the assistance to the group which it has created. International governance structures are poised to assist when either a national government refuses to or cannot assist its own people. Their connections with member states and their negotiating power provide a forum to discuss, create policy, and implement agreements which have a much broader scope than individually negotiated regional treaties. While helpful, these bodies still face institutional and political constraints. Their ability to incorporate those displaced by climate change into current structures depends on political will as well as the flexibility of their mandates. This book presents a qualitative case study of the UNHCR, IOM, UN OCHA, and the UNFCCC's Loss and Damage work program. It provides a historical account of the development of each intergovernmental organization (IGO) from the beginnings of its regime to formal institutionalization, how and why each has eventually expanded, and how each has incorporated climate change into their work. A comparative structural analysis is then employed to evaluate the different institutional components which guide each IGO beyond their specific mandate. Finally, it will also question which, if any, of these IGOs are the appropriate places through which to govern such movement. It had been suggested that a case can be made for each IGO to be the one which *should* take on this new and growing challenge, but their abilities, desires, and appropriateness to do so are not equal.

This research represents a new foray into the study of those affected by climate change as a part of the global dialogue. This book will demonstrate that climate change displacement, as a form of forced migration, has yet to be brought into mainstream research and will pose a significant challenge to current migration/displacement frameworks – specifically frameworks that relate to governance. Analytic frameworks are fluid and tend to work well for academic inquiry. They can change over time with new information, but governance is different. Governance of such an issue needs concrete and thorough information as it is derived from policy and international cooperation. To govern an issue such as this takes governments, IGOs, and regional/local coordination. Policy which can connect these points needs to be concrete and systematically outlined with specific agreed upon responsibilities to those the governance is for. In this case, not only has no current migration/displacement-based IGO stepped up to take responsibility for this phenomenon under its current mandate, neither have individual governments. Those being affected have called for action, but the international community has not decided to make this issue a priority. In some ways, doing so would force nations that have not wanted to commit to high levels of emission reductions to have to do so; admitting that their outputs are displacing people would indirectly force them to have to take responsibility for what they are causing. Additionally, this would be an expensive endeavor; thus, if a current governance structure were to take up the task, the IGO would need significant financial assistance to do something, but the responsibility would then be indirect to individual states using the IGO as a conduit.

In some ways, this is not a new challenge. When major environmental shifts happen, people have always had to choose whether to stay or to go. However, modern immigration policies have developed with closed borders, external processing centers, quota systems, and traffickers to sneak around all of these. Current policies make it very difficult to cross an international border. But moving within a nation is not necessarily an easier or safer endeavor. Many of the world's mega cities have significant slum areas being developed by individuals seeking better economic opportunities after leaving poor agricultural conditions. Not everyone chooses to migrate, and not all people have the resources to do so; individual choice is situated at the nexus of social, environmental, and economic conditions. When the impacts of climate change are increased, they weigh heavier on both the social situation and economic conditions of individuals and communities and its interaction effect on both is also larger. This interaction is important to keep in mind, since the decision to migrate or the reasons for displacement are never clear cut. When the phrase "climate induced migration" or "climate induced displacement" is used, either in the title or throughout this book, the implication of the phrase(s) is not intended as simplistic; any and every time these are used, it is under the consideration of other complex factors such as social and economic considerations. Individuals do not simply move because the climate is changing; they see the need to move because larger storm surges keep destroying ones' home or because changing monsoon patterns can no longer support ones' necessary crop yields. These short examples are not meant to be exhaustive, but to demonstrate that reasons to move are more complex, and this book (and its language) takes this into account. There are hot spots around the globe where these choices have already taken place. Their explication in the next chapter highlights some of these overlapping complications as they apply to each example. The book will proceed as follows:

Current state of affairs

Chapter 2 brings to the forefront a few of the examples of specific countries where people are already being displaced, highlighting both internal and external displacement. Examples include short case studies of the Carterets in Papua New Guinea, both Kiribati and Tuvalu in the Pacific, and the Maldives in the Indian Ocean. Each represents the difficulties many are facing in the attempt to navigate acquiring new land. The Carterets have been in the process of seeking a solution to their disintegrating islands for decades, and their process has been stalled due to a lack of funds, lack of land, and a looming vote for autonomy; these islanders are seeking to internally migrate. While it has already been shown that most movement due to climate change will be within national borders, the case of the Carterets demonstrates that internal migration should not be considered synonymous with easy migration. The cases of the Pacific and Indian Ocean islands will eventually necessitate the crossing of an international border. Similar geographies create some convergence between cases, but when these are layered over on top of other cultural and development-based issues they begin to exemplify the difficulty faced when an entire nation needs to relocate. Additionally, resettlement solutions will question how sovereignty can still be exercised when a nation is possibly nested within another. Lastly, the chapter will provide a brief overview of additional areas that are vulnerable to displacement, highlighting locations in both the developing and developed world.

Hyperbole versus fact

Chapter 3 outlines how well-known concepts and definitions are being challenged by this new phenomenon. In the media, those who are already being displaced by climate processes and those who will be are described in very colorful language. Being touted as the "canary in the coal mine", and representations of the "lost city of Atlantis" and "sinking islands" are becoming commonplace. However, these representations only serve to skew the much more complicated realities that most people face. This chapter begins with an evaluation of these commercial frames, to move from the overly dramatic characterizations to a realistic version of events. Each concept above has come to epitomize a certain level of futility and concrete proof of climate change. Their use in normative discourse evokes vivid imagery and some spectacle, but is not useful. The chapter deconstructs this idea and moves the reader beyond hyperbole and into the true thorniness of this phenomenon.

This serves two purposes: to disassociate the reader from any oversimplifications that journalistic accounts tend to provide, and to show that theatrical simplifications can do more harm than good. Once that is accomplished, the chapter demonstrates the larger implications of such lines of thinking. Poor characterization leads to a misunderstanding of human security issues as well as minimization of long-term adaptation measures. Finally, the chapter suggests a different language to discuss these vulnerable areas in a plural fashion that does not degrade the seriousness of their situation.

Academically understood context

The field of migration studies, both voluntary and forced, has a way to further deconstruct and classify movement due to climate change. Though helpful, they also serve to demonstrate many more levels of complication. Academic fields beget more specific subfields, and those who will need to relocate based on climate-induced phenomena can fall into many categories and yet still – in other ways – fall through the cracks. Chapter 4 situates climate induced migration and displacement in the field of migration studies, forced migration studies, refugee studies, and the subfields of environmental migration and survival migration. It also attempts to distinguish the different scenarios in which the agency of an individual can shift this interpretation. If one chooses to leave a location that is inevitably uninhabitable, is this voluntary migration or forced? In this chapter, climate change and its effects are seen as an additional layer over current understandings of migration and displacement, but one which challenges normative and legal understandings of causation. This culminates with both a descriptive and legal analysis of the label "climate refugee". By the end of this chapter, the reader can see how otherwise fairly demarcated concepts can overlap when new scenarios challenge our current understandings.

Institutional expansion

Chapter 5 situates climate induced displacement in the realm of governance. How a phenomena is labeled and conceptualized can affect how it is governed; Chapter 4 delved into these labels and understandings and Chapter 5 explains their implications. This chapter begins with an introduction to global governance; what it looks like and how it functions without a world government to enforce it. It provides a discussion of governance at the meta-level and then proceeds to incorporate specific structures at subsequent levels. Governance is also a type of international cooperation at the highest level; this chapter also provides a short overview of how cooperation works between nations, as told through traditional international relations literature. Beyond cooperation, institutions of governance and their mechanisms come from specific mandates that eventually expand if the institution is to grow or change over time. This chapter provides several institutional expansion theories, from the general sense to more specific theories of neofunctionalist spillover and firm theory. It provides theoretical and functional explanations for institutional expansion.

Lack of expansion

Currently, there are three IGOs that govern several forms of migration and displacement – all of which have expanded over the years when the situation has demanded it. They are the UNHCR, the IOM, and the UN OCHA. Chapter 6 tells the story of their corresponding regime development, refugee, migration, and humanitarian, respectively, as well as the development of each IGO as the solution to a specific international problem. Subsequently, each has also gone through an institutional expansion beyond its original mandate when new situations demonstrated a further need. It also provides examples of how each IGO has related itself to the topic of climate induced migration and displacement and how each has significant challenges to additional expansion to govern this new group of displacees. This chapter is organized by each IGO to be examined.

Filling the governance gap

Chapter 7 contextualizes the institutional analysis of Chapter 6 into the topic of global governance. Reluctance to expand on the part of UNHCR, IOM, and UN OCHA has left a large gap in the general international governance framework. Alternatively, the gap is slowly being filled through the UNFCCC, which seeks to govern the mitigation and adaptation measures relating to climate change at the global level. In recent years, it began a new work stream to assess loss and damage beyond what both mitigation and adaptation can prevent. This chapter outlines the development of the modern climate regime and introduces the Loss and Damage mechanism as an alternative to the established IGOs that have been discussed until this point. The work stream has been extended multiple times and was codified as the Warsaw International Mechanism (WIM) at the COP 19 in Warsaw in 2013. However, where it is going is still in question, since its focus continues to be lopsided; it is supposed to consider both economic and non-economic loss, but economics has tended to be more of a focus thus far. The chapter will evaluate what kind of mechanism it is, how it has developed until now, and its potential to fill the governance gap that currently exists in this realm – comparing this new emerging apparatus to the IGOs previously mentioned. The chapter also provides both a structural and political analysis of the three IGOs under investigation (against the WIM) in order to demonstrate where substantial challenges to expansion will lie and their implications.

Conclusion

Chapter 8 concludes with an overview of the main premises provided throughout the book, such as revisiting how the case studies reflect the issues brought up in earlier chapters. It evaluates if either of the expansion theories help explain each IGO's expansion to date and if these theories can assist in understanding why they have not expanded into the issue of climate induced migration and displacement. This chapter will also delve further into insufficient political and theoretical reasons for the main IGOs to be averse to expanding. Additionally, it will consider which, if any, of the governance institutions under consideration should be the space in which this issue is handled. With the slow loss of land and habitability in many areas simultaneously, there is need for concern in the realm of governance. Most considerations of migration and displacement have taken place under the umbrella of security and conflict. Without adequate measures to curb climate change, its progress will not slow, and without governance or assistance to those on the front lines, there is potential for security theorists to be right. However, sufficient measures to govern these consequences as they come can assuage this. The conclusion will suggest several policies and their global implications.

2
Current State of Affairs

The current state of affairs for many around the globe is becoming challenging. Those feeling the effects of climate processes are already seeing the need to migrate as a way to adapt to their changing situations. Spontaneous and planned relocations are being considered across the globe. In the larger climate change discourse, displacement is slowly coming to the forefront. More scholars and policy makers are acknowledging that migration can be a positive way to adapt and that, in many cases, it is the only option. The temporal element also varies over this fragile geography: for some, this need to move is now; for others, their short-term future is in jeopardy. But the long-term future is bleak for many. This section will highlight several of the geographical areas in which people and communities are already being displaced, both within and outside national borders. Each example represents the difficulties many are facing in the attempt to acquire new land through complicated negotiations with local and national governments as well as NGOs. Cases which highlight the need for displacement solutions include the similar situations of the Carteret Islands, Kiribati, and Tuvalu in the Pacific Ocean and of the Maldives in the Indian Ocean. These islands and their situations are the best known and detailed in the media as well as in academic writing. In these cases, similar geographies create some convergence between cases, but when these are layered over on top of other cultural and development-based issues, they begin to exemplify the difficulty faced when an entire nation needs to relocate. There are other areas where displacement will be necessary as well; this chapter will conclude with a short overview of those areas and their future struggle for relocation.

Migration as adaptation

Migration has slowly come to be seen as an adaptation strategy to extreme environmental stresses, and now to climate change. McLeman and Smit (2006) detail several situations of environmental difficulties, such as the African famine (in the mid-1980s) and Hurricane Mitch, as specific evidence for migration as adaptation. The authors also offer a model that provides a basic framework of the process as it applies to climate stressors, which demonstrates that decisions to move are made at the household level. Banjeree et al. (2012) also provide evidence of migration as adaptation caused by climate-related vulnerability. Their report for the Foresight Project outlines three areas of vulnerability: dryland margins, mountains, and low-elevation coastal zones. The authors argue that in each of these cases, there is evidence of the effectiveness of migration as a form of adaptation in terms of its persistence as a strategy adopted by those facing deteriorating or extreme environmental circumstances. Yet across such vulnerable locations, there is a lack of specific empirical studies on the role of migration in the context of adaptation to environmental variability and change. Thus, while the greater body of academic work is still lacking in empirical studies in this area, migration as adaptation can still be seen. It is also occurring in two ways; adaptation that takes place prior to the impacts of climate change and migration that takes place afterwards are referred to as anticipatory adaptation and reactive adaptation, respectively (McCarthy et al., 2001). *After* a climate-related environmental disaster, households may have lost their belongings, valuables, and homes; those moving in reaction to this type of situation will need aid and assistance. It is pre-planning and governance that can assist with anticipatory migration, paving a way for those in most need to have access to a place in which to relocate and the means to travel there. Until such time as this is universally implemented, each vulnerable location, community, and household will have a different experience. How this is currently unfolding will be described below.

Making the move

The Carteret Islands, a territory of Papua New Guinea (PNG), is only 1.5 meters above sea level and is already being inundated with salt water, which is destroying crops and contaminating freshwater wells. This has left the inhabitants with a diet of rainwater, coconut, and fish, facing chronic hunger (Lateu, 2008). The residents of the Carterets are, unfortunately, already living with the most serious results of climate change and represent the first organized relocation. The local population created their own association to tackle this issue. Called Tulele Peisa, which translates to "sailing the waves on our own", its purpose is to advocate for conservation, culture and identity, reloca- tion, and sustainable livelihoods for its people. Its founder is a woman named Ursula Rakova, a native of the Carterets. Chosen by the Council of Elders, Rakova has worked with local and regional NGOs since 1993 and is considered a pioneer of the environmental movement in PNG (Tulele Peisa, 2008). The Carterets Integrated Relocation Program is a proposal to assist the 3,300 residents of the Carterets who are losing their homes due to sea level rise and to integrate them into three exist- ing communities (Tinputz, Tearouki, and Mabiri) on the neighboring island of Bougainville. As early as 2001, the Bougainville government was discussing the relocation needs of the nearby Carterets; the Council of Elders made the final plans to form the local NGO in 2006, when it became apparent that they would need their own organization in order to implement a planned, staged program to relocate its people (Tulele Peisa, 2008). Official preparations to evacuate began in 2008 (Loughry and McAdams, 2008).

The most dramatic images depicting the necessity for migration as a form of adaptation have come from these islands. This was chronicled in the documentary *Sun Come Up: The Story of Climate Change Refugees*, pro- duced by Jennifer Redfearn and nominated for a 2011 Academy Award. The film follows a group of young Carteret Islanders as they search for land in Bougainville, an autonomous region of PNG, 50 miles across the open ocean (Big Red Barn Films, 2010). While it did not win the big prize, the film continues to tour different festivals to raise awareness and money to assist in the relocation.

On December 11, 2008, Tulele Peisa's resettlement initiative was presented at a meeting organized by Displacement Solutions (DS), a Swiss NGO which focuses on land rights and resettlement projects. In attendance were representatives from Bougainville, AusAid, UN Habitat, the government of Tuvalu, OXFAM, Mantle Group, the International Commission of Jurists Asia and Pacific Office, the University of Florida, the Australian Centre for Peace and Conflict Studies, Tulele Peisa, the government of Kiribati, a Maldivian climate change expert, the UNHCR Pacific Regional Representative, and DS (Displacement Solutions, 2008). This meeting demonstrated the overlap of national, regional, and inter- national interests in the relocation process. Those present also dis- cussed issues of responsibility and how to fund the project. Participants expressed the need for the PNG government to earmark funds to pur- chase land on Bougainville and to compensate those forced to resettle;

this would coincide with its legal obligation toward its citizens. In addition, the government of Australia added a request for additional aid to come from the international community (Displacement Solutions, 2008). One of the biggest challenges discussed was the identification of land for resettlement. The islanders felt it was important that they be sustainable in their new home and needed sufficient land for each family in order for them to earn their livelihoods. This was decided as 5 hectares per family. The Catholic church donated 81 hectares, but the negotiators still needed 1,400 hectares more for all of the families to be able to move. The Carteret islanders did not have the financial resources to purchase all the land necessary, and it appeared that the PNG government lacked the political will to purchase it for them or expropriate the land. There are several layers of land ownership to contend with: traditional owners, the government, the title holder, and the user. A final barrier arose at the meeting; the political status of Bougainville (Displacement Solutions, 2008). A referendum for independence has been in the works for a while, which could complicate not only land rights, but also political will and any monetary agreements with the PNG government, as well as change internal negotiations into international negotiations. The PNG government's window to hold the referendum is between 2015 and 2020 – which potentially leaves years of uncertainty (*Radio New Zealand*). As of September 2014, seven families had already been relocated (*Huffington Post*), but many more need to follow.

This short discussion of the Carteret resettlement plan demonstrates the complications of this type of planning among the many stakeholders with varying degrees of commitment. Although the PNG government is ultimately responsible for the safety and wellbeing of its people, it has been largely absent in the planning and meetings held by Tulele Peisa. If one looks at the sources of funding for Tulele Peisa, the PNG government is absent again. Because of the layers of barriers and stakeholders, a global governance structure would have better reach as an arbiter than the small NGO that began the process. While both UN Habitat and UNHCR had representatives at the resettlement meeting, they have not taken a leadership role either.

Resisting a move

There are several other islands that are slowly seeing the need to plan for relocation like those in the Carterets, but which have more time. Tuvalu is arguably the most researched set of islands in this group of vulnerable countries and consists of nine coral atolls. Located in Oceania, its highest point is 5 meters, with an estimated population total of 10,472.[1]

King tides, the highest of the year, have been increasing and lasting longer than they ever have in the history of the islands. This flooding has hurt crops, caused in-migration from outer atolls to the capital Funafuti, and, in turn, has caused overpopulation and a strain on resources. The inundation of sea water has leeched into the drinking water and has also damaged the already small amounts of arable land. King tides do not only roll in from the sea, but also bubble up through the sand, affecting anything growing within it. Because of this phenomenon, some Tuvaluans now grow crops in tin cans instead of the ground (Price, 2003). The former prime ministers of Tuvalu have been outspoken on the matter and have argued that the industrialized nations need to do their part to mitigate the damage they are doing to these islands due to their carbon dioxide (CO_2) emissions (Ielemia, 2007). In 2002, the former prime minister announced a plan to sue the USA and Australia in the International Court of Justice (Allen, 2004). Though the case never went into litigation, the next prime minister, Apisai Ielemia, still said that he will keep the option open (Ielemia, 2007). Tuvaluans, either at home or abroad, also participate in this debate using chat rooms, blogs, and letters to the press (Farbotko, 2010).

Internal ecological destruction in addition to sea level rise creates a process which erodes an island's ability to continue to sustain human habitation. In the case of Tuvalu, climate change is affecting where people live, and thus one driver of vulnerability is overpopulation. This is likely to be a process which is the most pertinently destructive. A move from one island to another forces not only more stress on a strained ecosystem, but on the economy as well. Ecological destruction leads to economic destruction, as environmentally based economies are very fragile. When fishing grounds, agricultural land, and tourism are simultaneously being destroyed, the chances of economic improvement are nil. Most importantly, with sea level rise, the concern is the irreversible salinization of water resources. Contaminated wells affect drinking water supplies and cannot be used for agriculture. While foodstuffs and water can be imported, it becomes cumbersome and expensive if this becomes led by aid versus regular economic conditions.

Kiribati, also located in Oceania, consists of 33 islands, 21 of which are inhabited. It has an estimated population of 99,482.[2] Kiribati has also been highly researched academically. Kiribati's population lives at a subsistence level where most people are actively involved in fishing and farming. Two-thirds of the workforce is employed by the government, with about 14% employed as seafarers on German and Japanese fishing vessels. Remittances are a significant source of money for extended families and communities, especially those in rural islands with little development opportunities, infertile soils, and long-distance markets (Borovnik, 2006). Dense population growth and high poverty exacerbate the human pressure on its small landmass. Of most concern is the use and management of Kiribati's freshwater, which is highly vulnerable to salt water intrusion and pollution (Storey and Hunter, 2010).

For Kiribati, and especially Tuvalu, their internal environmental issues have raised questions about whether climate is really the impetus for their problems or development. Locke (2009) argues that the influxes of population movements to urban central islands have changed the socioeconomic structure of these small island developing states. His work focuses on both Kiribati and Tuvalu and demonstrates how overpopulation strains resources and makes people less healthy. He has observed that Kiribati imports more and more processed foods to make up for poor agricultural production and increased foreign aid and remittance money. The population spike has also led to poor sanitation and inadequate sewage and garbage disposal. Similar circumstances prevail in the capital of Tuvalu, Funafuti. Much of the capital is built on water and garbage-filled pits. It also imports poor quality foodstuffs which has hurt the Tuvaluan death rate. Allen (2004) describes these issues, comparing Tuvalu to a small planet; its poor environmental stewardship is no more egregious that that of bigger nations, but because of its fragile, remote, and resource-poor landscape it has less room for error than other countries. However, these internal problems have become a barrier to outside help. Tuvalu and other islands have been implicitly and explicitly encouraged to resolve what is seen as their own "development issues" (by the developed world) before neighboring nations will seriously consider additional migration schemes (Connell, 2003). Loughry (2009) explains that the populations of both Kiribati and Tuvalu deal with overcrowding, unemployment, poverty, pollution, and modernization. Climate change not only drives these issues, but also multiplies their effects. Sea level rise has forced this initial internal migration from smaller atolls to their overcrowded capitals. Thus, these nations' adaptation capabilities have become extremely challenged already due to climate induced internal migration.

The Maldives is a series of 1,190 coral atolls with 80 used as resort islands; its highest point above sea level is 2.4 meters. It has an estimated population of 395,650.[3] Situated in the Indian Ocean, its low lying nature has already made it vulnerable to intense cyclones and storm surge. The Intergovernmental Panel on Climate Change (IPCC) predicts that most of its low lying islands will be submerged by the year 2100. Concerned by this prospect, former President Nasheed announced he was starting a fund to relocate his entire population; this was even before he was sworn in as president on November 11, 2008. News of this plan circulated through major news editions and networks such as the *Financial Times, Guardian, Telegraph, BBC*, and *CNN* on November 10. The plan involves earmarking a certain percentage of tourism revenue to purchase land in neighboring Sri Lanka, India, or Australia. While *The Telegraph* reported that Nasheed found the nations he approached to be "receptive", the *Financial Times* adds that the Director of DS, Scott Leckie, questioned the logic of this plan suggesting that it has not been thoroughly thought through. By the time this news story ran, rumors had already spread that Maldivian officials had begun purchasing land in Sri Lanka.[4] Another quote from Leckie poses an essential question for understanding the specific complications for Maldives' climate migration, "Are they actually asking to re-establish the Maldives elsewhere?" (Evans, 2008). Nasheed was looking to reestablish their cultural and national integrity within this process. Since the Maldives produces 0.001% of global greenhouse emissions (Climate Lab, 2011) and yet faces the brunt of the total damage, why not ask to be totally restored? To do so requires a new interpretation of international law. If the Maldives were to buy land in Sri Lanka and move its population, would it be autonomous there or would it be subject to rule by the Sri Lankan government? These are not questions easily answered, but necessary evaluations which could set a precedent for peoples in the Pacific Rim as well. This line of questioning can be posed in the Carteret case as well if Bougainville does secede from PNG.

The Maldives (like Tuvalu and Kiribati) is threatened with the eradication of its entire landmass; in this circumstance not only is out-migration a necessity, but it makes the idea of purchasing a new homeland less crazy. The Maldives has a unique culture which has spanned the rule of European and regional powers, its own language, and is an Islamic state. It should be no surprise that the president voiced such a strong plan for his people this early in his tenure, as it is one of the places most vulnerable to climate change. Over 90% of government tax revenue comes from the tourism industry, which can be very fragile. Tourism has been a developing industry which, after implementing a more liberal foreign investment policy, has boomed through the "one island-one hotel" scheme converting each resort island into its own sustained enterprise (Domroes, 2001). However, it is an industry which creates much solid waste, increasing pollution, and uses large quantities of the Maldives' limited fresh water. President Nasheed had been adamantly voicing the Maldives' concerns to the world. In 2009, he and his cabinet held a meeting underwater in scuba gear in order to bring attention to his nation's plight (Buncombe, 2009; Omidi, 2009). The event sparked many news stories again, but little sincere action. He was the main character in *The Island President* (2012), a movie about his role in the Copenhagen Climate Change talks in trying to garner international assistance for their vulnerability to climate change. International cooperation over this issue has been slow moving and will be outlined in later chapters. After President Nasheed was ousted in a coup in 2012, the country has not had as outspoken an advocate for their relocation.

As a group, these nations have much in common. All are low lying, have environmentally based economies (either tourism, seafaring, or agriculture) and have governments which are keenly aware of these issues and how it will affect their people. The simple geological similarity of being atoll islands explains other parallels. Islands are, by nature, restrictive environments of limited sustainability. Any kind of economic base is structured within this limit. Island nations already understand the difficulties in sustaining a growing population or economy on scarce resources. They will be damaged more quickly than larger land areas because they have comparatively lesser ability to deal with climate change. For example, the development of industry, individualized products, and disposable packaging creates mounds of garbage all over the globe. However, the small land area of an isolated island leaves less room for disposal. This phenomenon has already been mentioned for Tuvalu and Kiribati. It can be understood as an unfair bind for small islands; the developed world pushes for open commerce and for purchase of their products, only to criticize those that acquiesce, but who cannot feasibly handle the unintended effects. These examples demonstrate the complex set of issues facing small islands and their drivers of out-migration. Internal movement, overcrowding, and pollution signal the need to move – that this homeland cannot sustain early adaptational methods. But this tends to be held against them by nations which are not as vulnerable as they are. Conventional adaptation measures pose a long-term question of adequate fit when it comes to nations of smaller landmass and capabilities. Thus the less conventional idea of migration as adaptation needs higher consideration in these situations.

Additional areas vulnerable to displacement

While the islands detailed above are some of the most identifiable areas of concern, there are many more areas – some in just as precarious situations – with the need to migrate even more quickly. The Asian

Development Bank's (2012) assessment finds that population mobility has already grown in East Asia, Southeast Asia, South Asia, Central and West Asia, and the Pacific, being driven by varying economic forces. However, climate change will aggravate this region in four ways: through sea level rise and storm surge, cyclones and typhoons, riparian flooding, and water stress. The assessment also identifies that some governments in the region have already begun to plan for the resettlement of vulnerable populations. Bangladesh in Southeast Asia, situated in the world's largest delta, has been dealing with environmental challenges for years, but its increasing vulnerability to cyclones is forcing some to consider and some to begin migration within the country's borders. Kartiki (2011) studied this decision of the people affected by Cyclone Alia in 2009. With the scale and frequency of these events increasing, many are choosing to move somewhere else. The factors under consideration in such a decision were identified as both push and pull factors. Individual push factors included: destruction of livelihoods, destruction of households, landlessness, lack of cyclone shelters, dissatisfactory current living conditions, insecurity for children, lack of optimism for the future, and the threat to life. Pull factors were better pay in urban areas and friends/family in other areas. Thus in Asia, some level of rural to urban migration can be attributed to the increase of climate impacts, just like in the islands of the Pacific. Elliott (2012) confirms this analysis with respect to other areas in Asia. Investigations by both the US National Intelligence Council and the Asian Development Bank reinforce the prediction of large-scale movement from rural or coastal areas to cities in Indonesia, Thailand, Cambodia, and Vietnam, as well as the possibility of cross-border migrations in the area.

As it stands, there is plenty of research on the developing world (Global South) and its challenges to climate change and potential displacement, but much less has focused on the developed world, though it will be affected in much the same way. Upscale homes in places like Florida, Louisiana, and New York in the USA are under threat from increased storm surge. Coastlines in the Mediterranean face similar concerns in places like Spain and Greece. In Australia, the biggest concern is the eventual lack of water due to record heat (Fatoric, 2014). Bronen (2013) outlines the immediate challenges faced by Native Communities in Alaska. The author explains that the erosion of these communities is well documented, even by the communities themselves, going back to the 1980s. But it was not until 2003 that the US Government Accountability Office (GAO) issued a report to document the impact of flooding and erosion on Alaska Native Communities. After it concluded that 86% of the villages are affected by such flooding (several very severely), Congress authorized the relocation of specific communities at full federal expense in 2005. However, the US Army Corps used all of the funds provided to conduct studies to determine the viability of relocation and to assess certain relocation sites. These reports found that it would be more cost effective to invest in erosion control instead of relocation, even though it had already been authorized. To date, none of the villages that were identified as living in imminent danger have yet been relocated, due to governance issues; the GAO report also recognized that no government agency has the authority to relocate these communities and that no government organization exists that can address the planning or funding specifically designated for relocation.

The state of affairs related to climate induced migration and displacement spans the globe, and does not neglect the developed world, even though the plight of the developing nations has tended to be more severe. However, while there is the need for relocation in many areas, the governance to facilitate these plans is lacking. This chapter has demonstrated the layers of governments, local organizations, and NGOs working on this issue and how the process has still been slow, obstructed, and generally unsuccessful. The previous paragraph confirms this: if the USA cannot orderly relocate its own people under the threat of climate impacts, what chance does Tuvalu have? This highlights the necessity for global governance that can and will assist with independent migration and relocation and which can act as a go-to beyond the usual bureaucratic processes. Those in precarious situations which are only going to deteriorate further cannot cope long with conflicting layers of multi-level governance refusing to take their fate seriously.

3
Hyperbole versus Fact

A problematic discourse

While climate change displacement has been identified as a possible risk in Table 1.1, it *is* already occurring and has become a nascent international concern, even though few have noticed this (Monbiot, 2009). Although not widely known or understood, spontaneous and organized internal and external migrations due to climate change are occurring around the globe. Projected hot spots of movement include the dryer areas of Africa, regions near the delta systems in South Asia, the coasts of Mexico and the Caribbean, and the low lying islands in the Pacific and Indian Oceans. Beyond projections, movement is already occurring on small, low lying islands which are the most vulnerable to the effects of sea level rise. What is known about such movement has been described in colorful, exciting, and hyperbolic terms. While this has been mostly propagated by the media, it has encroached into academic writing on the topic as well. What is disturbing is that these discourses provoke spectacle instead of understanding, fiction instead of fact, and a macabre longing for the worst to happen. The three main discourses in which this is most evident are the "canary in the coal mine", "lost city of Atlantis", and "sinking islands". This chapter deconstructs these ideas and moves the reader beyond hyperbole and into the true thorniness of this phenomenon. It serves two purposes: to disassociate the reader from any oversimplifications that journalistic accounts tend to provide, and to show that theatrical simplifications can do more harm than good.

Canary in the coal mine

As a well-known metaphor, the "canary in the coal mine" is often used as a way to identify when something bad is happening that cannot been seen with the human eye. Literally used in mining, a caged canary would be carried in with the miners. If noxious, but scentless, gasses built up in the mine, the canary would die, and thus the miners would know to exit the mine immediately. A canary is more sensitive than the miners to these gasses, and thus its death would indicate that the miners were also threatened if they did not leave the area. Using this metaphor to describe a literal or metaphorical death due to climate change would suggest that the death or loss incurred by islanders would prove that climate change is real, its effects are real, and both are in fact dangerous. This metaphor is especially powerful in demonstrating the effects of climate processes for lay people. As Farbotko (2010) explains, for climate scientists, climate science is concerned with changes in the earth's systems seen through specialized measurements: "On islands that are disappearing from view, however, such (scientific) changes are apparently rendered visible, without complexities, to the non-expert and imagined as phenomena that can captured photographically in real time." An island disappearing can be seen by anyone and thus verified as a canary – without instruments or formal training. Even some academic articles use this metaphor to describe areas vulnerable to extreme climate deterioration. Bailey (2010) uses the phrase to describe the islands of the Maldives, but takes it from a BBC article. Others have noticed this as well; Connell (2003) mentions that even the more cautious social scientists see Tuvalu as the "canary in the coal mine", a true indicator of the seriousness of climate change. This terminology is much more pervasive when its usage in the general media is considered. Ayers and Forsyth (2009) also use the same phrase to describe the Arctic Rim communities in their article in *Environment Magazine*. A simple Google search of "canary in the coal mine", "climate change", and "islands" brings up over 12,900 results.[1] Farbotko (2010) argues that this metaphor is utilized as a way for the developed world to construct their anxieties about climate change and for newspapers to assign the people of (specifically) Tuvalu a label of victimhood (2005). They would have to be victims, because in order for the island or coastline to be the "canary", it has to disappear, and this implies the people and culture attached to it will disappear, as well. That is the importance of the canary; it is a sacrifice. For miners, it was a sacrifice for industrial capability of the world; to build and grow, it needed coal. Islands, islanders, and other coastal dwellers would also be the sacrifice (to development) if they are deemed the canary in this instance. The actual loss of these areas should cause more than anxiety for the developed world; it is unclear if Farbotko is suggesting that the anxiety is that the island of Tuvalu *will* have to disappear and only then will the developed world have no more excuses not to make large emission cuts. This would finally demonstrate that there is a big price to pay for inaction. However, the world's nations already know this – their participation in the COP process and their reliance on the IPCC's reports are evidence. But then again, would losing one island or coastal town be enough to force political will? Until or unless the canary dies, there is no way to definitively know.

The lost city of Atlantis

Using the "lost city of Atlantis" to describe both small low lying islands and threatened coastlines is also hyperbole and misleading. Using either literary or figurative descriptions of the Atlantis myth lead to different variations of the same macabre inevitability of complete loss. The literary description comes from Plato's *Critias*. In this work, Plato describes a vast island through the Straits of Gibraltar from the Mediterranean, "larger than both Libya and Asia put together" and thus deserving to be called a continent. While Plato's *Timaeus* provides an account of how Atlantis was populated and about its resources and livelihoods, it is *Critias* which accounts for the continent's demise. "But afterward there occurred violent earthquakes and floods, and in a single day and night of rain all your warlike men in a body sunk into the earth, and the island of Atlantis in like manner disappeared, and was sunk beneath the sea." In one line, a legend was born. Without archeological substantiation, this version of events is all that there is. As it stands, the literary version of the Atlantis myth is of earthquakes, floods, and rain. If this is taken literally and the idea that an earthquake is the main reason for destruction is considered, which, in turn, could have created a tsunami and a subsequent flood, then the loss of Atlantis can be attributed to a natural disaster. The figurative myth takes a bit of a different interpretation; it can be attributed to Plato's *Timaeus*. It explains how advanced the civilization was: abundance of wood for carpenters' work, temples, palaces, harbors and docks; manmade canals, bridges, statues of gold. The Atlantians built their intricate world and lived by a strict code of laws passed down by Zeus[2] and the bloodline of Poseidon. But even with their comforts and abundance, things eventually changed.

> By such reflections, and by the continuance in them of a divine nature, all that which we have described waxed and increased in them; but when this divine portion began to fade away in them, and became diluted too often, and with too much of the mortal admixture, and the human nature got the upper-hand, then, they being unable to bear their fortune, became unseemly.

The Atlantians had abundance, wealth, stability, and yet they were unable to keep it. This part of the myth implies that the Atlantians were responsible for their own demise in some way; they were unable to handle what they had built. Moving away from the literary, this has come to have a slightly different meaning in popular culture; Atlantis was a highly developed society with great advanced technology; they eventually destroyed themselves with that technology. This is not unlike the fears shared by many in the early years of the Cold War when global annihilation due to nuclear weapons was a very real possibility.

Scholars and journalists have also used this metaphor, but usually without any indication as to which description, direct literary or metaphorical, the author is implying. Jain (2014) uses the same quote above from Plato about Atlantis being lost in a day and a night and argues that a "substantially similar fate" is likely to befall the Maldives, Tuvalu, and Nauru. Additionally, the author uses a fictional state of "Atlantis" to be representative of this group. Similarly, Blitz (2011) suggests that a new Atlantis be reorganized as a colony on another state when discussing Melanesia, Micronesia, Polynesia, and the Maldives in their relation to climate induced statelessness. Shen and Gemenne (2011) call the plight of climate change refugees as portraying an "Atlantis in the making", not unlike Whitty (2003) and Price (2003). Wong (2013) suggests that "Atlantis-style" inundation of small islands is unlikely because many climate processes are of slow onset in nature, but the author uses the metaphor nonetheless. This process is again examined by Tol et al. (2006) though the use of the "Atlantis Project" which was to look at the extreme scenario of the West-Antarctic Ice Sheet (WAIS) collapse, or a 5 meter sea level rise in the next 100 years. Another crude Google search for "Atlantis" plus the same search terms as above provides an impressive 246,000 results.[3] It appears that this imagery is even more pervasive than the canary in the coal mine when considering the disappearance of small island states and potential climate refugees.

Comparing those islands most at risk to extreme degradation or to complete loss due to rising seas to the lost city of Atlantis is not helpful as either a visual or metaphorical characterization. In terms of the visual, the idea of an entire continent being lost to the sea in a day and a night is extremely dramatic. The size and magnitude of the earthquake (and subsequent hypothetical tsunami) which would be able do this would be well beyond what anything human experience has otherwise witnessed. And even Plato agrees:

Many great deluges have taken place during the nine thousand years, for that is the number of years which have elapsed since the time of which I am speaking; and in all the ages and changes of things there has never been any sediment of the earth flowing down from the mountains, as in other places, which is worth speaking of. (*Timaeus*)

While exciting, this visual is not accurate as to what small islands and coastlines will face. Climate change is a slow onset event, especially sea level rise. Even the IPCC's projected increase in cyclonic activity would not cause this type of destruction. In essence, the drama is unscientific and unfounded. Both the literary and metaphorical myth conjure up images of tragic victims who came to be extinguished – but in different ways. As victims of a natural disaster, Atlantians are nothing like those under threat from increasing climate stressors. Not only is climate change a slow onset process in nature, it is also not a natural occurrence. This literal literary interpretation implies that climate change is natural and out of any human control. In this way, using Atlantis as a descriptor for the effects of climate change has the potential to incite those who deny that the climate is in fact changing. Additionally, it can add fuel to arguments of similar sects of people who do believe that the climate is changing, but that it is completely natural. Either interpretation makes the destruction of islands and coastlines inevitable and suggests there is nothing that can be done. Those lost to such a tragedy were immortalized by Plato and thus will be fate of those lost in this time, but they have to actually *be* lost to fit into the mold of the Atlantians. The metaphorical interpretation also portrays islanders as victims, but the victims of the actions of the developed world. They become a bedtime story of the perils of self-induced technological overload again, inevitably lost to man's insatiable lust for development. Victimization in this way also strips away the agency of the people living in these vulnerable areas. If they are indeed doomed like Atlantis, fighting back is futile. However, adaptation efforts are not something these places have taken on in vain; they will sustain habitation in many places for years to come. Moreover, if they are again "doomed", the world's nations have little reason to actively put together an emissions reducing climate deal with any significant targets; if islanders are already fated to be lost, then there is only need to plan for a future without them. The Atlantis characterization, like the canary in the coal mine, becomes a self-fulfilling prophecy that strips individual peoples' identities and designates them to be a harbinger rather than sovereign.

Sinking islands

Often referred to as the "sinking islands", these places have tended to garner some attention as their predicaments have unfolded. The islands most popularly designated to this grouping are particularly at risk due to their small, low, and flat nature, which is typical of reef islands on coral atolls (Yamano et al., 2007). The "sinking island" is a concept that has become another well-known metaphor for the long-term consequences of climate change. The term is often used to describe those places which will be most severely affected by climate change; those which may be completely lost to rising sea water. These are islands whose highest point is only a meter or so above sea level. Unlike coastal areas, which have also been referred to as Atlantis (*Environment*, 2003) or canaries, only islands can completely "sink". These tropical islands conjure up images of idyllic palm trees, crystal waters – and imminent doom. It is an image of tragedy in "paradise". However, this image can be damaging to substantive research in that it detracts from serious issues that need to be addressed and refocuses on simple doomsday scenarios which are sensational, but disempowering not only to those who are personally affected, but to adequate research as well. It is scientifically unsound.

Sea level rise will not make an island "sink"; climate change is not a geologic process at work. While the structural integrity of islands do change over time, it is part of a process that includes reef integrity as well as the lagoon and island itself. As Kench et al. (2005) describe, coral-reef islands are accumulations of the sands and gravels that characterize the surface of atolls and other reef platforms. The islands' low elevation, small size, and reliance on locally generated sediments make them particularly vulnerable to the impacts of climate change and sea level rise. Thus the stability of reef islands is of major concern in atoll nations where such islands provide the only habitable land. Improved understanding of the depositional history of reef islands is required to better resolve their future stability. However, this stability is complicated. Coastline adjustments are just one complication. The "sweep zone" suggested by Kench and Brander (2006) proposes that alongshore reorganization of sediment characterizes siliciclastic shorelines. But there are alternative models such as the Bruun Rule – a simple geometric profile model which implies coastlines will migrate landward due to erosion and the relative extent of erosion is a direct function of the magnitude of sea level rise and the gradient of the coast. It is also a tool advocated by UNEP (Webb and Kench, 2010). Woodroffe (2008) explains that sea level rise in itself need not endanger all elements of atoll systems until a critical threshold has been exceeded. In addition, reef flats may be recolonized by coral, and can (eventually) provide more sediment to be transported to the oceanward shores. He also acknowledges that it is unclear what effect increased wave run-up will have on these islands. It may build the ridge crest higher or, alternatively, waves may run over the ocean-ward ridge and inundate the island's interior. And this inundation is exactly what most islands are feeling. Rising King tides have continued to make these islands more difficult to live on in the long term. Such flooding is now a regular occurrence in Tuvalu and the Marshall Islands, and has caused great damage to at least one northern Maldivian island (Simonelli, 2014). This chapter is certainly not meant to be a complete background on atoll geology, but serves to demonstrate the complex life spans and stability of small reef islands. Their shapes may ebb and flow over time and even retain some of their structural integrity through some sea level rise, but to "sink" or "disappear" is not the same as being unable to sustain human habitation. Severe coastal inundation makes living on a small low lying island difficult, if not impossible over time. An island does not have to "sink" for its inhabitant's long-term future to be threatened.

Using the term "sinking islands" suggests that Tuvalu, the Maldives, and other such islands are expendable – as are their inhabitants. It also advocates that there is no hope of saving them and thus no need for discussion on mitigation tactics – these islanders are simply doomed. Because scientific time frames are mere generalities, not only are island-ers doomed, but not knowing exactly when heightens the drama. Many news magazines and publications refer to this imagery as a dangerous paradise (Allen, 2004; Ede, 2002/2003; Lynas, 2004; Morris, 2009; Patel, 2006; Sheehan, 2002; Warne, 2008). This drives normative discussions about climate change and island nations into a place where the details on the ground do not matter; any island that is sinking can be inte-grated into this frame and delegitimized as an individual society. This also affects islands that are not sinking. As the discourse is overwhelmed by the conceptions of "sinking", it leaves no room for less dramatic, but necessary, adaptation policy. What is not fully understood is how damaging this label of "sinking islands" is to their actual plight. Not only does it suggest something that is inevitable, but it also implies a steady, continuous process. Because scientific forecasts provide long-term projections, we cannot know exactly when an island will "sink". However, only focusing on the time line for sinking ignores the fact that there are more problems associated with sea level rise other than the loss of land to stand on. Long before islanders will be permanently ankle-deep in the ocean, they will suffer losses that will make it virtually impossible to stay that long. The salinization of drinking water and agricultural land, as well as more frequent and severe tropical storms, have the potential to leave low lying island nations in an extremely vulnerable position – even without sinking. The difficult thing is finding another way to discuss this general phenomenon in terms that are easily understandable. When academics use this language, even for description (Cordes-Holland, 2008; Farbotko, 2005; Kolers, 2012; Kolmannskog, 2012; Prasad and Narayan, 2008; Simonelli Berringer, 2012; Wong, 2013), it can still cloud the discussion of what is actually happening in these places. The difficult thing is finding a suitable language to use. While academics may occasionally play with general semantics within a sphere of deeper understanding and sentiment, it is the thoughtless use of such language which is damaging. A last Google search for "sinking islands" and "climate change" produces another 7,660 results.[4]

Implications

It is apparent that low lying islands are not actually sinking and that most may not completely "disappear" either. But this discourse has a way of steering the direction of conversation away from the people living on these islands and toward a need to demonstrate physical loss for definite "proof of climate change". No one is sinking, nor are they Atlantians or canaries. Each hyperbolic phrase to describe climate change's worst case scenario depicts a largely Western way of conceptually understanding the results of what it has primarily caused. In some ways, it may make the situation more palatable – the suggestion that certain areas are already lost. However, there are hundreds of thousands of lives at stake – minimized by eager journalism and the human imperative to relate the past to the potential future. Finding short cuts in verbiage or drama in metaphors cannot simplify certain issues.

More important than sloppy interpretations are the issues of lost agency and human security. When a specific discourse takes the focus away from the individuals in a precarious situation and onto these islands in the "perverse hope" (Farbotko, 2010) that one day they will not exist, it not only minimizes the inhabitants themselves, but also what they can do about their own situation. Victimizing them in this way dismisses local means of adaptation and social resilience. If the outside sees their plight as simply inevitable, they will be blind to the determination of these people to save their homeland or at least find meaningful ways to cope in situ. Additionally, this may force the locals to have to stay in deteriorating conditions until the damage is *enough* to fulfill the prophecy of "doom" and allow their moving to be justified. If so, this will jeopardize certain aspects of these people's security. To stay in a place where higher tides are constantly damaging one's property would mean having to replace that property, and over time this burden can undermine previously financially secure people. Additionally, such conditions threaten health and environmental security, in that higher tides and flooding disrupts garbage fills in shallow earth. Standing water full of contaminants affects the surrounding environment and can breed water-borne disease. There is also the risk to physical security, not from the threat of others, but from the water. Walking, wading through, or bailing high water can overexert the elderly.[5]

The normative discourse that emphasizes "sinking", "disappearing", or Atlantian disaster as a potent hyperbole provides ample research space to explore climate change migration/displacement. The inevitability of "sinking" is exciting and fascinating, but leaves virtually no room for mitigation or adaptation projects which could extend the habitation of these and other islands. Migration is the only option. However, it remains in the future, which can force necessary research and governance intervention away in the meantime. But if, in this early stage, the focus can be shifted to direct attention to the many ways in which climate change will exacerbate migration, then its necessity as an adaptation mechanism can be seen as legitimate much earlier. Human migration has linkages with climate and other societal processes and can be better understood in its complexity, instead of considered an automatic response to a singular risk. Climate change causation for migration is thornier than "sinking"; it also entails adaptive capacity, which will vary from place to place (McLeman and Smit, 2006).

Is there an alternative to poor descriptions of long-term climate processes that are not hyperbolic and yet are able to explain the complications that these islands face? Simonelli Berringer (2012) uses the phrase "eventually uninhabitable islands" or EUIs. The author preliminarily defines this as: "those islands/islands chains that are geographically the closest to sea level, will lose their ability to support human habitation, and have already begun to deal with the consequences of rising sea water including frequent storm systems, larger storm surges, and tidal flooding." As a group, nations that can fall under this description have much in common; all are low lying geographically, have environmentally based economies (a combination of tourism, seafaring, and agriculture), and have governments which are keenly aware of these issues and how it will affect their people. With so many islands facing long-term damage to human habitation, this descriptor is getting closer to a realistic concept to work with. The acronym EUIs does not suppose any form of "sinking" or necessitate physical loss or disappearance. Furthermore, there are plenty of land areas that are currently uninhabitable by humans. While modern transportation and technology allow for the importation of food and water to places that do not produce them otherwise, there are still places where human habitation is difficult to impossible. The term also considers the long-term nature of such processes by only assuming that the process is "eventual". Until such time that the world's nations act to swiftly curb global emissions, the earth is locked into a certain degree of change no matter what is done, and considering the current damage to many Pacific islands, this process is currently moving in the direction of "eventually" being the case unless monumental changes take place. There may still be better language that can be adapted to describe long-term climate processes that will displace island and coastal populations, but for now the term EUIs will be used in the remainder of the book, when applicable. It will, however, be slightly expanded to include those islands/islands chains/ coast lines that are geographically the closest to sea level, which (according to current projections) will eventually lose their ability to support human habitation due to the consequences of one or any combination of climate processes. This includes but is not restricted to: rising sea levels, more frequent storm systems, more intense storm surges, increased tidal flooding, ocean acidification, and beach erosion. The adjusted term will be Eventually Uninhabitable Islands and Coastlines (EUICs).[6]

4
Academically Understood Context

The focus of this chapter is to adequately conceptualize migration and displacement in the context of climate change. There are many varying definitions of both migration and displacement which have changed over time and have been dependent on different contextual understandings. For the purpose of governance, it is imperative to contextualize the group under inquiry in this book. Legal and conceptual categorization of those being affected by climate change has been difficult and slow moving, and scholars in this field use inconsistent language. Many authors have begun to define those affected by climate change in terms of refugeehood, such as "climate refugee", "climate change refugee", "environmental refugee", "disaster refugee", and "ecological refugee". Legally speaking, the word "refugee" defines a very specific identification which carries with it certain rights and obligations; a concrete meaning and acquired privileges. These rights do not apply equally to all persons fleeing their homes simply because the term "refugee" has been presupposed onto their condition. Using legal terminology for purely descriptive purposes can and does confuse academic analyses. These "fuzzy concepts" can be understood in a multitude of ways – but for governance, concepts need to be codified, defined, or fixed. How can a phenomenon be governed if there is no agreement as to what it is? Governance and subsequent policy necessitates clear definitions, and thus far, there are few. The group in question is also referred to as "climate change migrants", "climate migrants", "environmental migrants", and "climate displacees". These inconsistencies occur because there has been no common academic or policy-based consensus of where this group fits into the current discourse on climate change, let alone migration. While cases can be made for many of these labels, their varying use has been problematic for accurately placing them under the most appropriate governance structure. This chapter will serve to comprehensively define these labels, properly place them in their respective subfields, suggest some alternate interpretations, and provide a specific definition that will be used throughout the rest of the book.

Migration as a field of study

Migration or migration studies is a broad field of study which encompasses many forms of movement. Definitions of migration also tend to be expansive in nature. The census definition of migration is a change in the address at which one usually resides (Hyman and Gleave, 1978), while others see it as the movement of any distance leading to a change in residence (Young, 2002). These common definitions can include those making cross-national journeys or those moving down the street. There are many conceptual problems with defining a "migrant". Petersen (1978) argues that this depends on equally vague criteria concerning distance covered, the relative permanence of the move, and its seeming importance. Is a person that changed residences within a mile radius as equally a migrant as one who moves overseas? Is anyone who moves out of their literal home of birth a migrant? Pronk (1993) argues that there is a little bit of nomad in each of us. For some it is a way of life, such as gypsies, pastoralists, employees of multinationals, or diplomats. For others it is a periodic escape, such as for tourists. In some cultures it is a requisite for adulthood and obtaining the right to marry. It is also a feature of seasonal economics. People move temporarily or permanently to improve their living conditions, to gain experience, to flee from oppression or persecution, or to seek adventure. The difficulty is to disentangle proper conceptual categories. If migrants are potentially everywhere or everyone, categorization is the only way to begin to differentiate between motives. However, data on migration is currently collected through legal and political definitions which have been argued to be too specific. This calls into question many other facets, such as how a "migrant" sees himself/herself. How do values act upon the attitude of the migrant in question (Mangalam and Schwartzweller, 1968)? Or should the criterion be more social in nature; whether a migrant crosses a cultural or societal boundary (Petersen, 1978)? Or whether a migrant crosses a national boundary? The field also considers internal and external migration, but suffers from a lack of consensus as to how to understand cross-national migration; frameworks and research assumptions have mostly been based on national intellectual assumptions and policy models. In an era of globalization, the study of international migration necessitates transnational tools (Castles, 2007).

Theoretical studies of migration have focused on economic push-pull factors and larger spatial models versus individual journeys (Anthony, 1990; Clark, 1986; Hyman and Gleave, 1978; Lewis, 1982; Petersen, 1978; Weidlich and Haag, 1988; Young, 2002). Demographic studies are attentive to the characteristics of migrants, their means for social mobility, the direction of migration, and their destination (Mangalam and Schwartzweller, 1968). However, demographics are purely descriptive and do not lead to any theory development without knowing more about the drivers of migration. Migration as related to social institutions, group coherence, and collective behavior has been relatively neglected for purely economic models (Petersen, 1978). These focus on labor migration and have dominated migration analysis with their emphasis on job opportunities, labor markets, and rising expectations. The sociological theories of migration study a much smaller unit of analysis, the individual migrant. They also argue that the economic assumptions about the individual being a utility maximizer are an inadequate basis for theorizing social action (Boswell, 2008). The sociological focus is on the choice of leaving or staying based on the advantages and disadvantages of the two alternatives. This focus can also have a strong tendency to be economically driven, with the exception that it also includes those escaping religious or political oppression. This literature is also very US-centric, beginning with explanations for the Irish potato famine and other large-scale westward European migrations (Petersen, 1978). Over the years, this field has amassed a quantity of knowledge which has yet to be connected by a general explanatory system. Because migration is such a broad issue of inquiry, developing a framework that can interpret its diversity has been lagging. Migration theory tends to be time-bound, culture-bound, and discipline-bound. As a social phenomenon, it cannot be understood in meaningful terms without a comprehensive grasp of the interplay of demographic, economic, psychological, and other dimensions that converge in the process of migration (Mangalam and Schwartzweller, 1968).

Human migration has been around much longer than any economic or sociological analysis. Scientists date large-scale human migrations out of the African continent as far back as 130,000 years ago (Balter, 2011). This assumes that early human ancestors migrated great distances to follow big game and eventually occupied all the continents. No dominant species had ever spread so far, so fast. Early civilizations also migrated with the rotation of crops as well as across open water with the advent of capable sailing vessels around 4000 B.C., became pastoralists, and began to expand by direct conquest (McNeill, 1984). Human history is almost entirely based on migrations. The English today are not indigenous to England, neither are the Malays to Malaysia, nor the Turks to Turkey (Sowell, 1996). What is interesting is that considering it has been a natural activity of all times and places (Pronk, 1993), migration has become a topic of international debate. The advent of the national border, the international search for jobs during the Industrial Revolution, and the post-World War I (WWI) refugee flows changed the way in which migration was seen. Until this time, migration had been conceived of as an exercise of individual decision and choice. Before WWI, passports and official regulation of migration were thought of as improper infringements on personal freedom. However, a mass of refugees threatened to put a strain on industrial societies (and their social welfare systems) and became a potential threat to native born citizens (McNeill, 1978). This opened the door to using migrants as political pawns; irrational and inaccurate opinions have found great influence (O'Brien, 1996).

The politicization of migration has continued and strengthened in recent years. The 1990s saw political discourse in the richest countries that immigration was out of control (Papademetriou, 1997/1998). Much of this can be attributed to a misunderstanding of the dynamics of migration. As Mangalam and Schwartzweller (1968) argue, migration is not a random event, it cannot be understood by approaching it like bird migration, it is a social, not an individual behavior, and while each case of migration can have superficial differences, patterns can be connected between movements. Each nation tends to see its own unwanted immigration in isolation. Thus, there has been a global tightening of borders for legal and illegal migrants as well as asylum seekers. A new legitimizing ideology has developed to justify this inequality. The hierarchization of the right to migrate can be seen as a form of transnational racism which posits the "naturalness" of violence in less developed regions and other perceived cultural incompatibilities with non-Western peoples (Castles, 2007). Even the field of migration studies has been driven by political considerations. Research questions and even some findings have been pushed by government officials who can undermine the scientific nature of investigation in this area and has isolated migration studies from broader social inquiry. Politicians believe that if they can work out the root causes of international migration, they can reduce it. This attitude suggests that immigration is a bad thing that ought to be stopped (Castles, 2009).

Voluntary migration

The study of modern migrations is most often derived from ideas about economic utility. It is based on immigration patterns and the idea that people move by choice for better economic or living conditions. This movement is usually classified as voluntary in that the move is a tacit choice based on conditions that either pull or push them out. However, there are distinct differences between push and pull factors; those include the lack of economic opportunities, jobs, land, and freedoms, respectively, as well as political repression (Belton and Morales, 2009). One is pulled out by a better job opportunity, while one is pushed out by not being able to pay one's bills. While both are economic-based, there is a clear difference between the two. Much of the voluntary migration literature which emphasizes economic pull factors relates to globalization and the individual desire to improve one's economic condition in the world. Shaw (1975) explains that this approach is guided by the idea that man is economically rational, an economic maximizer, and that he will perceive and evaluate migration options from this point of view. This is an opinion which is posited from the outside in that the migrant is not consulted to understand if that was indeed his/her motivation. Stark and Taylor (1989) reinforce this view by providing evidence that international migration is influenced by both relative as well as absolute income considerations. However, their research demonstrates that migration motives have more to do with one's relative income based on his/her peer's versus a basic determination based on poverty. This adds support for other theories in the field, which argue that it is not the poorest that migrate, but those who have the means to do so; this is a consequence of globalization. Contending literature, however, argues that if migrants are asked about their motives, a different picture will prevail. Winchie and Carment (1989) demonstrate that non-monetary career reasons can also be important. Their research shows that having existing familial relations overseas can be an equally strong pull factor as the desire for economic mobility.

Migration in this fashion is considered voluntary in that one wants to improve his/her lot and thus moves in order to do so. However, migration theory often omits those voices which oppose capitalist globalization or heavily critique it. Pull factors like wanting economic mobility suppose the decision to migrate is purely selfish; one is currently economically secure but chooses to find a way to acquire more. However, capitalist development often raises some while disaffecting others. Migration is an instrument of the capitalist work economy and the exploitative economic and development policies by dominant states which affect weaker ones (Belton and Morales, 2009). Capitalist development causes both pull- and push-based migration. Globalization essentially means flows across borders of capital, commodities, ideas, or people. National governments remain suspicious of the latter two (Castles, 2007). This body of work, however, does not consider the larger sphere of situational influences that can affect a potential migrant. A subsistence farmer who has a poor growing season is not necessarily looking to move into a new career as much as to supplement a current short-term difficulty. The same can be said for the same farmer whose business has dried up due to trade agreements. NAFTA has been a disaster for small farmers in Mexico, increasing rural poverty. An estimated two million Mexican corn farmers have been forced out of business by cheaper, subsidized US imports (Belton and Morales, 2009). These examples and their implications begin to question the extent to which migration is purely voluntary in the sense of economic maximization or is forced based on prevailing outside influences. This discrepancy will be discussed further in the next chapter.[1]

Forced migration as a field of study

Forced migration studies is a subfield of migration studies. It is concerned with the types of "push" factors which drive migrants to leave their homes. This also includes studies on displacement types, such as disaster induced displacement, development induced displacement, environmental displacement, and all those labeled refugees.[2] The main debate within this subfield is whether refugee studies should be part of forced migration studies or be a separate field of study. Hathaway (2007) argues that marrying refugee studies with forced migration studies will take away from the special circumstances of refugees and encourage work on the phenomenon itself instead of on refugee rights. While DeWind (2007) agrees that refugees are a special category of forced migrants; he believes that Hathaway overemphasizes the effectiveness of the international community and underappreciates the positive contributions of forced migration studies. He argues that practitioners have a difficult time distinguishing between refugees and forced migrants in cases of human rights abuses; there are conflicting ideologies of legitimacy for legal rights. These distinctions demonstrate the way in which the field has tried to incorporate the ideas of practitioners who deal with these conflicts every day. There are also other authors who disagree with Hathaway. Adelman and McGrath (2007) see his ideas as puritanical; Hathaway presents no evidence that forced migration studies will pose a risk to the study of Convention refugees. Cohen (2007) adds that not all refugees are Convention refugees; many flee generalized violence. In addition, Hathaway ignores internally displaced people who do not enjoy the protections of their government and yet have not crossed an international border in order to receive assistance. Cohen's main problem is that academics should not argue over priority; a better response is to work toward protectionary needs. What has initiated much of this debate is the way in which the asylum paradigm has changed over time. Crisp (2003) explains that there is more and more pressure for migrants to be managed, and there is a growing unwillingness to admit and provide for asylum seekers. Therefore, keeping the fields separate may be a tactic to lessen the erosion of current protections; lumping refugees in with other forced migrants may exacerbate this policy process.

Refugee studies

The refugee regime is arguably the most developed in terms of literature, governance, and protections. Asylum is one of the most ancient institutions, dating back to the Mediterranean civilizations. It was based on the guarantee of liberty and protection against oppression. This norm was accepted as one of "minimal standards", which meant that refugees in another land should be accorded the same treatment as nationals (Krenz, 1966). The evolution of a formal protected status, however, took a bit longer. This began after WWI and proceeded in three distinct phases, argues Hathaway (1984). The first phase emerged around 1920. The concern during this time was with refugees as a member of a group which had no freedom of international movement because its members were deprived of the formal protection of their government. This remained the theme concerning refugees until 1935. From 1935–1938, there was a move away from preoccupation with state protection that saw refugeehood as encompassing those who were victims of broad-based social and political upheaval. Finally, from 1938–1950, there was a move back to understanding the relationship between the individual and the state. After WWII, mass movements of refugees through Europe necessitated governance and attention. The newly founded UNHCR passed the 1951 Convention on the Status of Refugees, which outlined a specific definition of a refugee as well as their legal status and protections. A refugee is a person who,

> owing to well-founded fear of persecution for reasons of race, religion, nationality, membership in a particular social group or political opinion is outside the country of his own nationality and is unable

or, owing to such fear, is unwilling to avail himself of the protection of that country; or who, not having a nationality and being outside the country of his former habitual residence as a result of such events, is unable to or, owing to such fear, is unwilling to return to it. (Article 1, section 2)

It also set the legal principle of non-refoulement, which states that no refugee should be returned to any country where he/she is likely to face persecution, ill treatment, or torture. States have endorsed this principle, but have looked to define its limits. Others have extended protections, such as the Organization of African Union (1969) and the Cartagena Declaration (1984) treaties, which add circumstances such as events of external aggression, occupation, foreign domination, and events seriously disturbing the public order (Goodwin-Gil and McAdam, 2007). While the term "refugee" has a specific legal meaning, it is often still used as a general concept which can vary in meaning. Shacknove (1985) argues that refugees should be seen as persons whose basic needs are unprotected by their country of origin and who have no remaining recourse than to seek international restitution for their dilemma. This definition would open up the label to many more people around the globe, including those displaced by climate change.

Refugee studies focus on many of the legal issues with refugee processing, who can and cannot be considered an asylum seeker, issues of resettlement, reconstruction, peace building, aid, and protracted crises. The most contentious issue may be bureaucratic labeling, which can blur the lines between refugees and other groups of forced migrants. Many academics and policy makers use language that implies refugeehood, such as "environmental refugee", "economic refugee", and similar terms. This language confounds the important distinctions between those forced to move because of these issues and the fact that even though they may need assistance, there is no legal precedent for individual nations to have to provide it. Politically, this bureaucratic label can also be used as a tool for marginalization (Zetter, 1991). All migration labels are weighed against the Convention; however, more labels have been created as the world has sought to restrict its protections (Zetter, 2007).

Environmental migration

Development induced displacement and disaster induced displacement are common and widely discussed forms of forced migration. Both are generally considered part of the larger sphere of environmental migrants. There is general agreement on three causes of environmental migrants:

natural disasters and environmental or industrial accidents, planned or unplanned relocation due to development, and health-related effects due to inadequate resources to maintain life (Cardy, 1994). This definition has been influenced by El-Hinnawi (1985), who specified that the first category encompassed temporary displacement because of earthquakes, cyclones, or environmental/industrial accidents; the second includes those who are permanently displaced due to man-made changes to a habitat, such as development projects; the third are those who migrate temporarily or permanently because their original habitat can no longer support them, such as due to drought and crop failure. Direct and indirect displacement will likely span all categories. As more frequent and increasingly stronger hurricanes, cyclones, and drought occur, environmental migrants from group one will increase. If a government decides to erect improved sea walls or divert water into drought areas, an increase of migrants in group two will be possible. Finally, if people begin to move due to the inability to sustain their lives and livelihoods, they will fall into group three. This includes those living on coastlines which are being lost to rising seas or agricultural lands that have been ravaged by desertification. These examples are certainly not exhaustive, but offer a glimpse as to the way that climate change can exacerbate known groups of environmental migrants and complicate current labels. Dun and Gemenne (2008) argue for a better definition of environmental migration in that it is often difficult to isolate environmental factors from other drivers of migration. Environmental factors are challenging to differentiate from other drivers, as they are often underlying and not necessarily seen by those affected by them in the same way that they are viewed by those studying the event. The environmental damage of human habitats can initiate a chain of events which affects peoples' lives and livelihoods. The driver is not specifically the environment, but its effects. People do not migrate simply because of drought, but because they cannot produce food anymore. They do not move because of a cyclone, but because the cyclone has eradicated the industry which provided the community with jobs.

Development induced displacement and disaster induced displacement have been classified, but still need to be defined more specifically. Robinson (2003) provides a thorough description of both. In the 1950s and 1960s, development was seen as the way to westernize traditional societies. Robinson explains that large-scale capital-intensive development projects in developing countries accelerated the pace to a brighter and a better future. Uprooting many along the way was seen as necessary for the majority to benefit. These projects include transportation, water supply, urban infrastructure, energy, agriculture, parks and forests, and population redistribution. Development induced displacement and resettlement can also be thought of as a form of state induced displacement. The state, as a sovereign power over its land, can and often does sanction infrastructure projects to provide benefits to those it considers part of the in group of society. Hammar (2008) explains two understandings of sovereignty which justify this practice. The first is the right to own and protect one's territory, which can be legitimized by expulsions by non-citizens; the second relates to the authority to define distinctions between worthiness and unworthiness and the power to define who is an insider who an outsider. These concepts can be used to understand forced displacements and replacement in places like Zimbabwe and Israel, where governments have forcibly moved those who they see as outsiders and replaced them with those who are loyal to the sovereign group. This is also seen around the world, where governments have displaced indigenous groups in order to establish environmentally protected areas (Dowie, 2011). Literature on development-induced displacement falls into two categories. At one end of the spectrum is a category of scholars who consider displacement to be the inevitable, unintended outcome of development, and at the other are research scholars to whom displacement is a manifestation of a crisis in development (Dwivedi, 2002). The first category considers development as a given, while the second considers it a catastrophe. Concerns of the first group include minimizing the adverse consequences of continued development. Concerns of the second include the political and negotiation rights of the people being displaced. Group one seeks to reduce negative effects, while group two seeks new ways of doing development.

The first of the two main development induced displacement and resettlement models which see development as a given is Scudder and Colson's four stage model. It attempts to explain how people and sociocultural systems respond to resettlement and was later applied only to supposedly successful cases. The stages include recruitment, transition, potential development, and incorporation. Many cases failed to go through all four steps, and a new theory became necessary to explain this tangled process. From here, Michael Cernea's *The Risks and Reconstruction Model for Resettling Displaced Populations* (1997) has relatively monopolized this field. This model, also referred to as the Impoverishment Risks and Resettlement (IRR) model, resides in category one and utilizes economic methods. It is a conceptual model that is built around eight risks of impoverishment; landlessness, joblessness, homelessness, marginalization, increased morbidity/mortality, food insecurity, loss of access to common property, and social disarticulation. Cernea also articulates four steps to use this tool in practice: carry out a risk assessment in the field, design targeted responses, engage proactive responses and participation of the population at risk, and establish transparent information and communication between planners and the at-risk population. One main reason for the specifically outlined model is his refutation of the traditional risk-response pattern: the cost benefit analysis (CBA). Cernea explores reasons why this method is inadequate. He concludes that the true costs of displacement are typically not included and accounted for fully. This perpetuates situations where some people share gains while others share victimization. Massive personal costs are paid for by the projects displacees, and thus this approach minimizes what compensation is directly connected to property loss and not livelihood loss. Those who will be moved are often seen as calculated for the benefit of the masses. Their compensation is also calculated haphazardly and without long-term consequences of the disruption that displacement will cause to current livelihoods or the education of the young. Thus the CBA approach to development induced displacement and resettlement accepts the cost of the lives and future of potential displacees for the convenience of the masses.

While Cernea (1997) is widely cited, Dwivedi (2002) takes issues with some of his conception of risks from the movementist tradition in the second category. The Cernea conception is considered managerial, as it seeks to manage risks. Because it accepts that development will still occur as it has, the only durable solution is to manage the damage. Dwivedi has four concerns with this framework. First, risk perceptions are constantly changing; a resource valued by one community may not be valued by another. A risk assessment may undervalue a resource or overvalue a resource, depending on the perspective of the person making the assessment. Second, the model is bereft of any systemic aspect or the global economic processes that cause displacement. Third, it neglects an understanding of the sequential nature of risk; risk is not a singular phenomenon, and it can unfold in a complex sequence of events which show that variables used in the IRR model cannot be isolated from one another. Finally, the model adopts a mechanical strategy for problem resolution in it that it assumes that land can be substituted for more land, as jobs can be for more jobs – things that, upon resettlement, are rarely equitable trades.

The descriptive literature on development induced displacement and resettlement also exposes the shortcomings of the managerial approach, however. Heming et al. (2001) and Stein (1998) discuss the involuntary resettlement policies of China concerned with the Three Gorges Dam. While affected peoples assume the state will take responsibility for their transition and compensation, this does not guarantee that managerial decisions made by the state will be fair or efficient. Heming et al. (2001) find that increased poverty was common in Chinese reservoir resettlement areas. A main reason for this is a low rate paid for lost assets, which failed to be sufficient to rebuild new homes and/or restore original living standards. Stein also finds that failures also occurred in not involving local people in resettlement plans; no new employment options gave way to high unemployment, with 60% of resettled residents living below the poverty line. Similar findings appear in India from the Nagaon Paper Mill project (Bharali, 2007). The result of the Land and Forest Allocation Programme (LFAP) in Laos also shows a shortage in draught animals after relocation due to the need to sell them to buy rice. Farmers were not given quality information about their new environments to adequately farm and thus were unable to do so (Vandergeest, 2003).

Beyond development, disaster induced displacement as a driver of environmental migration is a broader phenomenon. It includes natural and man-made components, but needs to be considered carefully. Not every fire, earthquake, drought, epidemic, or industrial accident constitutes a disaster, only those which exceed a society's ability to cope and where external aid is required. Robinson's dissection of the term identifies two types of disasters and separates them into several subcategories (Table 4.1).

Disaster-displacement issues appear to be handled as a form of relief rather than a more comprehensive rebuilding or resettlement strategy, domestically and internationally. Unlike persons displaced and relocated

Table 4.1   Man-made and natural disasters (as per Robinson, 2003)

Types of disasters		
Natural	Sudden impact	Flood, earthquakes, tidal waves, tropical storms, volcanic eruptions, landslides
	Slow onset	Drought, famine, environmental degradation, pest infestation, desertification
	Epidemic diseases	Cholera, measles, dysentery, malaria, HIV, AIDS
Man-made	Industrial disasters	Pollution, spillages of hazardous materials, explosions, fire
	Complex emergencies	War, internal conflicts, and natural disasters in conjunction

domestically due to development, environmental migrants usually have no rights to compensation for losses due to natural disasters (Heming et al., 2001). Without legal protections entailed in crossing an international border, those who lose their homes and livelihoods due to natural causes have to rely only on short-term help to survive and possibly rebuild. Lautze (1996) explains that international relief resources are to be used to return communities to the status quo prior to the emergency. In essence, international aid is used to manage the situation. Cernea (1997) does suggest his model is a possible option for natural disasters, but it is unclear if it has been used as such. For natural disasters, relief and rehabilitation are different from redevelopment. The aim in a disaster is to alleviate human suffering. For the USA, funding for relief is based on lending a helping hand when others are in need, but development or redevelopment is still an individual nation's domestic concern. Additionally, both development induced displacement and disaster induced displacement are defined in terms of internally displaced persons (IDPs). IDPs share many of the same difficulties as refugees, but have no defined legal status. Persons displaced by dams or cyclones are usually displaced within their country of origin. IDPs are a broad classification of those who could be considered refugees if they had crossed an international border (Robinson, 2003). Although development and natural disasters are thus cast as domestic problems, they are both sensitive to international influence. Many development projects are underwritten by the World Bank, and disaster assistance is leveled by global resources.

Survival migration

The newest conception in forced migration is called "survival migration". Conceived by Alexander Betts (2010), it is defined as those persons outside their country of origin because of an existential threat against which they have no access to a domestic remedy. This threat has been interpreted as environmental/climate change, livelihood collapse, and state fragility. The definition has three specific elements; people are outside of their home country, they face a threat which includes the right to dignity,[3] and they cannot/have not been able to remedy this situation within the domestic sphere. What makes this a relevant new category is that it identifies deprivation of socioeconomic rights, which may make many of those currently considered economic migrants be designated as survival migrants. While the literature usually depicts international migration as a simple dichotomy between refugees and economic migrants, this conception gets at the complications and multi-causality that can be attributed to migration. It also recognizes such gaps in protections, in that those who are survival migrants may need assistance but currently cannot attain it as "refugees" under the 1951 Convention.

Climate change complications

As mentioned before, climate change displacement will likely overlap the various categories of environmental migrants. The IPCC suggests that long-term variation in mean temperatures will only exacerbate short-term issues that already disturb the public order. Modest projections for what Norman Myers calls "environmental refugees" from all causes could, by the year 2050, amount to 1.5% of the world's population (Cardy, 1994). This estimate has been since considered wildly alarmist in nature, but the reality is that not every person migrating can be interviewed; thus it will be extremely hard to assess how many would legitimately fall into this particular conception. With no academically agreed upon definition, many may be lumped in with other categories of migrants. This would include victims from every category of environmental migrants as well as many survival migrants. If each of these scenarios is confined to domestic spheres, and the previously noted inadequate planning (concerning development) and short-term resources (concerning relief) continue, these circumstances can lead to conflict. If so, actual refugees will be produced in this process as well. Therefore, it is likely that millions of people driven by the same factors will be treated differently, based on how their individual situations play out. Treatment will be (and most likely already is) uneven. There is a need to apply a label to this situation which can adequately define its intricacies, can disentangle this group from others (as best as possible), and can be attached to governance policies that are specific and equitable.

The "climate refugee"?

Labels matter. They recognize a process of identification or identity that has been independently applied and chosen. Bureaucratic measures seek to prevent access to the (refugee) label, and ever decreasing numbers of people are afforded full refugee status (Zetter, 2007). The choice to use the term "climate refugee" is purposeful and imposes an identity. It suggests that those affected by climate change are victims of circumstances beyond their own control and thus are deserving of international protection. However, deserving a legal entitlement is not the same as having one.

The refugee label can reinforce alienation and divisions within society as well. It can cause an exacerbation of welfare issues as refugees are afforded protections and assistance. More specifically, this label assumes a set of needs and a distributional apparatus (1991) such as food, shelter, protection, and a way to receive them. But "refugee" means much more. Zetter also considers the broad ramifications of refugee labeling. The label originates within the confines of an extreme situation, but over time it becomes a permanent status. How long should a refugee be considered a refugee; how long does one have to be resettled or assimilate to be considered a citizen? Many protracted refugee situations encompass generations of "refugees". Zetter's work on Greek-Cypriot refugees demonstrates generations of assistance even after resettlement. While the label is necessary for assistance, when it becomes significantly longer lasting it can seem like a burden to a host society. In addition, many refugees may not want to be viewed solely as victims in need of "international charity" (Robbins, 1956). Though somewhat controversial, the refugee definition has been argued to be the starting point for every discussion on international refugee law, though often the UN treaty definition is not adequate to meet today's realities (Helton, 2002). There are many labels which fail the test of the Convention for legal status, some of which articulate very desperate situations: tsunami refugees, development refugees, environmental refugees. No matter who is called a "refugee", all forced migration labels are ultimately tested against the Convention (Zetter, 2007). Can those being displaced by climate change legitimately receive refugee protection based on the Convention's established legal mechanisms and expanded operations? Two short analyses will clarify the situation.

Convention analysis: strict definitional

The Convention's definition of a refugee is structured around the concept of persecution, with the only other clearly identified stipulation being that a refugee must cross an international border. A refugee is one who

> owing to well-founded fear of persecution for reasons of race, religion, nationality, membership in a particular social group or political opinion is outside the country of his own nationality and is unable or, owing to such fear, is unwilling to avail himself of the protection of that country; or who, not having a nationality and being outside the country of his former habitual residence as a result of such events, is unable to or, owing to such fear, is unwilling to return to it.

What needs to be established is if "climate change refugees" are being persecuted. The general conception of persecution is an individual threat. It can be a threat to one's person because of who he/she is, or a threat to one's safety because of an inclusion in a specific group. Are "climate change refugees" a definable group? They are peoples living in the many places where the climate is shifting. This can include coastal communities, forest villages, and/or urban areas. It will affect people of varied cultures, societies, and economic conditions. Because climate change affects the entire globe, there will be few, if any areas unaffected. However, the need to flee or become a "refugee" is only apparent in areas most severely affected. Therefore there may be pockets of displacees; the only commonality among them will be a deterioration of living situations due to environmental degradation. Thus the persecution could only be considered impersonal, as human environments are indiscriminately threatened in different ways. Climate change does not choose who to affect, but some areas are more vulnerable than others.

If impersonal persecution is acceptable, the question becomes, does an Act of God translate into persecution by a non-state actor? In this case, for what reason would nature persecute? This line of thought is obviously extreme, but the point is simply to show that persons already uprooted by famine and flood are not included in the UN definition (Robbins, 1956), and the Act of God explanation is therefore insufficient to offer any significant international protections. Can a case be made for impersonal persecution by the developed world? Persecution in this case would be equitable to negligence on the part of industrialized nations. It can be argued that as soon as they understood the damage they were causing, the developed world was complicit in such negligence. However, the 1951 Convention and 1967 Protocol do not conceive of refugees being created by the invisible emissions of industrialized nations. At this point, all nations directly contribute to climate change through development and are complicit when buying products produced in damaging ways. Carbon emissions span the globe, and it is unlikely that any particular refugee can be sure of whose carbon caused their predicament. Again, the persecution would still be impersonal and direct causation of any identifiable persecution impossible. It is clear that the predicament of those displaced by climate change cannot simply fit under the legal mandate of the UNHCR as it is currently written. However, UNHCR's humanitarian approach has expanded its reach to protect and assist those in "refugee like situations".

Conventional analysis: chain reaction

In addition to the legal documents UNHCR relies on, it also provides a Handbook on the Procedures and Criteria for Determining Refugee Status under the 1951 Convention and 1967 Protocol Relating to the Status of Refugees. This document outlines definitions of terms used in each of these agreements and how to interpret a person's situation into a status. As previously mentioned, the important thing to decipher is whether "climate refugees" are being persecuted. The Handbook states, "There is no universally accepted definition of 'persecution', and various attempts . . . have met with little success." Because there is no universal definition, UNHCR workers in processing interviews have some room to interpret individual situations and can decide if the reasons for the persecution feared is met. The Handbook also states that persecution is "normally related to action by the authorities of a country". Therefore, if the national government cannot protect its citizens from persecution, there is a case for refugeehood. In the case of climate change, and considering an earlier example, is the Maldives at fault for not being able to protect its citizens from the effects of India's industrialization? It may be impossible for any country to protect itself from the combined emissions of the world. Developing nations will also feel the effects of climate change before others; thus are they inadvertently persecuting their own people by not being able to protect them? Would all those living in regions susceptible to the worst damage from climate change be allowed prima facie group status as refugees due to this protection inability? This is a difficult case to make, because there would have to be agreement as to which areas are most at risk and what, if anything, a country would be expected to have done to protect its people.

However, the Handbook does provide one last way to include "climate refugees"; it is the concept of cumulative grounds:

> In addition, an applicant may have been subjected to various measures not in themselves amounting to persecution . . . in such situations, the various elements involved may, if taken together, produce an effect on the mind of the applicant that can reasonably justify a well-founded fear of persecution on cumulative grounds. . . . Needless to say, it is not possible to lay down a general rule as to which cumulative reasons can give rise to a valid claim to refugee status. This will necessarily depend on all the circumstances, including geographical, historical, and ethnological context.

Cumulative grounds can include the ways in which climate change will affect the lives of many; its complications, chain reactions, and refugee-causing catastrophes. Recent academic work is beginning to discuss these linkages. There are currently at least 40 case studies in which environmental resource scarcity has been cited as a contributing factor leading to violent conflict; environmental scarcity acts as an indirect cause of conflict by amplifying or triggering traditional causes of conflict (Martin, 2004). Global climate change will impact a region's ability to produce agricultural goods, will expose more people to floods and drought, and threaten the integrity of certain island chains. A chain reaction analysis demonstrates how climate events can/will trigger many types of societal responses.

These event chains are long term in nature, however. Many of them will happen slowly and ultimately redistribute natural resources. Arable land and current sources of drinking water will have new geopolitical owners. This can easily create struggles for power and incite violence. Martin argues there is a growing concern that scarcity induced insecurities can contribute to the amplification of the perceived significance of ethnic differences. There is a natural progression of events which can cause people affected by climate change to become legal refugees. However, they will have endured much hardship before that point. These are general events which can take on complicated processes as they play out. These conceptions under a chain reaction analytic frame can be applied to any situation where a climate event is threatening to or has already created a deteriorating state of affairs. At the end, there will be refugees, but is it not particularly humanitarian to make those in need go through much hardship before assistance can be had. Thus an argument can be made that "cumulative grounds" could be considered for climate refugeehood, depending on the processor of the asylum application. However, the person making the application may not be aware of the many years of circumstances that have led up to their displacement. It may take more than a generation for certain tensions to build, and it is possible that the applicant will not be able to piece the story together in a way that encompasses every step in a way that gets the whole chain of events correct. This consideration will also exclude assistance to those who leave before any violence erupts, but who still could use it. There is also the consideration that most of those being displaced will be inside their own nation's borders. However, UNHCR has created *The Guiding Principles on Internal Displacement* which explains that IDPs cannot be granted a special legal status like refugees, which would exclude refugee-like assistance to those who have not crossed an international border. Refugees are offered special international protections because they have lost the protection of their own county, and those still remaining in their own nation are considered a domestic issue. As per the Guiding Principles, IDPs are

> persons or groups of persons who have been forced or obliged to flee or to leave their home or places of habitual residence in particular as a result of or in order to avoid the effects of armed conflict, situations of generalized violence, violations of human rights, or natural or manmade disasters and who have not crossed an internationally-recognized border.

It is within this legal paradigm that the climate change "refugee" is considered. There are also great challenges to repatriation; those affected by climate change will not have the opportunity to overcome many of these. Those who leave early as the economy begins to irreversibly slip will not want to return, as they are already aware that their livelihood conditions will not improve. Those affected by recurring disasters will necessitate repeat emergency assistance and may also decide not to go home and face their destroyed personal belongings again and again. Not only will it become undesirable for the residents to return to an environmentally vulnerable area, but it also poses questions of whether the international community or the national government will continue to fund such actions. When the redevelopment of an unstable area becomes a burden, a nation may decide not to intercede and look to resettle its residents in a safer area. Those who experience conflict over dwindling natural resources may be in a similar position as those who leave for economic reasons; they know that the situation will not improve and that they face hardship if they return. Options for durable solutions for climate change "refugees" are not the same as for Convention refugees. With repatriation impractical due to continual environmental degradation, the only available option is resettlement. The difficulty lies in the extent to which settlements will need to shift. In the case of sea level rise, entire populations will need to be resettled, which includes those who some host nations may find undesirable. The 1951 Convention provides for exclusions for individuals who are unworthy of refugee status: those who had committed war crimes, those who had committed a serious non-political crime prior to their admission as refugees, or those who are guilty of "acts contrary to the purposes and principles of the United Nations".[4] These exclusionary categories can complicate negotiations between sending and receiving nations in situations where it is necessary for everyone to be relocated.

Because of the preceding evaluation, the term "refugee" should not be used to reference those potentially displaced by climate change. Regardless of the good intention of those using it to describe a desperate situation and to imply that individuals displaced by climate should be covered by the same protections as those fleeing systematic violence, they are very different situations. Considering the academic labels as referenced above, "climate change" or "climate" instead of "environmental" migrant or displacee differentiates those who face irreversible habitat deterioration and those who may face only temporary displacement. Specific language can also suggest those who will need to migrate or those who will be pushed out. The word "displaced" proposes that an event is occurring which moves someone involuntarily. Economic migrants are often pulled out of their communities with opportunities from abroad, but this is often coupled with unsatisfactory living conditions at home. Using "climate change displacee" suggests a push out instead of a pull, which can decouple those who choose to leave a deteriorating situation early from general economic migrants – the difference is layering climate processes onto a deteriorating economic situation. People will be displaced because their living conditions will only continue to erode – something which cannot necessarily be assumed to be the case for economic migrants. Being "displaced" does not imply a new location for movement. Using "migrant" suggests that the person or group is actively moving, while being "displaced" suggests a passive process. Understanding one label as active and the other as passive can more clearly identify the processes at work in international or national movement. Migrants choose to leave, while displacees are forced out. This distinction is an important one. Those who will need to move to Bougainville are not doing so by pure choice; it is the best of many other poor options. Using "migration" to discuss the effects of climate change can suggest that many may not have to leave, but wish to; this is an incorrect understanding which can hurt relocation efforts. Politicians and IGOs that are already reluctant to extend themselves to assist this group can use such a label to make it more difficult to attain help. If these people are perceived as choosing to leave rather than being forced to, they may be left to manage the process in their own ad hoc fashion. If they are considered to be forced out, there is a better chance their situation will be seen as necessitating humanitarian assistance. Labels matter, and once institutionalized it is much harder to change their meaning. The islanders of the Carterets did not contribute to the changes that are causing their imminent displacement. Most of those who will be affected will be situated similarly; thus displacement also implies that there is a causality which is beyond their control.

It should be clear that much of the academic literature on different forms of migration intersects, conceptually and in practice. While the word "refugee" is often used in conjunction with many other forms of forced migration, it should be apparent that this is not only inaccurate, but is creating a general misunderstanding. In the established academic community, the debate over the "climate refugee" is and has been over for some time. However, new academics to this field, journalists, and the general public still perpetrate this confusion. For the remainder of this book, the terms "climate displacement" or "displacee(s)" will be used to describe those who will be forced to leave their current homes due to the continual environmental deterioration and secondary concerns (those affecting their livelihoods or having other economic and social impacts) from the processes of climate change, migrating inside or outside of their home country. There is no expectation that all academics may eventually agree on this definition, but it is useful for this analysis to untangle these overlapping conceptions and specifically outline a definition for the group whose governance potential will be examined.

5
Institutional Expansion

The focus of this chapter is to situate climate induced displacement into the realm of governance. With climate induced displacement being a newer phenomenon, it has no governance structure of its own, but there are IGOs which govern migration and displacement in alternative contexts. Before the constraints on such organizations can be understood, governance structures must be situated within the broader international relations literature. While their unique inner workings and expansion will be discussed in the case studies in the next chapter, this chapter will serve as a conceptual background as to how governance at the inter-governmental level is currently understood (in the broadest sense) and what should be expected from it. Additionally, it identifies the functions of international governance organizations, how they develop, and why they are used. Finally, the chapter provides an overview of several institutional expansion theories and elaborates on how they can and do creep into various additional issue areas beyond their original mandates.

Governance at the global level

Governance at the global level is about the rules of world order, the agents who participate or are excluded, and the discourses about them (Mittelman, 2010). At the global level, governance is thus characterized by the prevailing centers of power which not only provide a platform to develop new rules, but also the ability to allow and disallow participation. These centers can be IGOs, non-state actors, and informal institutions (Kahler, 2013). It entails multi-level and networked relations and interactions for managing and facilitating linkages across policy levels and domains; it consists of formal and informal arrangements that provide order and stability in a world full of constant flux: it provides a range of international cooperation without a world government (Thakur et al., 2014) or, as Ruggie (2014) describes it, governance in the absence of government. There is no world government, and thus the overlapping systems of policy development can assist the state where it cannot go alone. More so, it reflects a capacity of the international system to provide government-like services at any moment in time in the absence of a world government (Weiss and Wilkinson, 2014a). It entails complex processes of social patterns of formation driven by rule-following, management, and self-organization in a world of turbulence, flux, fragmentation, disequilibrium, and uncertainty (Brachthauser, 2011). Ultimately, governance can be identified as a system of steering, coordination, or control that occurs at various spatial scales (Cadman, 2012). For the purposes of this inquiry, it is this steering of policy at the global level by states, non-state actors, and interest groups alike to tackle issues beyond the scale of an individual state alone and is based on interdependence between those actors and the necessary cooperation to achieve results.

Ikenberry (2010) argues that the generally recognized model of international governance is an American led model built on a Western foundation and entails complex notions of sovereignty and interdependence. It is also hierarchical, with the USA positioned at the apex. This developed into the bipolar environment of the Cold War; in the 1990s, the international order began to expand and American dominance was contested by new states seeking influence and changes in security interdependence. This hierarchical "old governance" model has limited utility in dealing with many of today's most significant global challenges (Ruggie, 2014). The development of hierarchy in international governance is also discussed by Abbott and Snidal (1998). They characterize the changes in IGOs over time as "Old Governance" and "New Governance". Old Governance is not unlike Ikenberry's description of governance after the Cold War; it is state-centric, centralized, and distinguished by bureaucratic expertise and consists of mandatory rules. This model provides a space in which hierarchy, designed by the USA, can be facilitated and upheld. It has a centralized regulatory authority and views societal actors as self-interested and unaccountable, and thus in need of rules. It can also be considered statist in that all formulation, implementation, monitoring, and enforcement of societal rules occurs directly or indirectly through state or interstate relations (Scholte, 2004). Scholte (2004) argues that governance in the more global world of the twenty-first century has become distinctly multi-layered and cross-cutting – rules for global companies, global finance, global communications, global ecology, and other matters are administered through non-governmental arrangements. Thus, in a world run by globalization, the "New Governance" theory rests on the premise that the state cannot by itself do all the work necessary to meet all of the pressing societal challenges it faces and so it needs to engage other actors to leverage its capacities (Ruggie, 2014). It is still state-centric in part, but the state plays a role as orchestrator rather than dictator. It is also decentralized, has dispersed expertise, and functions through soft law. In this model, the state promotes and empowers a network of outside institutions which are encouraged to create self-regulating activities. In addition to being self-regulated, these are also argued to be more participatory and democratic as institutions. International organizations like the European Union (EU) and the World Bank as well as leading grass-roots movements have pioneered new governance arrangements leading a pathway to more inclusiveness (Swyngedouw, 2005). This is one major critique of governance literature, that effective governance achieves not only efficiency and order, but also participation and accountability (Scholte, 2001), and that considerations of participation and democratic accountability tend to stop at the national border. Additionally, whose consent is necessary and whose participation justified in decisions concerning acid rain, AIDS, or the use of non-renewable resources (Held, 1995)? These are global concerns in which one democratic state can assert its participatory decisions and have an effect on others. In this context, it is important to question who makes such decisions in the international realm and how they are accountable for them. In this literature base, technocratic criteria receive more attention than democratic standards (Scholte, 2001). Nonetheless, technocratic criteria are one of the main subjects of inquiry in this book, as they do not question the level of legitimacy or democratic representation of such organizations. This analysis is more state-centric because the institutions under investigation have resulted from the consent of member states and the subsequent political inquiry describes the interplay among these states.

Both Old and New Governance are ideal types, but they can represent the changes that international governance has seen from the post-Cold War era to today. Furthermore, even if there is a shift to New Governance and states are not necessarily authoritarians in this realm, this does not mean that all states are considered equal. The assumptions underlying this argument is that countries in the global South are underrepresented in global governance, and governance cannot apply to countries with a minimum of institutionalized public order. Thus the global South is pitied for being excluded from world affairs (Overbeek et al., 2010). There are certainly many reasons to substantiate this claim; smaller nations do not necessarily have the resources or personnel to participate at the same levels as larger and richer nations and are not seen as equally influential. Governance actors wield authority across borders as states exercise authority over other states; it is cooperation within a hierarchy (Lake, 2010). Regardless of whether it's considered a specific hierarchy – such as the one proposed by Ikenberry – or a diffuse one where many states agree to use their authority over others, governance is still about the authority in relationships. This is especially important when dealing with climate change. Those most affected by this process are the smaller nations which do not have the global reach to insist that their concerns be addressed by the larger and significantly higher carbon emitting nations. Kahler (2013) argues that sheer economic weight and increasing military prowess do not directly translate into capabilities that provide bargaining power in negotiations or influence over the institution of global governance – that market size and the ability to open and close a market plays a larger role for some developing states. However, developing market shares have only compounded issues in the global climate arena and helped to create rifts instead of cooperation.[1] An alternate model comes from Ramachandran et al. (2009). These authors consider a system of global governance as needing a balance between people, economics, and nation-states; Ikenberry and Abbott and Snidal have already illustrated that balancing smaller nations can be a difficult task. Ramachandran et al. see a necessity to treat all nation-states as equals; however, they also add that there should be a requirement to include representation by economic resources which would reflect pragmatism. The authors' idea of pragmatism is already shown in the way in which minor states are currently treated in governance structures; those with small or developing economies do not receive the stature or have the influence that states with larger economies do. In terms of climate governance, it may be less than pragmatic to isolate the nations that are most affected, if effective solutions are the goal. However, if the goals of governance are guided by the powerful main actors, "effective" may not be the ideal if these solutions hurt their already strong economies. Whether pragmatic or not, relative economic power does shape global governance structures.

One way to define these structures is to use the term "architectures". This term has been used to describe the broader institutional complex in international relations such as security, finance, trade, and environmental protection; however, it has no clear, commonly agreed upon definition. Biermann et al. (2009) define global governance architectures as the overarching system of public and private institutions that are valid or active in a given issue area of world politics. The system comprises organizations, regimes, and other forms of principles, norms, regulations, and decision making procedures. It can be described as the meta-level of governance. Meta-level governance is a bit abstract, but as the authors describe it, the term focuses on the overall environmental setting in which distinct institutions exist and interact. Additionally, there are degrees to the amount of fragmentation of an architecture; this description of global governance structures emphasizes the layers of governance that can make them more global in nature. If some states/ regions are not directly active within the highest level of organizational entity, there may be alternative ways in which participation is still achievable. Similarly, Eberlein and Newman (2008) describe the development of what they call "incorporated transgovernmental networks" in their discussion of the EU. This form of international governance is comprised of national regulatory authorities who are embedded into the supranational policymaking process; transgovernmental actors guide the process of integration and harmonization. This can also be seen as a type of architecture, as layers of national governance harmonize with supranational structures. A step below this meta-level would be the international regime which tends to have distinct institutional elements of the larger architecture (Biermann et al., 2009). Regime complexes, argue Van de Graff and De Ville (2013), have components that are loosely coupled, but these writers draw on Keohane and Victor (2011) to elaborate that regime complexes can offer advantages such as greater flexibility across issues and adaptability across time versus legally integrated regimes. Orsini et al. (2013) argue that regime complexes have several distinct properties: the constitutive elements of regime complexes are regimes in their own right; they are composed of at least three elemental regimes; their focus is a specific subject matter, often narrower than an issue area; their elemental regimes must overlap at least partially, but seldom entirely; any set of three regimes does not automatically constitute a complex, and policy makers/stakeholders must see the simultaneous existence of the elemental regimes as problematic. Thus, while regime complexes are not as nebulous as governance architectures (as per Orsini et al., 2013), what is under consideration is a wider governance space than an institution developed to govern one particular regime.

International governance structures

The emphasis on international governance structures for the purpose of this academic inquiry is the IGO. This structure would be one of the more concrete forms of governance within an architecture. The literature on IGOs focuses on the specific structural attributes of an organization, not on a conceptual entity like global governance scholarship. Early efforts to develop international governance have attempted to parallel domestic governance forms – especially federalism. Abbott and Snidal (1998) argue that replicating domestic governance is difficult in the anarchic structure of the international system. IGOs may be governed and created by their members, but are in essence not state-centric. They are member-centric, and powerful member states often exercise substantial and disproportionate influence over IGOs. Centralization is also limited, with many decisions made by consensus. IGOs are important centers of bureaucratic expertise, but rarely adopt mandatory rules, instead relying on individual states to ratify any treaty before it can take effect. The shortcomings of these organizations have given way to an alternative bodily configuration. Deemed "Emerging Transnational New Governance", this updated form differs in that there is little state orchestration, it is highly decentralized, its expertise is disbursed, and agreements have become voluntary.

Going one step further, Ingram et al. (2005) define IGOs as organizations that meet regularly, are formed by a treaty, and have three or more member states. Similarly, Bernauer et al. (2010) recognize such an organization as one with a permanent secretariat and one that holds regular meetings. Minnich (2005) argues that IGOs should be defined as those associations established by governments or their representatives that are institutionalized sufficiently to require regular meetings, decision making rules, and a permanent staff and headquarters. Additionally, they can be distinguished by their formal organization, purposeful activity, bureaucratic design, and legal personality. These definitions move closer to a description of what an IGO is as a physical entity. They argue that the policies of an IGO with capabilities – such as effective mechanisms of communication, coordination, and dispute resolution and enforcement – should have more of an impact than those bereft of these attributes. Rey and Barkdull (2005) define them as formally recognized, permanent institutions created by a treaty among nations. Thus an IGO is multilateral, with an enduring character, headed by a secretariat which holds regular meetings, and backed by some sort of international legal standing. However, the more one can define an institution's operations, the better chance one can have of understanding its effectiveness. Volgy (2008) distinguish between IGOs and formal intergovernmental organizations (FIGOs) which have a comprehensive operational system. These authors take a nuts and bolts approach to identifying such organizations through structural criteria. They define IGOs as entities created with sufficient organizational structure and anatomy to provide formal, ongoing, multilateral processes of decision making between states, along with the capacity to execute the collective will of their members. In doing so, they also offer 11 specific criteria and the respective thresholds of those criteria which classify FIGOs (Table 5.1).

These explicit criteria provide a complete view of what an IGO consists of above and beyond conceptual explanations; the structures examined in the subsequent case studies can be categorized as FIGOs, but will be referred to as IGOs. IGOs also vary as institutions by membership rules, scope of issues, centralization of tasks, rules for controlling the institution, and the flexibility of their arrangements (Korememos et al., 2001). They are structures which come to exist in a rational and purposeful manner. Much of this section thus far has focused on how IGOs are viewed, conceptually as well as structurally. It is important not to overlook the way in which they function. IGOs are independent from states, because they control information and expertise. They have an authority derived from their member states to act independently and thus do not necessarily mirror state decision making. IGOs bargain over turf and funds, are constrained by individual state preferences, and create a ritualistic behavior which can be disconnected from the outside (Barnett and Finnemore, 1999). Beyond structure, there is also the

Table 5.1 Classification threshold for a FIGO

Thresholds for FIGO criteria

Criteria	Threshold
Number of states	Three or more
Mix	Predominantly states, no veto by non-state members
Representation	Representing central government or its subunit
Rules of governance	Specified in its charter
Meetings	Routinized and meeting at regular intervals and at least every four years
HQ secretariat	Permanent
Staffing presence	Non-symbolic, more than two, paid by the IGO
Staffing independence	Independent of any IGO
Budget amount	Sufficient to cover minimal staffing and operation
Funding mechanism	Routinely identified and regularly available
Source	Majority funding not controlled by another state or IGO
Source of information	Varied, including direct contact with IGOs and their websites, news reports, and original documents

internal functionality of IGOs. There can be no denying politics, which often conflicts with the functionality within these structures, and this will be further examined in the next chapter. One of these political factors is a strong Western dominance.

A type of cooperation

The IGO can also be understood as a form of international cooperation. While IGOs can conduct themselves in a somewhat disconnected manner from their member states, they are representative of the way in which states cooperate in the international sphere. The international relations scholarship on institutions and cooperation is broad; it will be reviewed with consideration to its relevance to the research at hand. The international sphere is generally characterized by a state of anarchy where there is no super state which can wield supreme authority over the rest. It is also argued that considerations about relative gains and concerns about cheating prohibit cooperation (Mearsheimer, 1995). These concerns from those in the realist community have been a challenge to their liberalist rivals, who identify not only that cooperation does occur, but that it does not have to be coerced. Realists are mainly concerned with power, how states guarantee their own survival, and how to maximize their relative power over others. Axelrod (1984) challenged this notion using game theory and realist assumptions about self-interest using a tournament of Prisoner's Dilemma (PD) games. He finds that cooperation can evolve from those who base their cooperation on reciprocity, a strategy which can thrive over many more protective strategies. The tit for tat strategy of reciprocity is nice, but also retaliatory when necessary. While this game represents cooperation in a sphere of anarchy, it is usually used to demonstrate whether states go to war. Deciding to act on climate change is a different matter. This situation is a "tragedy of the commons" (Hardin, 1968), where the freedom of each state to pollute with carbon emissions is harming them all. It is a problem of overdevelopment of common pool resources (Ostrom et al., 1994), not war. Thus the imposition of PD on this situation may not best represent a strategy for cooperation. All states know that climate change is a bad thing, that no one is immune from its consequences (good or bad), and that they would be better off with an agreement to stop it. Krasner (1993) would describe this as a coordination problem best represented in the Battle of the Sexes. The disagreement lies in what is to be done, not that something needs to be done. There are many Pareto points to consider, depending on the issue area; coordination on emissions limits will have different options.

Neoliberal institutionalism explains cooperation through transnational institutions and regimes (Milner, 1997) and is thus is a better theoretical location for this research than basic realism or liberalism. The term "institution" can mean various things; Keohane (1984) describes institutions as formal NGOs, international regimes, and conventions. Norms and regimes have the ability to develop into more formal organizations, depending on how widely accepted they become. Finnemore and Sikkink (1998) describe this process as a life cycle where new norms emerge, cascade, and then become internalized. This process can create what Milner (1997) calls an epistemic community – where many states share certain thoughts about a particular issue on which they agree and have adopted the same stance on said issue. Cooperation is more likely when these communities exist. Inversely, Dorussen and Ward (2008) argue that those who participate in IGOs are exposed to norms which generate a type of social capital which creates network links among nations. States within these institutions develop ties to other states and learn how to bargain with each other; thus, rather than seeking asymmetrical advantages through coercion, states are functioning in a realm of reciprocity. Membership in an IGO is in itself a distinct form of cooperation, as well. Coordination can be found in their formal structures, such as agreed upon goals, established decision making procedures, and coordination of policies (Minnich, 2005). In addition, IGOs can assist outcomes and alleviate fears of unequal gains (Keohane and Martin, 1995). They create credible commitments (Morrow, 1999), reinforce norms, mediate conflict, reduce uncertainty, aid problem solving, socialize actors, and contribute to identity formation (Rey and Barkdull, 2005). IGOs are also purposefully designed with membership rules, a scope of issues, and with centralization of tasks, internal rules, and certain flexibilities of arrangements (Korememos et al., 2001). Constructed as such, IGOs are specifically crafted, not haphazard. They are a form of cooperation which is tacitly agreed to in order to form a concrete structure, form, and purpose.

Why do states agree to join IGOs?

According to Abbott and Snidal (1998), states agree to create and prefer to institutionalize certain arrangements because they can generate centralization and independence. IGOs have the propensity to contribute much to the international community. These organizations enhance efficiency by economizing transactions costs and resemble governments more than business firms; they can thus mirror the activities of governments and achieve cooperation in ways that other institutions cannot. There are additional functions which IGOs provide as well. They (seek to) provide a neutral, depoliticized space to make arrangements, which is essential for productive negotiations. They equalize power among nations through rules; while not always successful, rules are necessary for smaller nations to be able to have their say against those which are more powerful. IGOs have an administrative apparatus which allows them to continue their work on a day to day basis – not just when heads of state or government are available. They manage operational activities and pool risks that individual nations may not feel comfortable tackling alone. Lastly, they push negotiations forward by facilitating cooperative relationships and agreements. While democracies tend to join more IGOs than do non-democracies, this also varies by democratic institutional structure. Rey and Barkdull (2005) find that those with more competitive party structures and multiple legislative chambers join more IGOs. Boehmer and Nordstrom (2008) add to this, demonstrating that dyads of nations that are economically dependent and/or that are democratic and enjoying peace join IGOs at higher levels than those which are not; development and alliance also increase IGO involvement. It could be argued that these results represent the effect of the epistemic community and that developed, democratic, and peaceful societies share a similar international outlook which increases their propensity to join international governance structures. The realist tradition suggests that IGOs only represent the current and prevailing power centers, but if others begin to buy into the values which these centers represent, then there can be an agreement between the two schools of thought in which resulting cooperation can stem from varying perspectives on the same phenomena.

In order to be able to fulfill the tasks listed in the section above, states need to be able to relinquish some level of sovereignty. Sovereignty within the sphere of international cooperation can be a complicated issue. It is the justification for domestic rule, but sovereignty becomes more difficult to maneuver in the international arena. The anarchic nature of the international realm is characterized by the lack of any entity which is sovereign over the rest. Even in the case of a powerful hegemon, realist explanations for peace between nations lean more toward alliances that support a balance of power rather than takeovers (Haas, 1953; Morgenthau, 1985; Waltz, 1979). In understanding international cooperation and how and why individual states would give up some of their sovereignty in order to join such an organization, there is a need to consider the benefits gained from becoming a member.

Abbott and Snidal (1998) provide many reasons why it can be advantageous to join an IGO. While the general purpose of such an organization is to create rules which constrain its members from choosing policies which are negative for the other members (Heinmiller, 2007), this does not have to seem detrimentally confining. Referring back to the epistemic community, there is some room to reconcile sovereignty issues. Rules are only constraints if a nation does not see them as in their best interest. If one is a member of an epistemic community consisting of a set of shared values and is doing what is in one's best interest, this can also be in the best interest of the community as a whole. Shared values should facilitate choices and regulations which will be more agreeable to the entire community.

Institutional mandates and enforcement

IGOs can be individually identified through their mandates and also consist of mechanisms to enforce agreements which facilitate the enacting of such a mandate. As Korememos et al. (2001) remind us, institutions are rationally designed to solve specific problems. While this may change over time, the original structure is built for a purpose. Mandates outline the reason for and justify the existence of an IGO. In order to create an IGO, the question of what will be governed needs to be answered. The mandate creates the initial image of the IGO, states what it will stand for, and distinguishes its activities.

IGOs govern through differing types of international law. Such agreements often vary between hard and soft law. Additionally, these mechanisms are multi-layered in that agreements at the IGO level must also be incorporated at the state level. Since WWII, there has been no shortage of instruments which states have consented to and formally share under the UN and other governing bodies. They have developed from common concerns and normative principles and rules that originate in regional and domestic law (Cottier, 2009). Again, the epistemic community provides a basis for the development of law as cooperation. It is important to refer back to the development of IGOs to understand the use of these mechanisms. The period of Old Governance was rooted in "hard law" which was legally binding and mandatory. Hard law rules are uniform across regions, enforced by legal procedures and backed by civil regulations. State compliance is monitored by other states. In contrast, New Governance relies on flexible norms and procedures (Abbott and Snidal, 1998). These rules affect a state's compliant behavior. Governments make commitments to further their own interests and comply to preserve their reputation (Simmons, 2000). However, hard and soft laws pose different challenges and incentives. Abbott and Snidal (2000) explain why there is the use of both. Legalization enhances credibility while codifying rules and consequences. Hard commitments are often used when the benefits are great but opportunism costs are high. When compliance is difficult to detect, hard law can increase the credibility of commitments. Sincerely committed states will also use them to symbolize their seriousness. In terms of climate change, it will be difficult to detect changes in others' emissions, and the risk of free riding is high. Conversely, soft agreements are argued to be more effective in that they are easier to achieve and allow actors to learn about the impacts of these agreements over time, which fosters compromise and cooperation.

Though it would make sense that hard law be used at the outset to deal with this global challenge, what has happened is that states with the least ability or desire to commit to such treaties simply do not. Soft agreements on this issue get widespread participation, but leaders do not seem interested in compliance (Von Stein, 2008). This is the tragedy of the global commons: exploitation of natural resources in support of economic growth and energy consumption which has lacked responsibility. Efforts to combat climate change have been dysfunctional and driven by these national interests. Long-term vested interests loom large (Cottier, 2009). Power does play a role in regulatory outcomes; these are not simply sterile technocratic processes (Shaffer and Pollack, 2010). Hard and soft law and agreements are not specifically alternatives, however. They can complement each other, but can also be antagonists. They can and usually are discussed in binary terms, but the usage of both can and often does lead to inconsistencies and conflicts among complementary norms. They are choices along a continuum. Hard and soft laws also interact; non-binding or soft law can lead to binding hard law, and hard law can be elaborated through soft law instruments. In the presence of distributive conflict (which can be understood as the winners and losers of climate change), the interaction of hard and soft law is often seen as the strengths of each regime being weakened by the other (Shaffer and Pollack, 2010). Those states that prefer one type of agreement over the other may also be involved in this antagonism, arguing for one's preferred competing jurisdiction over another.

Institutional change (general)

Lane and Ersson (2000) explain why institutions matter by classifying their importance into two distinct categories: intrinsic and extrinsic.

Intrinsic importance means that institutions matter for their own sake because they are interesting and affect overall social outcomes; extrinsic importance means that they are important because of their consequences or outcomes. For the purposes of this book, the focus is on their extrinsic importance. There can be no doubt that institutions have value as social instruments. Regimes and norms have continued to become institutionalized throughout human history. In recent years, many more have become structures which, in turn, are dissected by political science. The field has a need to understand how they come into form, how they function, and when and how they change. This type of inquiry is certainly interesting for the purposes of intrinsic value. Whether as a norm or structure, institutions have many beliefs, rituals, and actors to investigate. Change or expansion can also be seen in this light, but not for the same reason that an institution has extrinsic value. Institutional change, as intrinsic value, speaks more to its survival than outcomes. An institution is not one if it dies; thus if change is what is needed to survive, there is value in change for the sake of simple perpetuation. This is not an assessment of the institution's quality of inputs or outcomes, but an observation about the institution as an object in and of itself. Because climate induced displacement is a pressing issue that will only get worse over time, this book is interested in investigating the outcomes of IGO expansion. Institutions do grow and change over time, often evolving under the pressure and demands of outside forces. This section will demonstrate how and why these things happen in order to appropriately situate the IGOs to be evaluated in the next chapter. If IGOs expand and evolve over time, it is logical to assume that the ones under examination will also have done so and/or be in the process of doing so to address climate induced displacement. This book's concentration on expansion and change stems from the role of outcomes of institutions – in the form of IGOs. Institutions act for many purposes, but IGOs act under the auspices of governance which seeks to manage or create solutions to global problems. In this case, expansion and change equates to problem solving. The implication is also normative; expansion in order to take on global challenges is a good thing. IGOs represent the strongest form of cooperation ever attempted as a way to improve international relations. The institutions which will be examined here were conceived in the aftermath of WWII, when international agreements were viewed as essential to correct the injustices perpetrated during the previous wars and to protect the world from new conflicts. Three of the four IGOs to be researched are a part of the UN system proper and fashioned under these norms.

While change and expansion can be considered normatively positive as a way to tackle newly recognized global problems, there are two main forms of constraints which can work against such progress: political will and structural path dependence. Political will describes the salience of new issues to the majority of member states involved, while path dependence describes the internal workings of the institution, such as bureaucratic culture and rules. Most literature on institutions and IGOs focuses on the latter rather than on the former. Shanks et al. (1996) explain that while bureaucracies have been described as "practically indestructible" by Max Weber, they do grow and have developed differently over time. The total number of international organizations has grown significantly since 1981, but only two-thirds are still active, suggesting that without growth many have become stagnant or insignificant. From 1981 to 1992, most IGOs were created by other IGOs with common goals. Large-scale cooperation is now important to IGO growth as well as state objectives. Within these institutions of common goals, arrangements can generate regularities that may become taken for granted, as Clemens and Cook (1999) describe. In this case, change occurs when such an organization is no longer perceived as inevitable. Connecting to Shanks et al. (1996), a third of IGOs which have fallen out of favor could have suffered this fate as either their mandates were no longer necessary or another IGO appeared and became more credible on a particular issue. Clemens and Cook also offer another reason for change: learning. Internal actors modify institutions in order to solve new problems or increase efficiency. This reason implies the actor involved is some sort of bureaucratic employee, as the writers' discussion does not touch on member states. And even if a bureaucrat can initiate change from within, such change needs to align with institutional tradition, as it must still be compatible within certain models of behavior.

IGOs have a propensity to create their own specific activities and behaviors. Meyer and Rowan (1977) describe this as myth and ceremony. Institutional rules function as myths which organizations incorporate to gain legitimacy, resources, and survival prospects. These myths can be isomorphic and affect the formal structure of an organization, which is distinct from its day to day activities. These institutionalized myths define the organization's domain of rationalized activity. The flexibility of the myth can either assist or deter expansion. Certain mythologies can be very confining, as an organization sees this as their identity, which, in turn, is internalized by its staff, who will protect it. One possible deterrent to change may be that in terms of climate induced displacement, there may be no prevailing myth yet which can be absorbed.

An institutional mythology can act as a way to legitimize territory; a way to ensure survival. If IGOs can lose their significance, a strong mythology can perpetuate an organization's existence. However, it can also create conditions which are less conducive to change. A mythology needs to be protected in order to ensure an institution's survival; changing it could be seen as a threat to the organization itself and to the people it employs. Organizations are not mechanical tools doing the work of their creators; they are alive in that they interact within their environments and maintain personnel who try to use the organization for their own ends. IGO secretariats spend time and energy in ritualized conferences, establishing agendas, co-opting state representatives, developing data, and generating resolutions (Ness and Brechin, 1988). There is also research which confirms that IGOs provide the negotiating space which member states use to make the case for their own interests and to facilitate cooperation. Bearce and Bondanella (2007) find that IGOs make member state interests more similar over time; this remains the case even considering the levels of inequality between nations (Beckfield, 2003), and the results are stronger within global IGOs than among their regional counterparts (McCormick, 1980). These outcomes suggest that if a member state saw the opportunity, it could use an IGO to make the case for expansion on an issue if it thought it was important to do so. This is not to suggest that the process would be easy or guaranteed. What the research does not answer is whether interest convergence happens through cooperation or coercion; thus smaller member states may not be the ones initiating a discussion to bring something important to the forefront or, even if they do, they may not find the convergence going in their direction.

This chapter has discussed the theoretical propensity for institutional change as somewhat detached from the individuals who initiate it. Much of this literature treats institutional bureaucracy and even member states as if they were autonomous of any sort of human design or control. However, there is a subset of neoinstitutionalism and neofunctionalism which identifies Eisenstadt's conception of "Institutional Entrepreneur" as important when discussing human agency; it designates individuals and groups who adopt leadership roles in episodes of institution building. Eisenstadt maintains that institutional change is partially contingent on the activities of such entrepreneurs, and while they may still rely on institutional myths, they do so only in ways to legitimize the changes they seek. Institutional entrepreneurs find innovative ways to articulate what they want, even using some of the confining aspects of the institution in their favor (Colomy, 1998). However, institutional arrangements often reflect the ideas and goals of the most powerful system actors (Seo and Creed, 2002); thus changing the institution can also mean standing up against those who would benefit from the way it currently works. This makes the task of the entrepreneur difficult, but also reflects the need for persuasion. These institutional entrepreneurs can be seen as the personification of Finnemore and Sikkink's norm entrepreneurs; once they begin to make progress and hit the necessary (and non-specific) tipping point, the norm cascade will bring others within the institution along.

Institutional expansion theories (specific)

For the purposes of this book, it is not sufficient to explore what an IGO is and does, but also how it acts over time. This means looking at IGOs' propensities to expand. As mentioned before, many international governance structures are initiated with the purpose of aiding a specific issue area. In recent years, however, governance has expanded rather than remained static. Thus, the expansion of migration governance to cover those affected by climate change would not be without precedent. Below are two different theories which can assist in understanding the impetus for governance enlargement. They are from both international relations and business literature.

Neofunctionalism is the brainchild of Ernst Haas, from his seminal work *The Uniting of Europe* (1958). It is a theory about the growth of governance of the EU. It began as a departure from two earlier works about the development of the EU: transactionalism and functionalism. Transactionalism refers to the amount of economic and human capital-based transactions across European borders. The more transactions, the more integrated nations become (Puchala, 1970). Transactions are a description of integration; they do not cause it. Functionalism has emphasized a union of neutrality which is suggested to be apolitical and based on regional institutional building (Mitrany, 1948). Haas' departure was moving beyond functionalism's vision of simple technocratic governance to offer a utilitarian approach to the fulfillment of interest (Rosamond, 2005). With neofunctionalism, actors matter, and they need to be comfortable acquiring the new loyalties of the developing governance organization. European scholarship has identified nested identities as an important way in which this process is facilitated, much like Russian Matruska dolls (Risse, 2005). This is important, as some have argued that neofunctionalism suffers from a macro bias and does not give enough credence to human agency which provides the leadership for institutional change (Colomy, 1998). It can also be extrapolated that national leaders can work as entrepreneurs to push for institutional change. Additionally, integration is a sporadic and conflictual process, but through democracy and pluralism, national governments will find themselves devolving more authority to these regional organizations, and thus citizens will expect more of them as well. This is the process of "spillover" (Schmitter, 2005). The better the system functions, the more will be expected from it. The process of spillover is optimistic, but also somewhat paradoxical. For the process to begin, existing states need to come together on some relatively non-controversial and separable issue area where tangible gains from cooperation are sufficient to give up autonomy. However, if this issue were so non-controversial, there would be little reason to necessitate expansion to handle it (Schmitter, 2005). In the case of climate change, the issue is much bigger than any one nation can tackle on its own and is also controversial in terms of blame. But, because the problem and its consequences are so large, it necessitates governance and cooperation beyond the national level. There is the need to provide functioning apolitical governance in order to tackle such a complicated issue. An important facet of spillover is that (at least in the case of the EU) it does not specify a time line for its occurrence. Contemporary authors have found this difficult, as the development of the EU stalled for many years. This could also be the case with climate change; there may be little movement now to expand migration governance structures, but this need not be the case forever.

Another set of literature which outlines organizational expansion comes from organizational behavior and firm theory. While firms or corporations do not function identically to IGOs, there are some theoretical similarities which may assist this particular investigation on institutional expansion. Organizational theory and firm theory tend to complement each other because a firm is a particular type of organization. Organizational theory assumes that organizational forms are effective in that they promote the survival of the organization. In addition, the job of a firm is to economize transaction costs. There are also political theories of the firm which argue that the basic problem facing organizational actors is to create a stable world so that the organization can continue to exist (Fligstein and Freeland, 1995). Stability is one thing, but expansion requires a large amount of resources (Hu et al., 2008) and a full evaluation of immediate operating needs, the competitive environment, and dimensions of management, finance, and macroeconomics (Kumar and Waheed, 2007). However, there are good reasons to expand. Taylor explains that expansion is often driven by clients' requests or demands, attractive economic options (2005), and when clients are already based in an alternate geographic area (Felts, 2005). Bringing IGOs back in, once established, they have bureaucratic structures which employ many who do wish to promote the survival of the organization, if for nothing other than to keep their jobs. Continuing to exist is important, but so also is the issue of continued relevance. This is complicated by overly effective governance of the issue area at hand; is governance necessary after all member states actively integrate IGO regulations into their own domestic policy? With nothing more to govern, there is no need for the institution to survive. However, expansion can be the answer to stagnation and irrelevance; thus firm theory can be extrapolated here. Expansion for IGOs is just as resource-intensive as it is for a firm; similarly, there are the requisites of office space, personnel, and sufficient revenue. In terms of climate displacement, there is also a growing demand to expand from those who are currently affected as well as from academics and practitioners.

This chapter is not meant to be fully comprehensive, but to provide some theoretical and pragmatic background on how and why institutions expand over time. International relations literature is flush with work on cooperation and calculating how and when states do so, to the extent that to be thorough it would necessitate a book of its own. This chapter's focus has been on IGOs in international relations as they form, develop, and expand, which is the basis for the in-depth case studies to follow. Neofunctionalist spillover and firm theory are two possible frameworks through which to see this process. These will be referred to again later in the book as a way to reintroduce, more specifically, a structure for understanding the options for each of the IGOs being researched. Most importantly, this chapter has shown that IGOs do expand in order to facilitate cooperation on new issue areas, and thus those IGOs which currently assist with migration and displacement do have theoretical and practical precedent to draw from. The next chapter will follow the development of three influential IGOs and their expansion over time, as well as their current involvement in the issue area of climate induced migration and displacement. They will be evaluated against each other and their structural constraints, political constraints, and theoretical frameworks to expansion in this issue area will be discussed in the final chapter.

6
Lack of Expansion

There are three IGOs that govern different forms of migration and displacement – all of which have expanded over the years when the situation has demanded it. UNHCR, IOM, and UN OCHA each assists migrants in a variety of contexts and will be discussed in this chapter. UNHCR is arguably the most capable and successful institution when it comes to protecting and assisting forced migrants. However, its potential expansion to deal with this newly identified type of migration cannot be seen as a given. Unlike refugees, the bureaucratic term "migrant" represents a much broader group of immigrants and does not clearly implicate a regime. While migration has been a continuous part of human existence, the development of any such governance in the international sphere, such as by IOM, has been relatively recent, leaving border governance to the individual state. Humanitarian agencies such as UN OCHA often deal with those displaced by environmental disasters, and the effects of climate change will only exacerbate their work. Thus, the development of humanitarian structures and governance is crucial to determining if this regime and its corresponding IGO are yet prepared to deal with the additional strain of climate change. This chapter will explain the development of the refugee, migration, and humanitarian regimes, their respective IGOs, and their current expansion to date. It also provides examples of how each IGO has related itself to the topic of climate displacement and how each has significant challenges to additional expansion to govern this new group of displacees. The chapter concludes with the idea of reluctance to expansion and how these IGOs have responded when confronted with outside demands for this expansion in recent years.

The refugee regime and its evolution

The modern legal designation of "refugee" originated in ancient times. As a form of hospitality, Mediterranean and Near East civilizations often granted asylum to those fleeing violence. It was an early humanitarian gesture recognizing the human desire for liberty and protection (Krenz, 1966). This first form of international protection had no special regulations, bureaucratic processes, or formalities; it was an informal civil pact of profound importance, and it continues to serve as one of the oldest international norms. Krenz explains this as the rule of "minimum standards" which, in short, grants foreigners the same treatment as nationals in cases of conflict. Goodwin-Gill comments on this custom in his discussion of citizenship and the nation-state. He explains that in the seventeenth century foreigners were not to be denied local protection if they came within the territory and jurisdiction of a government not currently at war (Goodwin-Gill, 1989). These customs turned into law as early as 1685 and established a more modern system of asylum in Europe (Grahl-Madsen, 1966). The World Wars created a burden on the old ways far more onerous than ever before. This period also saw more exclusionary immigration policies by individual nations, leaving millions of the displaced in limbo. Rubenstein (1936) calls this situation an "exodus" which created political, legal, social, and humanitarian problems. Hathaway (1984) describes this phase as having three periods; juridical, social, and individualist. They represent a changing definition of international refugees. The juridical perspective period (1920–1935) was primarily concerned with the refugee as a member of a group that has no freedom of international movement because its members have been deprived of the formal protection of their government. This relates to a nation's drive for a homogeneous homeland and their use of what Adelman (2001) calls a population swap (or a less violent form of ethnic cleansing). These swaps allowed Turkish Christians to come to Greece and Muslim Greeks to flow into Turkey. The social perspective (1935–1938) shifted to encompass victims of broad-based social and political upheaval – regardless of their legality. Finally, the individualist perspective (1938–1950) moved away from group disenfranchisement and toward a consideration of the relationship between the individual and the state. Essentially, it was concerned with a fundamental incompatibility between the citizen and government. This was the prevailing understanding of refugees as WWII came to a close. Several refugee governance institutions came into and went out of existence during this time, leading to the development of UNHCR. They will be briefly outlined in the time line below to demonstrate the structural evolution of refugee governance.

1921: In the aftermath of WWI, the Office of the High Commissioner for Russian Refugees opened in 1921 with the express purpose of helping those who had become refugees due to the Russian revolution. Headed by Dr. Fridtjof Nansen, it provided travel documents, sought employment opportunities, and delivered aid to displaced Russians and those living in the Ottoman Empire. It was an arm of the League of Nations and would be replaced by the High Commissioner for Refugees in 1938 (UNHCR, 2005).

1933: As a partner to the Office of the High Commissioner for Russian Refugees, the addition of the Office of the High Commissioner for Refugees Coming from Germany became necessary when Jewish refugees began leaving. In two years, the office resettled 80,000 refugees, mainly to Palestine. It was also replaced by the High Commissioner for Refugees in 1938 (UNHCR, 2005).

1938: Formed in 1938 as a conglomerate of the previous two offices, the High Commissioner for Refugees played a very limited role until 1946 (UNHCR, 2005). The Intergovernmental Committee on Refugees was formed in the same year, with the responsibility for those who emigrated out of Germany and Austria on account of their political opinions, religious beliefs, or race and those within this group who had not yet settled (IO, 1947a). This organization supported a very specific mandate and resettled 240,000 refugees (defined as such) before the outbreak of WWII. Still in existence after the War, an initiative to expand came from the governments of the UK and the USA. In 1944, refugees included all persons who had to leave their homes due to the events in Europe. The organization's second expansion came in 1946, to include those persons who were considered "non-repatriable" refugees from Germany, Austria, and Italy (IO, 1947a). Additionally, the Committee worked with the UN on a draft constitution of the International Refugee Organization (IRO) which would take over its current functions (IO, 1947b). Its Executive Committee arranged for the transfer of all office equipment, vehicles, and stocks to the IRO. While under liquidation, final agreements allowed for the resettlement of refugees to Peru, Brazil, and Venezuela (IO, 1947c).

1944: Established by the Allies in 1944, the United Nations Relief and Rehabilitation Administration (UNRRA) was to provide emergency relief to the displaced. It organized the return of millions to their homes, but was not designed or prepared to handle those who refused to go back (UNHCR, 2005).

Permanent institutionalization

The IRO, established by the UN, took over for the Intergovernmental Committee for Refugees to provide a permanent solution for Europe's refugees. Its primary task was defined as repatriation, followed in other cases by resettlement of those who could not return to their countries of origin (IO, 1947b). The agreement and subsequent constitution it produced was far more comprehensive than that of the previous organization. It was also very clear about who would be considered a "refugee" for the purposes of its mandate. The IRO defined as "refugee" any person who "has left, or who is outside of, his country of nationality or former habitual residence, and who, whether or not he had retained his nationality" is a "victim of the Nazi or Fascist regimes or of regimes which took part on their side of the Second World War, . . . Spanish Republicans and other victims of the Falangist regime in Spain, . . . persons who were considered refugees before the outbreak of the Second World War" (IO, 1947d). It also specifically outlined those who would be excluded, such as war criminals, traitors, those who had assisted the enemy, ordinary criminals, ethnic Germans, those who already received financial support, those who had attempted to overthrow their government by armed force, and those who were currently in the military or were members of the civil service of a foreign state. In terms of mandate, these specificities are obviously time- and incident-related. Additionally, the IRO was deemed a "non-permanent organization" (IO, 1947d), suggesting that after those displaced by WWI and WWII were resettled, it would be an unnecessary institution. While highly tailored for a precise function, agreement on this institution was not without member state politics. A divide emerged between countries of origin and countries from the West. The Soviet Union preferred a policy of resettlement and a strict definition of "refugee", while others emphasized widescale resettlement and a wider definition. The gap continued to be an issue for the institution's financial arrangements as well. Much of the expenses of repatriation were to be charged to the governments of Germany and Japan (IO, 1947d). This consisted of external and "heirless assets". Some would come from German financial holdings in international bank accounts which were promised to the Intergovernmental Committee on Refugees, but were eventually allocated to the IRO (Rubin and Schwartz, 1951). In its four-and-a-half-year tenure, the governance structure dealt with more than one and a half million dislocated people. Of this group, 1,038,750 were resettled and 72,834 repatriated, leaving some 362 cases without a satisfactory resolution (IO, 1952). It provided care and maintenance rehabilitation, legal and political protection, counseling, vocational training and employment (Davie, 1957). As Sir Arthur Rucker (1949) recounts, it did this work juggling three headquarters (in Geneva, Paris, and London) and while nations paid their contributions very late in the year. As an insider, Sir Rucker's speech at Chatham House in 1948 represents a frustrated view from the inside which observes that the organization's important work must have public support, more money, and nations willing to assist refugees.

The IRO began to fall out of favor by the late 1940s, but it was clear that there was still much to be done. A subsidiary body of the General Assembly, the UNHCR originated in January 1951. Its mandate provided for it to function for three years (UNHCR, 2005). The agency's mandate, drawn from previous experience, was as specific as possible as to the conditions which would cause one to be a refugee in Europe at that time, including "race", "religion", "nationality", and "membership of a particular social group or political opinion", which are antecedents of the period between the two World Wars (Gallagher, 1989). The Convention on the Status of Refugees also specifically noted a time frame: refugees would be those displaced by events occurring before January 1, 1951. As is, the Convention provides for WWII refugees and very little else. The institution itself had only a restrictive budget; it was also explicitly prohibited from raising its own money (Barnett, 2011). However the Convention does stipulate that any contracting state can extend its obligations further than what is specified in the agreement. Its original signatories (those who signed and ratified the treaty before and in the same year as its effective date of April 22, 1954) are Australia, Austria, Belgium, Denmark, France, Germany, Israel, Italy, Luxembourg, Monaco, Norway, Sweden, and the UK. It is clear that directly after WWII, UNHCR and its work was primarily the concern of the European continent.

Extensions of the refugee regime – national policies and regional agreements

Many state and regional agreements to govern refugee flows did spring up after the establishment of UNHCR. The Convention needed to be ratified by individual states in order for it enter into force, but those states had the right to extend such obligations, and many did. In addition, regional agreements came into force in the areas in which conflict continued to emerge. The USA passed the Refugee Relief Act of 1953, which also incorporated the reason of "natural calamity" under its definition of refugee (Wenk, 1968). The USA was not an original signatory to the 1951 Convention and decided to implement its own law. In doing so, it expanded the definition of refugee which it would observe. While this move could be seen as benign due to the occurrence of hurricanes and the need for America to assist its neighbors in the Caribbean, accepting a larger definition of those labeled as "refugee" began to open the door to later expansion of the legal terminology. Additionally, the Organization of African Unity (OAU) prepared a Convention in 1969, adding an additional category to the 1951 UN Convention: those fleeing their country to escape warfare or other man-made disasters (Grahl-Madsen, 1983). During this time, a man-made disaster tended to be understood as the consequences of decolonization. The 1960s saw a huge wave of nations calling for independence from their colonial rulers. Specifically, the OAU Convention states, "The term 'refugee' shall also apply to every person who, owing to external aggression, occupation, foreign domination or events seriously disturbing the public order either in part or whole of the country" is forced to leave his/her country of origin. Africa was extending its definition of "refugee" to apply to its changing landscape. Central America eventually followed suit. In 1984 the Cartagena Declaration on Refugees also extended its refugee definition to those fleeing "other circumstances which have seriously disturbed the public order".

UNHCR expansion to date

Academics have been arguing for expansion of the UNHCR definition since its inception. Grahl-Madsen (1983) explains the category of "de facto refugees", or a person not recognized by the Convention but who is in a similar situation. These are people who could eventually be successfully recognized as refugees or those who cannot be recognized as such under the Convention but may be allowed entrance into another country on humanitarian grounds. The meaning of "humanitarian grounds" can also vary from country to country, but allows for receiving nations to expand their assistance in circumstances which may not exactly fall under the Convention's explicit definition. These critiques did not fall upon deaf ears. While UNHCR's original mandate and funding mechanisms were fairly well constrained, its officials were able to open some space to grow. Signatory states had labeled the organization "humanitarian" not only to describe the work it did, but in hopes that it would be resolutely apolitical. Humanitarianism gave UNHCR moral authority. The body used this role to increase its influence in protecting the weak and vulnerable, extending its mission and principles to assist in "refugee-like" situations outside of Europe and geared toward events occurring after 1951 (Barnett, 2011). This initial expansion conceived by its own employees set the stage for further growth as international conditions began to change.

Expansion of the 1967 Protocol

The demands on UNHCR grew as the world continued to spar with the growing Communist threat in Europe and Southeast Asia and decolonization in other locales. Increasingly, people fled from Communist regimes in Eastern Europe and the USSR, while many more were being affected by the conflicts in Korea and Vietnam. In 1967, the Protocol Relating to the Status of Refugees was expanded, which formally applied the status of refugee to any person who fits the definition "as if" the date requirement had been omitted. This act had an enormous effect on expanding UNHCR. Before the Protocol, its mandate only required that it provide legal assistance to those displaced because of the events of WWII. The Protocol opened the door to assisting anyone displaced due to persecution and conflict. While nations continued to slowly ratify the 1951 Convention, those nations who became the first signatories of the Protocol demonstrated the acceptance of this humanitarian role. They include Algeria, Argentina, Cameroon, Cyprus, Denmark, Finland, Gambia, Ghana, Greece, Guinea, the Holy See, Iceland, Liechtenstein, the Netherlands, Nigeria, Norway, Senegal, Sweden, Switzerland, Tanzania, Tunisia, Turkey, the UK, and USA.[1] Many of these nations were dealing with large refugee flows – especially on the African continent, where the OAU would take this as a first step toward its own regional agreement mentioned in the previous section.

Expansion to IDP issues

Hakovirta (1993) explains that in the 1990s, UNHCR's concerns broadened and prima facie group determination of refugee status largely took the place of individual interviews. The circumstances under which it offered protection in these situations included persecution and insecurity as well as starvation and critical environmental conditions. During the Cold War, those searching for refugee status were often making claims on their own. Subsequently, UNHCR began to see the need to assist many groups even before they became refugees. This began with the displacement of the Iraqi Kurds during Operation Desert Storm. In the midst of this conflict, UNHCR became increasingly involved in providing assistance and protection to the degree it possibly could. The agency decided not to wait until the Kurds crossed an international border, but to proactively help those who were internally displaced (Hammerstad, 2011). IDPs share many of the same difficulties as refugees, but have a different legal status. UNHCR drafted a new document entitled *The Guiding Principles on Internal Displacement*. It explains that IDPs cannot be granted a special legal status like refugees. Refugees are offered special international protections because they have lost the protection of their own county. As per the Guiding Principles, IDPs are

> persons or groups of persons who have been forced or obliged to flee or to leave their home or places of habitual residence, in particular as a result of or in order to avoid the effects of armed conflict, situations of generalized violence, violations of human rights, or natural or manmade disasters and who have not crossed an internationally-recognized border.

Defined as such, IDPs are a broad classification of those who would be considered refugees *if* they had crossed an international border (Robinson, 2003). UNHCR was changing its bureaucratic agenda from specific legal assistance to a more "on the ground" aid disbursing agency which would be more central to the protection of forced migrants, not just legal refugees. This period also saw a shift in the types of employees the organization hired. The number of staff working around the world in conflict zones increased, while the number of lawyers decreased.

Climate change and UNHCR

The previous chapters have offered an overview of the ideological forces that have provided and contested the existing legal refugee framework. The 1951 Convention and the 1967 Protocol outline a specific legal form of governance which is internationally recognized. It is also a confirmed obligation which guarantees a certain set of rights and privileges. Unlike economic migrants/immigrants, refugees have the right to seek asylum, cannot be returned to their country of origin, and have a protected status. They also need care and maintenance, reestablishment, and legal and political protection (Malin, 1947). Refugees, as defined by the Convention and Protocol and accepted by their signatories, are a recognized humanitarian commitment. They are internationally justified in their migration and are deemed worthy of assistance (monetary and otherwise). However, not every desperate situation falls under the protection of the Convention and Protocol; this is why it is important to question the use of the label "refugee" on a certain group, as they may not have a legally valid claim to it. Those who already are or will be affected by climate change to the extent that they will need to relocate are often called "climate refugees". It is a term that is popular in the sensational vernacular of journalists and some academics (Biermann and Boas, 2008; Trent, 2009). Former President Nasheed of the Maldives had made a specific call for UNHCR to prepare a new treaty which would include those displaced by climate change as "refugees" (Biermann and Boas, 2008), though no progress has been made on this recommendation thus far. Others have made the case that it is not just the people, but entire islands that should be considered "ecological refugee states" due to the loss of their entire geographical territory (Nine, 2010). However, just calling such people "refugees" does not provide them with legal status as if they were – as mentioned in Chapter 4. There are many important distinctions between those who fall under the refugee label and those who have been implicated due to climate change. The most important distinction is the ability to repatriate. Generally speaking, after a war or conflict, refugees want to go home. The refugee regime, even before the advent of UNHCR, focused on repatriation as the primary solution. Refugees rebuild; they reconstitute their previous communities. But climate change impedes this. The migration literature explains current problems with repatriation and other durable solutions. The three classic durable solutions are repatriation, resettlement in the country of first asylum, or resettlement in a third country. Resettlement is the least used solution, as not all countries are equally open to all refugees (Stein, 1983). The decision to resettle is in the hands of the state and not the refugee. First, states can and often do return asylum seekers in an attempt to not overburden their own society. UNHCR dubbed the 1990s the "decade of repatriation", with an effort to return three million people to 21 countries. The agency faced many challenges with trying to return this volume of people, and asylum seekers were often returned to areas which were politically fragile, so that protection was still necessary after return. A second issue surrounded material conditions. Many returned to areas which were destroyed or where land mines were abundant. Third, refugees going back to agricultural production faced rival claims to arable land. Fourth, when generations returned with children who had known no other life other than a refugee camp, many necessary skills were lost to the community. Finally, repatriation has demonstrated the organizational gap between humanitarian efforts and development assistance (Rogers, 1992). In many areas disturbed by climate processes, these same problems may occur and leave no room for repatriation – only relocation. In an unofficial research paper on this topic published by UNHCR, Ferris (2012) evaluates some of the research and lessons learned from previous forced resettlement and planned relocation. The piece also evaluates the use of normative frameworks such as the *Guiding Principles on Internal Displacement,* emphasizing forced resettlement and planned relocation and not touching on the need for external relocation, although it does comment that organizations like

UNHCR and the Red Cross have not paid attention to this need even though they have relevant experience in this area.

Internal displacement continues to be a growing concern for UNHCR ever since it decided to delve into the issue. While the discussion of "sinking islands" is provocative, the vast majority of people displaced by climate change will be displaced within the boundaries of their nation of residence. As such, a more proper discussion of the way in which UNHCR could assist those displaced by climate change under their current structure would be through its expansion into IDP issues. In 2009, the UN representative of the secretary-general on the human rights of Internally Displaced Persons presented a report to the General Assembly which outlined the way they interpret the nexus between climate change and internal displacement (A/64/214). The report credits climate change with the potential for voluntary and forced displacement, highlights the issue as one of humanitarian concern, and outlines a framework of protection under the *Guiding Principles on Internal Displacement*. The Report also clarifies that there is no legal basis for the term "environmental refugee" or "climate refugee", and argues that this term should be avoided in order not to undermine the legal regime for refugees.

On the surface, it appears that without expanding, UNHCR has included those displaced by climate into their fold. However, a closer examination of the Guiding Principles shows that in doing so, UNHCR is shifting responsibility of this group to the state level. Finalized in 2000, the Guiding Principles identify rights and guarantees relevant to those who are displaced in their country of residence. They are based in international human rights and humanitarian law and reflect previously established norms. The document consists of 30 principles relating to the treatment of those in a situation of displacement due to violence, human right violations, and those affected by natural or man-made disasters. While the Guiding Principles are thorough and based in law, they themselves are not law. The Guiding Principles are not a recognized treaty obligation to member states. They are, as is stated in the document itself, to "provide guidance" and "should be disseminated and applied as widely as possible". The Foreword, written by Under-Secretary for Humanitarian Affairs Sergio Vieira de Mello, emphasizes this point. He explains that these Principles are to serve as an "international standard to guide governments as well as international humanitarian and development agencies in providing assistance and protection to IDPs". Guidance is helpful, but not obligatory. The Principles relate the needs of the internally displaced to their current rights if the states in which the displacement is occurring adhere to the Universal Declaration of Human Rights, which is also a non-binding document, even though it serves as the basis for human rights law. Thus, the Principles are more of a reminder as to how to nation *should* act, rather than an outline of new international law. The importance of its non-binding nature is that it allows each state to decide if it will chose to adhere to such principles on its own. Many indeed do, but as a soft law instrument, there is no mechanism for enforcement. In sum, invoking the Principles is a way of taking a stand on the issue of climate change displacement without offering material assistance, situating the issue outside of UNHCR's legal mandate, and allowing UNHCR to rest on its moral authority in displacement situations for guidance.

This stance was strongly reiterated at The Nansen Conference: Climate Change and Displacement in the 21st Century, held in Oslo, Norway, in June 2011. Antonio Guterres, the UN High Commissioner for Refugees, offered a statement to open the conference which parallels this assessment. Guterres posited that primary responsibility for the protection and wellbeing of those displaced by climate change rests with the states in which displacement is occurring and encouraged such states' responses to be consistent with the Guiding Principles. In addition, he stated that UNHCR has refused to accept any label such as "climate refugee" or "environmental refugee" as is will confuse UNHCR's efforts to protect those who are persecuted. Finally, he recognized that it will not be easy to establish a new binding international treaty; therefore UNHCR offers its assistance in developing a "guiding framework" on the matter (Guterres, 2011). To restate, UNHCR is offering no material assistance, is placing this matter outside of its legal mandate, but is offering more guidance. But it does highlight this guidance in its 2014 publication, *UNHCR, the Environment and Climate Change: An Overview*. The document promotes UNHCR's involvement in the Nansen Initiative and its additional recommendations as a part of the Advisory Group on Climate Change and Human Mobility in the UNFCCC process, the Interagency Working Group on Climate Change, and the Sanremo Consultation on Planned Relocation. These actions were also stressed in the *Note on International Protection* – a Note from the High Commissioner (A/AC.96/1134).

Concluding remarks

The UNHCR is a highly developed governance structure for the protection and assistance of refugees. Its mandate and the definitions which it utilizes have developed through the World Wars, and it has come to be known as an organization with great experience and impact. Because of this, it is easy to see why many have called for it to weigh in on the matter of climate change displacement. Its mission has not in fact remained static over the years; the definition of refugee was expanded in 1967, and it has occasionally stepped in to assist populations before they became displaced. Former staff saw a need to expand its protection mechanisms as the political situation in the world changed. However, when it comes to displacement due to climate change, UNCHR has made it a point (as has its High Commissioner) to posit this issue outside of its legal obligations – not to offer specific material assistance, but to provide its legal expertise when it comes to human rights. Its response has been hands off, and there is no indication of this changing.

The migration regime?

Migration is the story of humanity. From the earliest hominids to modern man, history follows mankind through migration; it is in our blood. People have never remained static – which is the point often lost in many modern accounts of migration policy. The earliest stories of human origin come from the Rift Valley in Western Africa and demonstrate humanity's amazing capacity for migration; traveling to every corner of the globe to inhabit desert, icy wastes, and small islands in the sea. Migration is also ingrained in many of the major religious traditions. The Judeo-Christian religions tell stories of Moses leading his people out of bondage in Egypt and then wandering the desert for 40 years; one of the five pillars of Islam requires for its followers to make a pilgrimage to Mecca once in their lives (if possible). The human history of migration is not always benign. It is important to note that the English of today are not indigenous to England, nor are the Malays to Malaysia or Turks to Turkey; migration and conquest put them where they are (Sowell, 1996). Migration has also occurred by force, regardless of whether it is through slavery, indentured servitude, or military conquest that then expelled the current inhabitants of a geographic area. The European age of exploration beginning in the fifteenth century demonstrates all of these types of forced migration. Whether discussing the triangular slave trade, an aristocrat taking his/her servants to the New World, or the destruction of indigenous populations, migration facilitated all of these.

Migration is defined as a permanent or semi-permanent change of residence across some type of administrative boundary. A person can migrate many times, for many reasons, over his/her lifetime (Wood, 1994). People migrate because of population growth or disparities in economic development, for salaries and living conditions, due to economic crisis, or because of poverty, political instability, ethnic conflict, and ecological deterioration (Farrag, 1997). In addition, there is the thrill of being somewhere new, taking on a challenge and adventure. Each of the motivating factors listed above can be seen as individual causalities for movement, but more often they overlap. These dual and multiple causal relationships have made governance difficult. Many of the labels used to describe and categorize migrants only infer a single causality such as "economic migrant", "environmental migrant", and (for the purposes of this book) "climate migrant". The difficulty for governance is that these factors are intimately entangled and cannot easily be separated. We know very little about how changes in the environment affect migration and lack the data to move beyond estimates (Laczko and Warner, 2008).

A major nuance that influences how migrants are viewed is the idea of personal choice. Shanmugaratnam et al. (2003) explain that while migration is often viewed in a typology, all concept-types flow into the voluntary/forced dichotomy. This dichotomy has emerged in most recent studies of migration and plays a role in this in this inquiry, albeit an inferred one. In real life, the line that divides the choice to migrate or not is most often blurred, but in academia it can be helpful in the conceptualization of movement. For the purpose of this investigation, refugees are forced migrants, while most others are considered voluntary; this is where certain labels can help and hurt such conceptualization. While the many who migrate in search of survival may argue that survival is not a choice, others do chose to stay behind to suffer hunger or violence. This "choice" to stay can also be guided by the lack of resources to actually migrate (Haug, 2003). The key factor is not necessarily the type of coercion applied, but the migrants' belief that they must flee to survive (Wood, 1994). IOM facilitates migration in many forms, both forced and voluntary, while UNHCR is only concerned with forced migration. However, this inquiry does not presume to use these terms lightly. Without taking the time and space to reconceptualize this dichotomy, it is used reluctantly. At a metaphysical level, a person always has free will, but there are many things out of the control of most individuals which limit choices in such a way that they can feel forced. The reader should keep in mind that these distinctions are not as clear-cut as the terminology suggests. There are other dichotomies which blur conceptual distinctions and policies as well: skilled or unskilled workers, permanent and temporary migrants. Individuals usually belong to one or more categories at the same time or move from one to another – ignoring this fails to do justice to the complexity of international migration

(Report on the Global Commission on International Migration, 2005). The simplicity of these dichotomies can be helpful when trying to make immigration policy, but does not allow for varied interpretations or complicated situations.

What becomes apparent when researching migration/immigration is the lack of a clear regime; national sovereignty remains the deciding factor in immigration policy, subject only to treaty obligations to refugees (Report on the Global Commission on International Migration, 2005). Governance in this area is still almost entirely controlled at the level of the nation-state, and jealously guarded, although most governments recognize that they cannot control migration unilaterally. As employers, smugglers, workers, agents, and individuals continue to defy national policies, governments are extremely reluctant to relinquish any formal regulatory authority beyond the regional level. However, states have never had full sovereign control over migration and have lost what little control they have had through the forces of globalization (Newland, 2010). In many other policy spheres, national leaders acknowledge and use the international realm to cooperate on issues that are too large to handle on their own. The question, as posed by the former Commissioner of the Global Commission on International Migration, is: "Why do we persist with national approaches to a phenomenon that is inherently transnational?" It appears that some governments find the global governance of migration intimidating and fear that it would involve the creation of a new supranational agency. There is a preference for soft governance and sharpening existing instruments in this area, although it has not yet resulted in any coherence (Marchi, 2010). While there may not be a conventional regime for migration, what has developed so far in terms of international governance will be outlined in the following chapter.

Governance development

The governance of migration has lacked a coherent institutional framework at the international level (Koser, 2010). Its development coincided with that of the refugee regime, but with different roots. Unlike the norms that accompanied refugees, any assistance awarded to a simple traveling stranger does not have the same moral pull as does someone fleeing conflict, and the current dichotomous bureaucratic terminology continues to divide those who migrate. From the beginning of humanity until the turn of the twentieth century, migration was mostly ungoverned. However, human history is dotted with ages of migration – from Greek colonies and Roman military conquests to the Byzantine and Ottoman Empires. Exploration and colonial territorial gains provided many places to migrate to. Many lands, even those occupied by indigenous peoples, were seen as virgin areas ready to be populated by European settlers. No matter how old the process, virtually no society seems capable of managing it effectively. It can be described as a paradox, in that without proper management, the receiving country's sense of identity and capacity to maintain its own laws leads to political turmoil (Papademetriou, 2003).

Even though some countries (such as the USA) began to regulate migrants as they continued to flow in, international coordination came after the dramatic population shifts caused by WWI. The International Labor Organization (ILO) emerged in 1919 during the Versailles peace settlement, with a mandate to promote social justice and human and labor rights for migrant workers (Kneebone, 2010). It also assisted in the movement of refugees until the League of Nations' High Commissioner for Refugees was established. Additionally, it facilitated a conference in 1938 to enable collaboration on bilateral migration agreements. During the meeting, the Permanent Migration Committee was established, which convened a meeting the next year on how these agreements could be financed. However, as WWII began, the ILO realized that the issue of migration would be much greater than employment and settlement. Orderly migration would be necessary to realize the peace and social justice needed after such a war. It suggested a plan which would establish an ILO Migration Administration and the constitution of a Migration Aid Fund (Karatani, 2005). The ILO saw the need to assist in migration not only across the European continent, but across other regions as well. But the proposal would not be accepted by the Americans during talks. The ILO's operation emphasized providing non-binding standards which recognized the sovereign rights of all nations to determine their own migration policies, but its strategy was to "sell" individual rights to states and bypass their direct engagement (Kneebone, 2010). The ILO could not sidestep the negotiations for its own expansion, however. The ILO's suggested programs at the Naples negotiations were seen as too expansive and international, but the plan backed by the USA and presented at the next meeting in Brussels was intergovernmental and had a much more limited mandate. The US backing established the Provisional Intergovernmental Committee for the Movement of Migrants from Europe (PICMME), which would ultimately become IOM (Karatani, 2005).

The PICMME was established in 1951, but began its activities in 1952 under another name, the Intergovernmental Committee for European

Migration (ICEM). Its constitution was adopted in 1953 and came into force on November 30, 1954. The constitution outlines its purposes and functions as well as its membership and organization. The mandate, as defined in Article 1 of its constitution, explains that the organization shall make arrangements for the organized transfer of migrants to countries offering to house them, to assist refugees and other displaced persons in the same manner, to provide medical assistance, language training, and assimilation services when requested by the states concerned for voluntary repatriation, and provide a forum for states and other international organizations to promote cooperation in the coordination of efforts and development of practical solutions. This operating mandate is quite broad. Although its name change implies that the organization only works on European migration, its constitutional operations do not specify a timeline for assistance, who it can or cannot assist, or in what region it can work.

IOM expansion to date

Operational expansion

At the outset, the ICEM was situated to facilitate any migrants anywhere around the world, and it did. While its first task was resettling those from WWII, this did not limit its work to Europe. In its first decade of operations, ICEM arranged for the processing and emigration of over 406,000 refugees and displaced persons from Europe to other nations overseas, such as Uruguay. During this time it also assumed responsibility for 180,000 Hungarian refugees. By the 1960s, it had already assisted over a million displaced persons. In 1964, the ICEM developed a program to place highly skilled emigrants in the developing countries of Latin America and organized the resettlement of 40,000 Czechoslovakian refugees from Austria (IOM). Unlike UNHCR, ICEM's mandate allowed for it to assist refugees and non-refugees who were also called "surplus workers" in Europe (Karatani, 2005). Any populations in the "surplus" are often considered an economic threat – especially at a time when Europe's economy was in slow recovery. ICEM's efforts expanded with the political turmoil of the 1970s. It began to resettle Jews from the Soviet Union, resettled 130,000 persons from Bangladesh and Nepal to Pakistan, evacuated Asians from Uganda, helped resettle 31,000 Chileans in other countries, and initiated a program to resettle Indo-Chinese refugees and displaced persons. By 1980, the organization had helped to transport and relocate over three million migrants (IOM). The expansion throughout the mid-twentieth century had more to do with operations and less with mandate. The agency changed its name to the Intergovernmental

Committee for Migration (ICM) to reflect this. ICM's responses continued to expand into what were called Migration for Development programs in Africa and Asia. By 1985, ICM had assisted four million migrants. It would change its name again in 1989 to the International Organization for Migration (IOM). In the 1990s, IOM would become involved in the repatriation of migrants stranded by the Iraqi invasion of Kuwait (including 800,000 Kurds), organize the return of the displaced from wars in Mozambique, those fleeing the Rwandan genocide, refugees from Chechnya, Hondurans needing assistance after Hurricane Mitch in 1997, and Kosovar refugees in 1998–1999. In the last decade, IOM has been there to assist with refugees and displacement in East Timor, India, Pakistan, Sierra Leone, Afghanistan, Thailand and Laos, Myanmar, Indonesia, Sudan, Somalis in Kenya, and many more. What is evident from this growing list of operations, which is only a part of IOM's work, is that as global crises increased, IOM has been there to facilitate the movement of the displaced. What is also apparent is its growth into aiding natural disaster displacees in addition to those affected by conflict; specifically, the Haitian earthquake and the Pakistani floods in 2010.

Institutional expansion

As referenced in the previous section, IOM went through several name changes to reflect updates in its work and mandate. Weiner (1995) explains that IOM initially focused on the movements of populations from Europe to North America and Latin America. But by 1980, its work had expanded worldwide. An amendment to the constitution in 1989 eliminated all geographic limitations and broadened the range of its activities. In its institutional expansion, there are some similarities to UNHCR; however, IOM's constitution does not activate the same type of mandate as UNHCR, in that its mandate is not legal, but functional. IOM asserts that its activities do contribute to the protection of human rights, and it also uses the language of humanitarian assistance to describe its work, which has troubled other agencies that have a longstanding association with humanitarianism. Criticism stems from the fact that IOM lacks the proper mandate to act in this area and that it engages in activities which violate the human rights of migrants (Andrijasevic and Walters, 2010). Assisted voluntary returns are facilitated by IOM and receiving countries which have denied asylum. It has often been described as a way to achieve justice for those who have been forced out by war but are not refugees. The assistance IOM provides is short term and piecemeal. It cannot reverse illegal expropriations or ensure that the returnee will be treated well upon return to his/her country of origin (Webber, 2011). In addition, there is only so much that is "voluntary" when a country threatens to forcibly repatriate; options are limited. IOM will not physically remove people, but it offers temporary assistance rather than forcibly deporting people. It is also in agreement with the many governments who view those who are rejected for asylum after the appropriate legal procedures as illegal aliens. Even though their bid for asylum was rejected, it does not mean that the situation back home has in anyway improved, and thus they may face hardship or poor treatment if returned. While this is a concern for IOM and many of its member states, there is no legal recourse for returning someone if they are now an alien (Weiner, 1995). As stated in the previous section, IOM came from a bargain which needed US backing; it was always intended to be an economic counter agency to the humanitarian UNHCR. They are neighbors, but serve different functions (Duvell, 2005). While IOM may be sympathetic to humanitarian interests, its institutional functions, even in its expansion, should not be misinterpreted. A main reason for such a misinterpretation is that although IOM is a major operator in the field of international migration, there is surprisingly little academic research on the agency itself. Migration scholars routinely use the research material it produces, but rarely is IOM the object of research (Andrijasevic and Walters, 2010). IOM's reports and policy recommendations have been used in this investigation, but research about the development and evolution of IOM has been hard to find. Andrijasevic and Walters (2010) argue that carefully interrogating this agency will lead to a better understanding about the ethos and rationality of international governance. IOM works in the realm of power relations, tactics, and maneuvers between its member states. If immigration and migration is so highly researched, so should be the institutions which promote standards and communicate norms about border controls. However, migration is not "traditionally an area of interstate cooperation and is not governed by a single applicable entity" (Ionesco and Traore Chazalnoël, 2015). It is important to keep this in mind as IOM can facilitate the development of frameworks in this area, but is not a treaty making body.

Andrijasevic and Walters (2010) use Duvell's discussion on IOM to explain how the institution sees itself and manages the role it has chosen. However, Duvell challenges its technocratic self-representation with its sometimes violent activity of deporting people while calling this "assisted voluntary returns". He has also pointed out that its main goal is to align the migration policies of the global South with the control norms of the global North. Its task is the sorting of mobile populations into streams of the useful and useless, admissible and returnable, and employable and deportable. IOM is not under any mandate, nor has it expanded to adequately take human rights into account. This is not to suggest that the organization does not care about migrants, but its work is more clearly understood as managing the processes of migration – not specifically protecting migrants. Its mandate is not normative. Ashutosh and Mountz (2011) provide one additional piece of research on IOM's brand of migration management. The authors argue that the organization maintains the role of nation-states by ordering global migration flows. It acts on the behalf of nations using the language of human rights as if it were working for the benefit of migrants, but it is ultimately to benefit the state. These institutional goals (orderly migration and upholding human rights) are not mutually exclusive, but are different. Andrijasevic and Walters insist that IOM's work shapes and defines the way in which states understand borders and create their policies; its institutional role has developed to become constructive and constitutive. In this way, its open mandate has allowed it to act independently and expand its efforts to become a player beyond a consultative/operational figure. One way that is has done this is through its commitment to what it calls "frontier strategies" which incorporate control functions to non-border settings and include the harmonization of travel documents. The authors describe the role IOM has taken on as that of the entrepreneur when bringing together states and other actors to negotiate, identify opportunities, and implement support programs. In these ways, IOM has expanded its role from interstate facilitator to the specialist in migration management.

Extensions of the migration regime

Even though there is not much of a migration regime, certain tools have evolved to protect those on the move. Koser (2010) divides these into two sets of instruments. The first set includes the core human rights treaties such as the International Covenant on Civil and Political Rights (ICCPR), the International Covenant on Economic, Social, and Cultural Rights (ICESCR), the Covenant against Torture (CAT), and the International Convention on the Elimination of All Forms of Racial Discrimination (CERD). Each of these offers many freedoms, such as life, liberty, freedom from discrimination, freedom to choose one's job, and the like. The second set consists of the 1990 UN International Convention on the Protection of All Migrant Workers and the Members of Their Families; the Protocol to Prevent, Suppress, and Punish Trafficking in Persons, Especially Women and Children; and the Protocol Against Smuggling of Migrants by Land, Sea and Air. Even with all of these conventions, the human rights protections of migrants are much less developed than those of the international refugee system. Moreover, these instruments are very broad and do not get into identifying migration other than for work purposes. Again, economic opportunities are but one of the overlapping push-pull factors for migration. One could argue that their breadth should be enough to cover many forms/categories of migration. Koser adds that the Convention on the Protection of All Migrant Workers and the Members of Their Families has only been ratified by 42 states, none of which is a major destination country for migrants. Thus even if the breadth of the instruments appears to assist the many, not enough states have ratified the agreements to bring them into force. As largely unratified documents, they are little more than aspirations to protect migrants in any form.

Climate change and IOM

Since 2008, IOM has been publishing research on how climate change can affect migration. These papers and books include studies, brochures and informational sheets, and the Migration Research Series (MRS). Primarily, there are two types of publications which form IOM's archives: those which IOM publishes but are written by outside experts and those which are official IOM documents. The reason for the use of a dichotomy to describe IOM's publications on climate change is that it sets up a lens through which to view what IOM endorses and what actions it takes.

Non-official publications

IOM's MRS "presents the findings of research projects managed by IOM's Research Unit in Geneva, and studies prepared by IOM staff and its field offices. The series is designed to bring the results of policy-relevant migration research to the attention of a broader audience more quickly than would be possible in academic journals and books" (IOM). The series covers varied topics on migration and presents new research. The series has also presented documents on climate change. Another type of research that IOM publishes and that is not always official are assessment reports. One was published in 2010, entailing environmental changes and vulnerabilities in the islands of Mauritius. It is also part of the IOM's repertoire on climate change and will be discussed below. Another significant publication that is not official and yet is written and funded by IOM staff is the *Compendium of IOM's Activities in Migration, Climate Change and the Environment* (2009). The publication was compiled through 32 country offices and missions, in conjunction with IOM's Migration and Policy Research, Emergency and Post-Crisis Management, and Migration and Climate Change Focal Point. Its Foreword is written by IOM's director general, William Lacy Swing. It presents IOM's role over the years in the areas of migration, climate change, and the environment, and includes country program profiles which also contain IOM responses in these countries along with project proposals.

The Compendium begins with an evaluation of the nexus between climate change, environmental degradation, and migration. It notes that even though predicting the details of climate change remains difficult, the probability is very high that there will be an increase in those migrating for environmental reasons. It also hails migration as a necessary strategy for adaptation, as it alleviates pressure on population and land use, and one that needs to be adequately managed so that large-scale movements do not lead to the overexploitation of resources in other areas. Much of this is echoed in the other publications. However, the Compendium also states that migration, as a coping strategy, is not open to everyone; this depends on resources, information, and social and personal factors. The most vulnerable are those who cannot move. While this is the case now, there is no indication that management, the way it is being used in this context, means finding ways to include those who cannot move without further help. The word "management" is continually used to indicate the role IOM sees itself having in climate change migration. Again, it is emphasized:

> IOM is making the case that migration in the context of climate change does not necessarily have to be a worst-case scenario. . . Yet, for migration to become a viable alternative – an adaptation strategy that increases the resilience of vulnerable populations – environmental migration needs to be managed, in particular with a view of enhancing positive and sustainable outcomes.

At first glance, this can give the appearance that IOM will assist those who will need to move and that they will be managed. Conversely, the approach outlined emphasizes IOM's objective that migration should be a choice. If so, then its work on climate change and migration is reactionary – IOM acknowledges that migration cannot be open to everyone; functionally, it has to wait for individuals to move before they are managed.

The Compendium also consists of regional and state-centric evaluations of IOM's activities in the realm of climate and migration. There are only two projects that have a direct connection to facilitating migration as opposed to responding to it: the voluntary relocation of vulnerable communities in Madagascar, and the Framework to respond to mass migration in Trinidad and Tobago. Madagascar, located off the east coast of Africa, is very vulnerable to tropical cyclones. Their increasing intensity motivated several communities to relocate – regardless of their attachment to their ancestral land. Two entire communities submitted formal requests for assistance to Madagascar's disaster bureau. In turn, the bureau solicited support from IOM. However, the project was first proposed as part of the global early recovery plan led by the UN Development Programme (UNDP). For now, the Pilot Voluntary Relocation of Selected Communities Affected by Cyclones and at High Risk of Further Flooding and Erosion in Madagascar is still a proposal, with an estimated budget of US$ 2,052,467. The Framework for Emergency Response to Standard Operating Procedures: Mass Migration Emergencies, however, has been formally completed. In the case of Trinidad and Tobago, IOM, in consultation with the government of Trinidad and Tobago's Office of Disaster Management and Preparedness (ODPM) and the Ministry of National Security has compiled a manual to guide the development of a framework for standard operating procedures (SOPs) to assist in the building of technical capacity for migration management. The manual is specifically tailored to improve the response after a natural disaster forces those from Grenada and Guyana to out-migrate. Trinidad and Tobago is a destination country and one of the four Subregional Focal Points in the event of a disaster. The manual is a framework for orderly migration in states of emergency, which means it is for a sudden impact event. While these projects are a starting place for IOM, they are still not clear examples of IOM taking on the challenge of climate migration head on. For Madagascar, IOM was invited to help, but the project itself was initiated by UNDP. In Trinidad and Tobago, creating a manual which produces a framework for an SOP is different from working on the actual SOP.

Migration and Climate (Brown, 2008) considers how climate change will affect forced migration, incorporates climate prediction, assesses implication for development, and recommends policy responses. This publication skims the surface of the connection between climate change and migration, but represents the beginning of such research at IOM. With this in mind, its policy responses do identify the gaps through which those affected may fall. However, Brown's last recommendation does not specifically deal with climate change. He states that the international regulation of migration, adaptation to climate change, and capacity building in vulnerable countries are intertwined. Because of this, migration will be used by individuals to adapt to climate change. He advocates for policies which promote workers to stay in their home countries, while not entirely closing the door on international labor mobility. While labor migration may very well be the first form of climate migration as certain livelihoods are eroded, there have been no connections made which implicate this in highly skilled occupations. It appears to be a reference to the brain drain problem experienced by developing countries as their most educated leave for better opportunities in more developed countries. He seems to be concerned that opening borders to those affected by climate change will accelerate this drain. As the last conclusion in a publication about climate change and migration, it seems a bit out of place. However, it can been seen as a political statement; it's a recognition that the implications of the report (i.e. the work and money necessary to fix the issues identified) are tied into current migration types.

Climate change, migration and critical international security considerations (McLeman, 2011) is authored by a geographer from the University of Ottawa. This publication also considers how climate forecasts can assist in understanding climate migration and discusses this type of migration as a phenomenon; in contrast to the previous publication, the author more specifically identifies the regions that are at most risk for environmental damage and subsequent potential migration. McLeman also briefly discusses the nexus between labor migration and climate migration, explaining that they are those on the lowest end of the socioeconomic spectrum – not highly skilled workers. The chapter on policy specifically focuses on what can be done to avoid distress using policy as a tool for management. Political will is identified as the main barrier to action – not technological know-how or socioeconomic necessity; developed nations have what they need to mitigate climate damage. This is also a criticism of the governments of developing nations in Africa who lease out arable land and fishing rights to Asian conglomerates which essentially strip-mine these areas, killing their long-term sustainability for local livelihoods. In addition, McLeman delves into the instruments which are commonly called upon for preliminary climate migration management. He disqualifies the use of the 1951 UN Convention on the Status of Refugees and points out that although the *Guiding Principles on Internal Displacement* explicitly include those displaced by climate change, signatories to the Principles are not bound to enforce them in any way. Finally, the author argues that national sovereignty, in a more exclusive sense, has taken priority over humanitarian principles; again, the main problem being political will. He also advocates for an internationally binding treaty to protect those displaced by climate change. His last point is a personal statement about his work and the urgency of this issue.

It should be apparent that there is a considerable difference in tone between these two volumes of the IOM MRS publications. They both begin to connect climate science to migration literature and evaluate what kinds of migration flows may be seen. However, their approaches to evaluating policy are very different. Brown's MRS paper evaluates policy through a state-centric lens, while McLeman's approach favors the international perspective. Additionally, McLeman's work is much more critical of the lack of action by individual nations and the international community. Brown's MRS report is more statement, while McLeman's carries an edge of advocacy.

Gemenne and Magnan (2011) produced an assessment report on the current migration issues due to environmental degradation in Mauritius. The report includes interviews and field visits to adequately evaluate the situation on the ground. It was funded by the IOM office in Mauritius, supported by its office in Geneva and the IOM's regional office located in Pretoria, South Africa. The study differentiates the impacts of climate change from other environmental changes, outlines vulnerabilities to climate change, and provides a thorough evaluation of the current ways in which those on Mauritius have had to migrate due to environmental changes. The focus groups conducted by the study reveal that the sea has reclaimed enough beach in Riviere des Galets to have affected the use of that area by locals and tourists alike. They are aware of the eventual necessity of relocation, but are reluctant and generally unwilling to do so. On the adjacent island of Rodrigues, the fishing has deteriorated and many have already thought about migrating to the mainland of Mauritius, but are reluctant to do so due to cultural differences. In Cite Lumiere, the government has already started the process of resettling residents out of the slum as its increasing floods have posed health risks and difficult living conditions. In concluding the focus groups, it is noted that some populations will need to be moved and others have already done so; resettlement schemes are ad hoc and do not apply the same standards evenly. Recommendations state that inter-island migration flows need to be better managed and harmonized. The report goes on to propose a framework for pilot projects for adaptation to current and future environmental changes. The main point is that there are no migration projects suggested in the framework. The 13 that are listed are explained as examples, but considering it has been noted that migration is already occurring, it is remarkable that none are for migration itself. The projects include sea salt production, mangrove restoration, eco-tourism, and roof-top gardening, to name a few. These are adaptation projects which will extend the time that the vulnerable areas in Mauritius are livable, deterring and preventing migration, not facilitating it. This report focuses on the development of the kinds of mitigation and adaptation projects preferred by environmentalists, not migration scholars. It could be argued that when it comes to climate change, IOM's early foray into independent research demonstrates an interest in preventing migration than in assisting it (2014 Outlook Doc).

IOM authored publications

The final set of publications related to climate change and migration, as per IOM, come from documents which can be considered official. *Disaster Risk Reduction, Climate Change Adaptation, and Environmental Migration* presents IOM's efforts to assist vulnerable communities affected by environmental hazards through disaster risk reduction (DRR) and climate change adaptation activities. It argues that migration and environmental migration need to be integrated into sustainable development strategies in order to be properly managed. The document itself is an informational piece for stakeholders and IOM members. The text openly acknowledges that IOM considers most current and developing environmental migration to be a part of a slow onset process and that in worst case scenarios, relocation, either internally or to a third country, may be needed. It also reiterates the point made in the non-official documents that migration is not an option open to everyone, that the most vulnerable are not able to move. IOM also points out that climate change is increasing the vulnerabilities of communities around the world and leading to increased migratory flows; IOM states that because of this it places high priority on addressing environmental migration. Here IOM demonstrates that it considers and integrates climate migration into the sphere of environmental migration; it is not addressed as a separate issue. IOM explains its response efforts as working to increase communities' resilience to risk factors and changes in their environment, with an emphasis on empowering local actors to develop capacity. This response is based on its migration management cycle, which consists of five steps: (1) preventing, (2) preparing, (3) managing, (4) mitigating, and (5) addressing migration. IOM conceptualizes this as a circular form, with each step leading to the next. The diagram emphasizes under step 1 that "IOM's foremost objective is to reduce unmanaged migration pressure, preventing forced migration while also ensuring that the migration taking place is managed." Management, the buzzword also used in the non-official documents, is used to denote addressing humanitarian needs, protection, mitigating the impact of migrants on destination communities, and looking for durable solutions. However, management, like in previous publications, is required and needed only after displacement occurs.

The document also makes an important point about IOM's function. It states that IOM's responsibility is to support states in strengthening capacity and institutions to respond to emergencies; however, it can and will substitute the state's role (upon request) in cases of "imminent or ongoing humanitarian emergency". This was the case after the devastating 2010 floods in Pakistan. Climate change, as it affects the slow deterioration of living conditions, is an imminent and ongoing humanitarian emergency. Considering it as such can put IOM in a position of taking the necessary proactive role of facilitating relocation, if it so chooses.

In 2013, IOM conducted a survey on environmental migration among its Missions worldwide. It acknowledged that migration related to climate change is an established reality in many countries. The respondents in this survey found the topic of climate induced migration important, but the survey only concluded much of what IOM has already written in terms of the need for better preparation and awareness for environmental migration as a whole. It also produced an info sheet titled *Capacity-Building Activities on Migration, Environment, and Climate Change* (2014), which outlines the trainings it has been conducting. It outlines five different programs and how IOM aims to respond to member states' needs for training tools and experts. Its main goal in this endeavor is to develop a corresponding training manual and build staff capacity. In the same year, it published the *IOM Outlook on Migration, Environment, and Climate Change*, a 144-page document containing 14 separate briefings on IOM's approach to this topic, legal frameworks, state of knowledge, as well as development and humanitarian policy. Brief 5 explains the state of knowledge on migration, the environment, and climate change. In all of its information, it uses the IPCC's reports to identify how climate change will affect the movement (greater frequency of sudden and slow onset events, changes in livelihood security, rising sea levels, and competition over shrinking resources). Additionally, it says that some changes may prevent migration (citing a 2011 study out of the UK Office for Science) and that when displacement occurs it may be short term, long term, and for other durations. Most telling is Brief 14 on IOM's operational responses to environmental displacement. This chapter cites several case studies, one of which was mentioned earlier in this book – the relocation of the population of the Carterets to Bougainville. As a part of IOM's operational response, they cite this slow onset event and state what they have done to assist. In this case, IOM has only recently become involved with assisting the Bougainville government (as PNG has only become a member state in the past couple of years) by doing a vulnerability assessment of the remaining communities to environmental issues, as well as the need for them to relocate – temporarily or permanently. Additionally, IOM will develop and test research methodologies to train researchers in the field to produce vulnerability and resilience maps.

Concluding remarks

IOM, the migration agency, has a mandate to assist migrants and manage migration flows. Consequently, it is much better situated to deal with migration due to climate change than UNHCR, whose main business is to legally protect refugees. Its membership has continued to grow over time, and currently stands at 157 member states. In its favor are increasing levels of expertise dealing with natural disasters; it has also taken an interest in migration due to climate change and sees it as a growing phenomenon which will need to be dealt with. However, its integration of climate change adaptation into DRR places its efforts in the category of short-term acute efforts that IOM is used to. IOM (and even its non-official authors) focus on building capacity for individual states to deal with migration. The majority of its activities serve as an alternative to permanent migration due to natural disasters, such as suggesting temporary and circular migration strategies to support seasonal livelihoods. However, these proposals are not viable for those whose land is eroding from under them due to sea level rise or desertification. DRR and management are important tools for sustainable development, but fall short in terms of the slow onset disasters that IOM recognizes as the majority of what is happening.

Additionally, IOM's usual focus on management, even in this area, is not adequate when slow onset disasters are being considered. Its reactionary response to managing displacement only involves IOM *after* something has forced people to migrate. Furthermore, it has emphasized that although it seeks to allow migration as a chosen adaptation strategy for climate change, it repeatedly states that not everyone can migrate, because of individual resources. Is migration a choice for those with money? If so, they are not preventing forced migration, but allowing many to be left behind. The language of prevention and management are reminiscent of Chimni's (1998) discussion of how the global North views refugees; excess or unintended movement can be controlled. This indicates that the values and political discourse of the North is guiding IOM and its current response to climate change; the goal is to keep unwanted migrants from coming at all, which can be achieved by investing there so they don't necessarily leave their home. This includes DRR, management, infrastructure, and capacity – exactly what IOM is advocating for and simply adding climate change adaptation into. Only McLeman, who is not attached to the organization, suggests that the problem of climate change induced migration may need a bigger international solution. Thus far, IOM has been engaging its member states in different policy forums but the impetus for action (proactive or reactive) needs to be initiated by members. This has accelerated since 2007 as some are seeing this topic as one that needs further discussions. This lead to the 2011 Internal Dialogue on Migration workshop which disseminated policy recommendations from IOM to its members and the decision to create the Migration Environment and Climate Change Division and its recognition in 2014 at the 105th Council session (Ionesco and Traore Chazalnoël, 2015).

The evolution of a modern humanitarian regime

Humanitarianism, at its core, does not decipher between who is on the side of right or wrong – it seeks to eliminate the majority of suffering along the way. However, it does not attempt to alter the order of things, which is the job of politics. Pure humanitarianism works to assist all mankind through several principles: humanity, impartiality, neutrality, and independence. These command attention for every human, separate humanitarianism from politics, and demand that assistance is based on need (Barnett, 2005). It is driven by human sympathy and the obligation to better the human condition and is guided by the mantra of "do no harm" (de Waal, 2010). Humanitarian aid is a function of compassion. It is also paradigmatically regarded as a state of exception (de Waal, 2010) or humanitarian space (Hilhorst and Jansen, 2010) where humanitarians can work without the interruption of politics or outside forces to provide aid and follow humanitarian principles. This space is metaphorical and physical, as it also marks the camps and tents in which aid is given.

Humanitarianism is also understood in two main veins: assisting those affected by both natural disaster and war. Much of the literature on humanitarianism intertwines the two scenarios in its discussion of "aid". They will be disaggregated here for analytical purposes, but it is an important point to keep in mind throughout the rest of the chapter. A natural disaster's impact is primarily the outcome of a physically uncompensated interaction between a natural event and a social system, while a complex human emergency is the outcome of an institutionally uncompensated interaction between a societal event and a social system (Albala-Bertrand, 2000). Complex human emergencies are very much the effects of war, when societal structures collapse and their reconstitution is a threat to a particular vision. This is usually a violent and long-lasting conflict in which there is an eventual political aim. War destroys infrastructure and services, security and safety nets. The outcome is an unraveling of the basic social fabric, which necessitates aid to rebuild individuals as well as communities, and sometimes nations. On the other hand, natural disasters are not caused by any social or societal impetus, but simply consist of patterned responses to changes in atmospheric or geological pressure and temperature. There is no guilty party, because disasters are not personal in nature. When climate change is considered, these two categories can blur together. IPCC predictions clearly show the impact of disasters will become larger and more frequent due to anthropogenic change. In turn, climate change can be seen as a form of complex human emergency due to indifference; large polluting nations are knowingly contributing to the destruction of vulnerable social systems by altering their long-term viability. Thus, while the literature does not always delineate the humanitarian intervention in a war zone from that in a hurricane, this is not problematic if climate change is understood as a function of the two disaster events.

Unlike the refugee regime which originated in ancient traditions, the humanitarian tradition, as an organized entity, is truly modern. The idea of doing something altruistic or philanthropic is not new by any means, but an organized effort to alleviate suffering is. It can be argued that the 1860s produced such a turning point. The work of business-man Henry Dunant, who wrote about the suffering he saw at the Battle of Solferino in 1859, contributed to the founding of the International Committee for the Red Cross and the 1864 Geneva Convention on the Amelioration of the Condition of the Wounded Armies in the Field (Leebaw, 2007). Additionally, this was at the same time as the American Civil War and when Clara Barton's organization of nurses eventually became the American Red Cross. These early humanitarian organizations were mostly concerned with treating injured soldiers – no matter which side of a war they fought on. They exemplified humanitarian principles, but it is important to note that in the beginning aid of this sort was directed solely at medical need. Slowly, two different versions of humanitarianism emerged: Dunantist and Wilsonian. Named for Henry Dunant, Dunantist organizations define humanitarianism as neutral, independent, and the impartial provision of relief to victims of conflict. These organizations are sometimes accused of being "high priests" of humanitarianism, which fear that the relaxation of their principles will endanger their purpose and effectiveness. Alternatively, Wilsonian organizations, named for Woodrow Wilson, believe that it is possible to transform political, economic, and cultural structures to produce peace and progress. Wilsonian humanitarianism seeks to attack the root causes that make populations vulnerable (Barnett, 2005). While both claim to be apolitical, the Dunantists would claim that the Wilsonian organizations are in fact political entities. The International Committee for the Red Cross is considered Dunantist, while Oxfam would be classified as Wilsonian. Ultimately, Wilsonian organizations see value in more than just temporary relief, and while this form of relief is necessary, it can also be a constant recurrence.

During the mid-twentieth century, it was the belief that the responsibility, will, interest, and capacity to assist individuals in a disaster situation were that of the national government of the affected area. Additionally, the significance of national sovereignty reinforced the separateness of each nation in this respect. International interventions occurred in the 1970s, but had proved uncoordinated and ineffective. At this time, international aid was not yet seen as a supplement to domestic aid (Kent, 2004). Kent describes the beginning of humanitarian aid as a "sideshow" to real political concerns; the crises of the 1970s and 1980s were not conceived of as having real political consequences. While there was considerable empathy for those affected in places such as East Pakistan, Guatemala, and Ethiopia, they were defined by the momentum of the Cold War. During the Cold War, the superpowers provided arms to various regimes, but did not intervene directly for fear of direct confrontation with the enemy. In this era, aid agencies had a real necessity to be neutral, especially when assisting those in conflict zones; they could not be seen as pro-Russian or pro-American (Vaux, 2006). Agencies at this time were also highly unrefined. There were relatively few agencies providing relief, they had few interactions, and they did not yet conceive of professional standards. Operations were staffed by individuals with little or no experience who believed that all they needed was a "can-do" attitude and good intentions. In a sense, humanitarianism was not much of a field; those who participated in relief work treated it more like a craft than a profession (Barnett, 2005).

It was only after the Cold War subsided that fragile nations, vulnerable to humanitarian crises, lost their resources and political support. International collaborative support deteriorated as key governments began to disengage; this reflected a lack of interest in continuing to work for the false harmony that existed in the bipolar Cold War world and the sense that unilateral action would be best for individual power interests. These tendencies undermined the role of humanitarian action – regardless of the intentions of the UN and other agencies – and allowed for them to become inadvertent instruments of post-Cold War politics (Kent, 2004). In the post-Cold War era, humanitarian action and space became politicized by several environmental factors. First, geopolitical shifts at the end of the war increased demand for humanitarian action; without state-sponsored aid, unstable domestic situations threatened to become large emergencies (Barnett, 2005). Additionally, state spending on humanitarian aid increased dramatically as nations began to show an interest in utilizing such aid in connection with political goals; it was also seen as a rationale for regime change (Leebaw, 2007). Second, these domestic breakdowns became "complex human emergencies" or conflict-related disasters, which involved a high degree of social dislocation and required a system-wide aid response. Third, there was the political economy of funding; private contributions increased, but not nearly as fast as official assistance, with the United States as the lead donor. Political motives fueled this increase in giving, and conditions were often placed upon such aid. Finally, there was also a change in the legal environment; the concept of state sovereignty was becoming conditional, based on accepted behavior to one's own people (Barnett, 2005).

Humanitarian aid began to be viewed by states as an opportunity. This was exemplified by the US intervention in Mogadishu, Somalia, in 1992. Provoked by the potential mass starvation of 500,000 Somalis, the US military prepared Operation Restore Hope to provide logistics, security, and support to relief agencies who were attempting to provide help in the chaos of civil strife. One specific battle in Mogadishu garnered the world's attention in 1993 as 19 US soldiers were killed. This loss initiated the Clinton administration's disengagement in the situation (Kent, 2004). Additionally, the United States' loss in Mogadishu served as a blow to humanitarian involvement/intervention by other nations as well. By 1994, no government was willing to step in and prevent the planned genocide in Rwanda, and over 800,000 Tutsis were slaughtered. In the aftermath, humanitarian assistance poured in and it has been argued that it was used as an apology for the international community's unwillingness to act. Additionally, assistance was used as an alternative to political action in the former Yugoslavia. There it was used as filler, to plug policy gaps when the major powers could not agree on a course of action. One UN official called this "containment through charity" – a true politicization of humanitarian aid (Kent, 2004).

Politicization was not solely an issue that developed in state-sponsored giving; it has also become a major driver of aid assistance from NGOs, as well. Donor nations can and often do use subtle, indirect methods to guide aid where they wish it to be directed. These include bowing to international pressures, using charitable giving in videos that favorably sell the war at home, to win hearts and minds; in this, the USA has donated much more than others and guides much humanitarian aid for its own interest (Barnett, 2005). Individual donors have specific motives as well. Donors want to know that their money is being spent in accordance with their intentions – no matter if these do not align with need. For example, the Asian tsunami of 2004 evoked massive public support and response, but such high levels of support are not seen for every humanitarian challenge. People suffering in situations which have a low media profile also get less help than others in the opposing situation, and thus aid is more closely related to donors' interests than wider need (Vaux, 2006); these interests include basic charity, while others donate to assist one side over another in a conflict. Donors wish to know that their money has been distributed in the manner they see fit, which means they donate to a particular disaster of interest, not humanitarian aid as a whole. Thus while the highly visible disasters garner the aid they need, other more serious situations can still struggle for funds. Humanitarian organizations do not survive on good intentions alone, but are eventually steered by resources controlled by others. Ultimately, the dissemination of ideas, allocation of resources, and implementation of projects all take place as subtle power processes (Hilhorst and Jansen, 2010).

Humanitarianism and human rights

As the humanitarian regime developed, the general conception of charity in which humanitarianism tends to be situated began to be questioned. The human ideal had been defined through the development of human rights instruments, which continue to fail to live up to the realities of the human condition (de Waal, 2010). Contemporary formulations of humanitarian intervention try to fuse the urgency and immediacy of rescue with claims of justice that are seen in human rights (Leebaw, 2007). Walzer (2011) describes the ancient Hebrew political tradition of obligatory charity; the word used for "charity" comes from the same root as the word "justice", which is suggestive that charity is not only good, but also right. He argues that if humanitarianism does not connect with justice, then it is not what it should be; that it would be wrong not to act in such a fashion, and in doing so it is more like justice rather than benevolence. Accordingly, intervening to assist those affected by a natural disaster or war is just, and the idea of the right to humanitarian assistance was within reach (de Waal, 2010). Rights-based programming is now used by many humanitarian organizations which highlight the degree to which a person is denied or enjoys their rights as a basis of vulnerability (Linde, 2009). Connecting justice and a right to aid reflects the Wilsonian view of humanitarianism; real material improvement can be attained if aid is used to ensure people's rights. Greenwood (2010) explains that humanitarian law is the older of the two legal frameworks; references go back to the Bible, early codes of Hindu law, and the Koran. Its original principles were quite primitive and only applied when you were fighting people within your own community. Its primary function was to provide guidance to the military as to how to protect human values in the most inhuman of environments – war. Although there are traces of human rights law in the early twentieth century, it is only since WWII and the Holocaust that a body of law has emerged that established how a state should treat its people. Both frameworks apply directly to individuals and impose obligations on them.

However, there is also a situational view of the application of humanitarian and human rights instruments. With the origination of the regime, international humanitarian law is often assumed to only be applicable in times of war, while international human rights law is also applicable in times of peace (Laucci, 2009). In opposition, Barber (2009) argues that while humanitarian law does apply in times of conflict, human rights law applies in times of peace and in times of war – it trumps humanitarian law. While this dichotomy in theory is the result of two extreme visions, the separation between them can be questionable as to the separate applicability of each. First, no one knows when war begins and ends anymore; formal declarations of war and peace treaties have fallen out of fashion. Second, the theory does not reflect relevant human rights treaties. Finally, the theory runs contrary to what international courts have stated – human and civil rights do not cease to exist in times of war. As one can clearly see, the debate between when to apply human rights or humanitarian law directly relates to the area of war, not to other situations of humanitarian need. If one considers disasters such as famine or an earthquake, legal frameworks are still relevant, but in a different way. There is no rift over which side has the prevailing moral high ground. The principles of humane treatment and basic rights overlap, although the standards vary; humanitarian assistance is provided at a lower level than aspirations for human rights would propose. Humanitarian assistance is also a temporary solution. Thus while this form of aid aspires to provide for need, rights become important after needs are met, if levels of deprivation are considered. For humanitarian aid to develop into providing the physical components expressed as human rights, there needs to be long-term cooperation with development, as per the Wilsonian view. Reflecting upon the numbers of people around the globe whose governments currently cannot provide them with their rights, there continues to be a need to implicate humanitarianism as a supplement to human rights. Thus they are intrinsically intertwined as legal regimes and in practice.

The development of UN OCHA

As previously mentioned, the development of institutions at the intergovernmental level to assist in the work of humanitarian efforts is truly modern. The success of the UN in other endeavors allowed some to question what more it could do. In the early 1980s, pressure began to mount on the UN to increase its capacity to deal with disasters and emergencies principally through coordinating humanitarian responses. This was in hopes of avoiding more of the irrational, ad hoc responses of the 1970s. The UN Charter provides for three responsibilities: peace and security, economic development, and human rights. The addition of a fourth pillar was originally considered a dangerous development in that it could seriously jeopardize the effectiveness in its core functions (Kent, 2004). With Cold War conflicts and their inevitable displaced populations, UNHCR was struggling to keep up. Additionally, as the agency grew and professionalized, it set a standard which others could see as useful in other areas, especially humanitarian responses.

This began to change in the fall of 1991. There was a growing recognition that the UN system needed a stronger coordination mechanism; duplication of efforts from additional agencies had proven inefficient, and yet humanitarian crises only got more complex. The political issue at hand was the right for humanitarian assistance to be delivered to individuals while still respecting national sovereignty. This ended in a General Assembly Resolution (46/182) which was adopted by consensus and which set out guiding principles for UN assistance for those affected by natural disasters and other emergencies, and fell outside the legal mandate of UNHCR (Helton, 2001). These guiding principles include a reference to national sovereignty, stating that assistance should be provided with the explicit consent of the affected country. The resolution also affirms that the first and foremost responsibility for disaster victims is of the state in which they reside, which has the primary role of initiation, organization, and implementation within its territory. Finally, it acknowledges that states whose populations are in need of humanitarian assistance need to facilitate the work of other organizations which will be implementing necessary assistance (Barber, 2009). The resolution provided for a senior official to coordinate relief efforts and states that humanitarian assistance should be provided with the consent of the affected country (Helton, 2001). To develop the leadership role, the UN decided to take on a separate department that was established within the secretariat – the Department of Humanitarian Affairs, led by an under-secretary general with the title of "Emergency Relief Coordinator".

With the mounting challenges of Bosnia and Herzegovina and the Great Lakes region connecting issues of humanitarian relief as well as growing numbers of refugees, the UN secretary-general proposed a reform to this system: to integrate the Department of Humanitarian Affairs into UNHCR. This would have made UNHCR the permanent lead agency for all humanitarian disasters. Many were opposed to this proposal, including the World Food Programme and UNICEF. Instead of creating an integrated institution, the secretary-general decided to keep the agencies separate, and the Department of Humanitarian Affairs was renamed the Office for the Coordination of Humanitarian Affairs (OCHA) in 1997. OCHA was to have three core functions: coordination of humanitarian emergency responses, policy development and coordination, and the advocacy of humanitarian issues (Helton, 2001). More specifically, in order to coordinate international response, its work includes contingency planning, such as consultation with the countries concerned to reach agreement on priorities. In terms of being an advocate, it is concerned with reflecting the need for recovery and peace building. The reform package was also a way to push back the mission creep that others perceived with the Department of Humanitarian Affairs even when it was filling gaps between agencies, as it was seen as being in competition with other similar institutions (Helton, 2001). Kent (2004) explains this political wrangling as a case of the UN becoming overly absorbed with its own domestic harmony rather than developing the leadership and coordination roles offered by the General Assembly. He also critiques the emergency relief coordinator for rarely challenging the donor community in order to provide more equitable and consistent relief. Ultimately, its development is mired in the basic drive for institutional survival.

Extensions of the humanitarian regime

While the humanitarian regime is still quite young, it has not developed without its share of growing pains. As previously mentioned, it began with a distinct dichotomy between simple basic humanitarian aid and humanitarian intervention. This changed as the Cold War thawed. No longer were NGOs kept at a distance from conflict situations and the high politics of dealing with such areas. Aid itself began to have political and partisan prerogatives, or it was at least so perceived to be. Additionally, NGOs began to look critically at how their aid impacted the areas it was intended to help. In many cases, such aid inadvertently exacerbated existing tensions and divisions between rival social/political groups (Bock, 2011). Another major realization with regard to humanitarian aid was that many large humanitarian NGOs encountered high staff turnover and frequent reassignment, which makes organizational learning difficult. Such disruptions affect institutional memory, especially when disasters pull people into emergency responses (Bock, 2011). This is exemplified through Messina's (2007) concerns with the Humanitarian Coordinator System. Humanitarian coordinators as well as regional coordinators are essential to the organization of aid responses at the top-down level. This system is officially developed under OCHA, but is essentially important to all UN aid. Messina argues for the need to develop an understanding of how NGOs and the UN system cooperate. Also, the author argues for an update to the format of the annual retreat which will allow peer to peer exchanges of information and experiences and for the integration of regional workshops for humanitarian coordination. Finally, Messina's department at OCHA will draft policy papers on key issues to further inform such employees. Humanitarian assistance has also become highly competitive and has grown as a percentage of development assistance. It has increasingly become the only form of support some nations receive. And as these budgets increase, enterprises such as gender sensitization and livelihood support get lumped under humanitarian aid. While not always humanitarian, though well intentioned, these additional projects can threaten traditional humanitarian projects in that they can stretch many organizations too far. Additionally, donors may have specific objectives, such as projects that do not directly fit into traditional humanitarian project work but can be funded through humanitarian budgets. This causes overlapping plans, duplication, and fissures where there should be coherence. Another issue that has come with growing aid budgets is that while this has been a great success, it has caused a demand for professionalization and well-rounded permanent structures which maintain their capacity between crises. This creates a circular dilemma; like a fire station, these structures need continuous money and resources to be able to be efficient (Kent, 2004).

Coordination is an important concept. In part because it is an imbedded feature of OCHA, the meaning of coordination has been implied as the organization of humanitarian efforts at the intergovernmental level. However, it can be more complicated than that. Helton (2001) explains that senior UN officials refer to coordination as the "C word" as is usually represents bureaucratic fights over money, personnel, and programs. It can also mean control over resources and programming or merely sharing information and consultation. Coordination has been a success, but one that comes with the risk of territorial disputes between UN entities and those on the outside. Appropriate and expedient responses require an active level of cooperation and coordination, even if the parties involved are not always amenable to each other's organizational whims. Coordination problems are not new to IGOs, and humanitarian aid is certainly not either. There is also a division between the objectives of those in offices and those in the field, the difference between strategic and operational coordination (Helton, 2001). Additionally, the Inter-Agency Standing Committee (IASC) was developed to support the Department of Humanitarian Affairs by facilitating inter-agency decision making. The IASC consists of the FAO, OCHA, UNDP, UNFPA, UNHABITAT, UNHCR, UNICEF, WFP, and WHO.[2] Coordination of so many agencies is bound to be complicated. At the headquarters level, OCHA has the dual responsibility as the undersecretary-general for humanitarian affairs and the emergency relief coordinator who chairs the IASC. In essence OCHA is the overstretched coordinator of all coordinators.

A last expansion of the humanitarian regime is the incursion of the military. Recently, foreign military have assumed additional responsibility for the distribution of disaster aid and emergency assistance. Kent (2004) argues that this creates three problems in the humanitarian context. First, the mixed role of the military puts in jeopardy the very principles that lifesaving aid should be provided to everyone in need and is perceived to be impartial. Second, the lack of distinction between impartial and independent aid workers and the military can create security problems and tensions. Finally, despite the huge increase in humanitarian funding in recent decades, the involvement of the military increases the competition for finite resources. The military and civilian groups are also structured in different ways, which can and often does result in a culture clash where the military sees any civilian as an NGO (Helton, 2001).

This can undermine IGO governance if the military, which can be seen as a threat to certain humanitarian victims, takes primary control where IGO and NGO coordination is preferred. Competition between these two groups is highly counterproductive in a situation of real emergency.

OCHA expansion

Because OCHA is a relatively new entity, it has expanded, but not nearly as extensively as the other institutions investigated. OCHA's mission allows for it to fill in assistance gaps to those who cannot receive international legal protections, but who do need temporary assistance. One more gap that OCHA began to fill was supporting IDPs. In 1996, the UN General Assembly tasked the emergency relief coordinator with a central role in the inter-agency coordination of assistance to IDPs. OCHA advocates IDP issues to member states, donors, and the media, ensures displacement issues are included in briefings to the Security Council, and works with the IASC to address gaps in IDP policy and institutional arrangements (OCHA, 2010). Its work with IDPs can be considered collaborative because UNHCR also heads and has developed IDP projects. However, this process garnered criticism, and thus, in January 2002, OCHA established its own unit for IDPs, renamed the Internal Displacement Division (IDD) in 2004 (McNamara, 2006).

OCHA and climate change

Like IOM, OCHA has begun to research and consider the implications that climate change will have on its work and on human migration. Several publications outline its recent work on the topic. While these papers are few and represent only a beginning, what they do demonstrate is a different attitude toward the impending situation at hand. A joint study by OCHA and the Internal Displacement Monitoring Centre is titled *"Monitoring disaster displacement in the context of climate change"*. The aims of the study were to provide an estimate of the number of people displaced by natural disasters in 2008, a methodology for ongoing monitoring of forced displacement arising from such disasters, and an indication of the resources required to implement the methodology. It does not seek to analyze how current levels of displacement will be affected by climate or what proportion of current displacement can be considered a direct effect of climate change. Instead, it seeks to inform discussion by providing an indication of the scale of displacement from which to start when considering the increasing influence of climate change. The report considers only hydrometeorological extreme hazard events,[3] those which force temporary displacement. The results show that 20,293,413 people were displaced in the 322 sudden onset climate disasters (including hydrological, meteorological, and climatological – excluding drought). Disasters associated with flooding and storms have been found most likely to be major drivers of displacement. Additionally, the mass majority of displacees came from Asia. While the study does not go so far as to attempt to predict how many people will be displaced by other drivers, it does indicate that sea level rise will be a significant driver in the future and highlights that 146 million people live in areas that are less than one meter above sea level. However, the study assumes that return to prior homes will be the most likely durable solution for those displaced by extreme hydrological events, but that resettlement will also be needed.

The Policy Development and Studies Branch presents research and papers which are written by OCHA itself and by its employees. The month before, a joint research paper with the Norwegian Refugee Council was published, OCHA's Occasional Briefing Series published "Climate Change and Humanitarian Action: Key Emerging Trends and Challenges". It is a short paper, but identifies several ways in which climate change will affect humanitarian efforts. The paper recognizes that climate change will redraw the world's maps of populations, wealth, and resources and will generate higher demands for disaster assistance. It also finds that the results of climate change will contribute to massive movements in populations, which have the potential to overwhelm state authorities and the international community and even threaten global stability. Because of the impending vulnerabilities and the consequences of complex interactions, it recognizes that carbon emissions may become a source of geopolitical tension. Finally, it recommends that humanitarian actors must become proactive in order to assist in mitigating this risk. A few months later, in January 2010, an additional paper on the topic came out with a disclaimer. However, the disclaimer is different from those used by other IGOs. It states that this is a "non-paper" and is produced primarily for internal circulation and with the intent of promoting further discussion on policy analysis and that its views are not necessarily the official views of OCHA. The difference lies in the fact that OCHA identifies them as a spring board for discussion, rather than simply backing away from its findings. Gelsdorf presents "Global Challenges and Their Impact on International Humanitarian Action", which addresses the fact that the humanitarian community needs to broaden its view of vulnerability; insecurity will stem from non-traditional threats, and there in an increasing need to integrate humanitarianism and development. The paper identifies climate change and migration as global challenges, and those implications for humanitarian work include caseloads which do not have the legal or policy frameworks to support them – such as climate induced migration and displacement. Finally, the author calls for the humanitarian community to be more proactive – a sentiment already presented by OCHA itself. A 2011 unofficial policy brief produced by UN OCHA outlined the potential roles for the organization in slow onset disasters. Desertification and sea level rise are slow onset in nature, but the brief does not mention either of these drivers, neither does it connect this brief to climate-based migration or displacement.

Concluding remarks

The humanitarian regime is new and has been growing in significance since its inception. It is in constant tension in two ways: between those who would assist anyone in need and those who do not want to contribute to those who create such problems, and between those who promote humanitarian action to mitigate complex emergencies and those who see military intervention as the appropriate fix in those situations. When it comes to climate change, it is the responsibility of each humanitarian organization to reflect on the consequences such change poses on its mandated work (Braman et al., 2010). This OCHA has begun to do. Its work on the subject of climate induced displacement is still in its infancy, but it does recognize that climate change will affect mass migration, and it is evaluating how such processes will affect its work. It stands out from the other two organizations in that its organization and its employees see the need for proactive measures. While these are not in place as yet, OCHA at least has more active messaging than the other two IGOs. Humanitarian actors are some of the first on any scene of great suffering – whether people are migrating yet or not. They appear to acknowledge that in this way, whether they like it or not, they are on the front lines of the response to climate change. However, as the joint research project demonstrates, it views much of climate induced movement as happening in the future. While more recent events may prove otherwise,[4] OCHA sees the need, but not that it is sufficiently urgent as yet.

7
Filling the Governance Gap

The expansion ability of any of the IGOs investigated depends on both political environment and the institutional apparatus they have put in place. To date, the three established IGOs discussed in the previous chapter have come up short even under increasing pressure to expand, thus leaving a governance gap. While each has produced original research detailing their interpretation of this gap, none is yet actively seeking to close it within their own organizations. Each of the previously investigated IGOs had originated for a specific purpose: UNHCR to protect refugees, IOM to facilitate migration and resettlement, and UN OCHA to fill in the gaps left by the other two. As it stands, this new challenge may necessitate an individual response of its own and, indeed, a new IGO. Why a new IGO? Because the intricacies of climate change adaptation and migration schemes as adaptation are too complicated to simply slip into another existing IGO. There is the need to identify hotspots where the environmental and economic systems are already deteriorating, negotiate resettlement sites for different peoples, and invest in alternative livelihood training for displacees and temporary assistance while this is taking place. An IGO is also appropriate because the problem will only continue to grow, necessitating a full international buy-in in order to produce suitable results; a global problem needs a real global solution.

Organizing such an IGO under the UN umbrella would allow for as many nations as possible to be involved. Arthur Helton (before his death) and Susan Martin have both proposed alternative visions for how the international community can better serve forced migrants who are not refugees. Martin (2004) went so far as to offer a UN reorganization of UNHCR under a new umbrella office for forced migrants. Such reorganization can easily add in an office which deals with climate change displacement. Without any significant changes in the UN structure, it will prove more difficult to initiate a new office. Concerns about migration/displacement due to climate processes have been getting louder and louder in the past few years during the UNFCCC climate negotiations, and this has slowly created a space for discussion and an eventual governance mechanism. This chapter provides a comprehensive background of the inception and development of what will eventually be called the Loss and Damage Mechanism (LDM) and its Warsaw International Mechanism (WIM). This chapter will evaluate what kind of mechanism has developed, its potential to fill the governance gap, and compares this new emerging apparatus to the IGOs previously discussed through a set of structural variables.

The climate regime

The climate regime, not unlike the humanitarian regime, is a new phenomenon originating in the 1950s but not culminating into a distinguishable source of governance until the 1990s. Its development began with the scientific acknowledgement of the problem and eventual large-scale buy-in by individual nations to form an IGO to further foster the governance of the issue and possible solutions. The estimated effects of increased greenhouse gas (GHG) concentrations has held some sporadic scientific interest from as far back as the mid-1800s, but it was not until the early stages of the Cold War that it became embedded within a durable, well-funded research program. During that time, two discourses emerged: the cycling of carbon between land, atmosphere, and the oceans, and how increases in atmospheric CO_2 influence the climate system (Andresen and Agrawala, 2002). Important contributions came from oceanography and meteorology in terms of weather prediction and the first carbon cycle model. This research was also fostered by scientific networks under the World Meteorological Organization (WMO) and the International Council of Scientific Unions (ICSU) (Andresen and Agrawala, 2002). At this point (during the 1950s and 1960s), one could not characterize research being done by multiple institutions as a coherent regime. This is because climate change was still being evaluated and had not yet come to the forefront of any discernible policy – national, international, or otherwise. However, Andresen and Agrawala (2002) argue that the development of UNEP in 1972 and its executive director, Mostafa Tolba, initiated the turning point from science to policy. Established as a result of the UN Conference on the Human Environment in Stockholm in June 1972, the UNEP approved a declaration of 26 principles which would guide the nations of the world in multilateral agreements to "effectively control, prevent, reduce and eliminate adverse environmental effects resulting from activities conducted in all spheres". Andresen and Agrawala (2002) contend that UNEP was instrumental in establishing climate change as a political concern by focusing on the societal impacts of climate variability. UNEP was influential in its funding of a five year international assessment of the causes and consequences of climate change, presented at a meeting held in Austria in 1985. Shortly thereafter, an Advisory Group on Greenhouse Gases (AGGG) was established in an effort to initiate the consideration of a global climate convention. The AGGG facilitated two workshops on climate change and policy responses which led to a larger event in June 1988 – the Toronto Conference. The scientists at this conference endorsed a timetable suggesting the need for a global cut of 20% of CO_2 emissions by 2005, relative to 1998 levels. While the scientists were acting as substitute policy makers (Agrawala, 1999), their efforts gave way to the development of the IPCC in November of the same year (Andresen and Agrawala, 2002). In just 30 years, a sparse group of scientists, with the help of UNEP to contextualize their findings, brought climate change out of the shadows and into the beginning of a series of institutional intergovernmental arrangements. Coming out of the intergovernmental nature of the IPCC was a series of sessions of the International Negotiating Committee (INC) in 1991 and 1992. Its focal point was a solution to stabilize GHGs. The text coming out of these meetings (specifically from a limited group meeting in Paris) was presented and negotiated at the fifth session of the INC and led to the adoption of the UNFCCC a month prior to the Rio talks (Andresen and Agrawala, 2002). This ushered in a new era of intergovernmental talks on policy to address the consensus science on climate change.

As intergovernmental talks began in earnest, efforts to address the mitigation side of the problem were the sole focus of negotiations for a decade, from 1991 to 2001 (Roberts, 2011). However, this would not go smoothly. While 132 nations did sign on to the UNFCCC, the original treaty avoided tough details. Southern countries were concerned about limits that could possibly be put on their efforts to develop, while powerful, industrialized nations refused to curtail their own excesses unless poorer nations did the same (Parks and Roberts, 2008). To balance such demands would require a new way of thinking and a change in international norms. Mitchell (2005) explains that the stabilization of GHGs is a lofty goal, and thus the developing regime will need to create a broadly held and abiding norm among governments and within global society that appropriate behavior requires significant and consistent efforts to reduce GHG emissions. And thus the first challenge to the climate regime was a legally binding agreement to stabilize GHG emissions agreed upon by all negotiating parties. This began with the Berlin Mandate out of the COP 1 that acknowledged the need to strengthen Annex 1 commitments beyond the year 2000, followed by the Geneva Declaration coming out the COP 2 calling for quantified legally binding objectives within specific time frames (Andresen and Agrawala, 2002). But things become contentious in the lead up to the COP 3 to be held in Kyoto, Japan. On the line was a treaty arising from concerns about GHG stabilization and an understanding that all nations needed to participate to meet this goal. But how to bridge the gap between the developed and developing world? The agreement that was negotiated was the 1997 Kyoto Protocol, based on grandfathering – or the notion that countries should reduce their emissions incrementally based on a baseline year, 1990 (Parks and Roberts, 2008). The Protocol was described by Andresen (1998) as a "genuine compromise" in the sense that the EU got the numbers they wanted, the USA got its institutions, Japan got the prestige, the JUSSCANNZ[1] countries got their differentiation, and the developing countries avoided commitments. More specifically, the Protocol established emissions reduction targets for the period 2008–2012. For nations in Annex A and Annex B, it offered the beginnings of an accounting procedure to establish compliance, and provided a general description of various mechanisms to allow flexibility to reduce the cost of compliance. These mechanisms included "bubbles" within which several countries could meet their obligations jointly, a facility for crediting emissions-reducing projects in other Annex B nations (Joint Implementation or JI), a Clean Development Mechanism (CDM) to generate credits for investing in projects in developing nations that had not assured constraints under Annex B, and a system of trading emissions permits among Annex B nations (Babiker et al., 2002). Of course, regardless of the deal struck, individual nations are responsible for the changes needed to actually lower global emissions. Domestic implementation thus requires each party to develop a regulatory compliance and review mechanism consistent with its own political, judicial, and regulatory structures (Babiker et al., 2002).

The Kyoto deal, while seen a positive step forward, did not end international tensions. Beyond the agreement itself, many details still remained to be hammered out – specifically the terms of implementation. The buy-in to the Protocol, including developing countries, has largely come out of the argument that getting a foot in the door is critical, even if its current targets are diluted (Najam et al., 2003). But the process needed to begin with a broad-scale agreement, and thus there was a concrete place to begin. COP 6 at The Hague began with 250 pages of bracketed text,[2] with the intent for the diplomats to clean up the text the first week and leaving key political choices to be decided by ministers the second week. However, the negotiations broke down. A compromise was reached in Bonn at the next UNFCCC intercessional meeting (Babiker et al., 2002). At the same COP, there was agreement on a number of funds, including the Climate Change Fund for capacity building and transfer technology and the Least Developed Countries (LDC) Fund to assist LDCs in climate change adaptation. While the intent of these funds is noble, it is difficult to place much faith in their potential, argue Najam et al. (2003), because: (1) they are voluntary, (2) they are to be managed via the still-controversial Global Environmental Facility (GEF) which inspired little confidence in the developing countries because its governance and agenda remains Northern-dominated, and (3) they remain poorly funded (Huq and Sokona, 2001). Moving beyond Kyoto meant eventually contending with the core of the climate negotiations which have boiled down to the differing perceptions of justice in the global North and global South (Roberts, 2011).

The South's concerns about the climate regime have evolved as the Kyoto Protocol has taken shape, but their longer-term interests have remained unchanged. The key interests of the South as a whole can be characterized within three categories: (1) the creation of a predictable, implementable, and equitable architecture for combating global climate change, stabilizing atmospheric concentrations of greenhouse gas emissions in a reasonable period of time, while giving all nations a clear indication of their current and future obligations based on their current or future emissions, (2) enhancing the capabilities of communities and countries to combat and respond to climate change, with particular attention to adaptive capacity that enhances the resilience of the poorest and most vulnerable communities, and (3) sustainable development as a central goal – at the declaratory as well as operational levels (Najam et al., 2003). More specifically, what is required is a binding treaty that covers all nations with operationalized goals and compliance, assistance to those countries and communities that face the most suffering, and a commitment to a sustainable future in all countries. All of these considerations became increasingly important as time progressed toward the deadline for the next treaty. During the decade leading up to COP 15, a social movement for "climate justice" had taken off, with growing numbers of academic and policy-making publications supporting increasingly vocal debates by activists and the governments of developing countries. The core of the idea of climate injustice is that those who are least responsible for the problem are suffering the worst impacts of climate change, with the least ability to address those impacts (Athanasiou and Bear, 2002; Roberts, 2011; Roberts and Parks, 2007). The global South has had sincere concerns about the levels of effort they had been asked to achieve, knowing that the USA had not even signed onto Kyoto and that the developed nations (as a whole) had held back on making the same development sacrifices they had asked of others. The USA and other highly developed nations have used path dependency as a crutch, arguing that it will be detrimental to their economies to do so. This, combined with concern over historical responsibility – and considering that by the end of the 1990s wealthy countries belonging to the Organisation for Economic Cooperation and Development (OECD) failed to honor their policy commitments (Parks and Roberts, 2008) – there was still much more to be agreed to through the UNFCCC's continued meetings. The Marrakesh Accords (COP7) in 2001 brought about a new focus on adaptation measures which began to address the growing needs of the global South. At this meeting, the National Adaptation Programme of Action (NAPA) was formulated to identify the urgent and immediate needs and priorities of the LDCs. In addition, the Special Climate Change Fund (SCCF) and the Least Developed Countries Fund (LCDF) were also created to fund NAPA activities. This was followed by the launch of the Nairobi Work Programme (NWP) on impacts, vulnerability, and adaptation to climate change at the COP 11 in 2005.

At COP 13 in Bali, the parties agreed to launch negotiations to adopt new arrangements for ushering a second commitment period under the Convention that would include binding emissions reductions for developed countries and new programs on adaptation for developing countries, deforestation, finance, technology transfer, and capacity building (Kakahel, 2012). This meeting also resulted in the adoption of the Bali Action Plan, which established adaptation as another pillar of the UNFCCC (with mitigation: technology transfer and finance). This COP also operationalized the Adaptation Fund (AF) under the Kyoto Protocol (Okereke et al., 2014). As the UNFCCC grew into a regime that governed both mitigation *and* adaptation, the pressure was still on to fully address deepening concerns over long-term damages. This seemed to boil over, coming into the Copenhagen talks in 2009 (COP15). The "climate justice" and "climate debt" concepts and discourse arose from rather peripheral circles in the early 2000s, being a part of some of the near-final version of the Copenhagen texts on Adaptation and Financing (negotiating texts of Tuesday, 15 December, 2009); some parties even called for a 1.5% GDP of wealthy nations to be earmarked for climate adaptation and mitigation support (Roberts, 2011). With tensions rising and pressure mounting, there was a failure to agree on ambitious goals, and a number of countries stepped back from actions they would have otherwise been prepared to take (Hare et al., 2010). Vast numbers of brackets in the text showed that there was no clear route forward. After heads of state arrived in Copenhagen during the second week of meetings, the BASIC countries (Brazil, South Africa, India, and China) and the USA simply set aside the text carefully negotiated up to that point. The EU, used to playing the role of climate leader, was pushed to the curb, as were smaller countries in the developing world (Roberts, 2011). On the concluding day, it was announced that the heads of state and governments of 26 countries had negotiated a two-and-a-half page document called the Copenhagen Accord. While this document was submitted to the plenary of the COP, the Accord could not receive the requisite consensus, and the COP merely agreed to "take note" of the document, meaning that it had no legal status or validity (Kakahel, 2012). Making things worse, the Copenhagen Accord's program of voluntary emissions reductions were projected by IPCC metrics to lead to a 4°C temperature rise. But the slightly brighter spot were two clear and fairly ambitious promises of finance of US$ 30 billion "Fast Start Finance" over 2010–2012, ramping up to US$ 100 billion per year by 2020 (Roberts, 2011).

Copenhagen failed to produce any semblance of consensus on a grand scale and was nicknamed the Copenhagen Discord[3] based on that experience. Consequently, there were very low expectations for the results of the COP 16 in Cancun, Mexico in 2010 (Cavazos, 2012). But, after two weeks of discussions, the negotiators at COP 16 arrived at a multilateral agreement that covered four of the five major topics on the agenda. It established:

- A *Deforestation Accord*, to prevent clear-cutting and create a framework to allow developed countries to finance others for reducing emissions (REDD+).
- A *Green Climate Fund*, to be managed by developed and underdeveloped countries to support adaptation and mitigation. This had been proposed in Copenhagen and was revived in Cancun.
- The *Cancun Adaptation Framework*, a guide for decisions to support adaptation in underdeveloped countries. An Adaptation Committee was established to provide coherence and implementation.
- *Technology Transfer*, to support developing countries with clean energies, technologies and capacity building (Cavazos, 2012).

The COP 16 talks could be considered a success, based on the fact that agreement was found on so many fronts. However, some contend that the developing countries were strong-armed into the agreement, leaving only one country standing strong against it. Kakahel (2012) explains that the agreements were primarily based on the positions voiced by the USA and other developed countries during the negotiations, as confirmed by the Chief US negotiator saying, "The reality is we really got what we wanted". The negotiator added that developing countries acquiesced to save the decision-making process itself; the package was a non-negotiable matter. The strongest voice in opposition came from Bolivia's president, Evo Morales, arguing that the agreement was totally inadequate (Simonelli Berringer, 2011) and his country would be the only one not to sign on.

The next round of talks seemed to fare better all around. Not only did the COP 17 in Durban, South Africa, add to the (nearly) universal agreement struck in Cancun, the event itself was not as contentious. The Durban Platform established a standing committee to govern and distribute funds from the Green Climate Fund. It fully recognized the Adaptation Committee under the COP 16 Framework and operationalized the Technology Mechanism also confirmed in Cancun. It made the Measuring, Reporting and Verification (MRV) system operational and confirmed the next negotiating period up to 2015, presuming the next legally binding agreement to be implemented in 2020 (Malla, 2012). The outcomes of the Durban talks were substantially less than what came out of Cancun, but it is important to note that the negotiations did move forward many of the previous meeting's decisions and confirmed the next time line toward a new legally binding treaty, which was needed as the previous commitment period for the Kyoto Protocol was about to end. When the world converged on Doha, Qatar, in 2012, the agenda was quite extensive. Doha would focus on five aspects of climate change: adaptation, mitigation, finance, technology, and loss and damage – with its main objective being a procedural way to streamline the negotiating process into the next Kyoto Protocol commitment period (Streimikiene, 2013). The first commitment period (for the Protocol) came to a close in 2012, and thus the Doha conference was the last opportunity to confirm a way forward. While important to approve the issues under discussion, it has been argued that Doha was going to be anticlimactic because of the nature of the agreements made in Durban (Roberts, 2013). There were well-founded fears that no formal agreement on a commitment period could be achieved, thus creating a gap. But the conference did adopt, by consensus, amendments to the Kyoto Protocol establishing a new commitment period (2013–2020, KP2) and providing for increased quantified emission limitation and reduction commitments (Bothe, 2014). Beyond a confirmation of the new commitment period, the rules for this period were also agreed upon and the discussions of the Durban Platform in Doha were broad and inclusive (Roberts, 2013). Streimikiene (2013) argues that although successful in general, the COP 18 did not deliver any improvements in mitigation ambition when it came to major emissions, and simply ignored emissions from international aviation and maritime transport. But there was substantial progress on the adaptation side, which was the decision to establish "institutional arrangements" for some kind of LMD. This will be elaborated on further in the next section.

Up to this point, the UNFCCC has succeeded in facilitating the development of several governance mechanisms to address the increasing amounts of GHGs in the atmosphere through multiple means *and* initiated initial evaluations of the needs of LDCs with corresponding funding mechanisms. But what does any of this have to do with displacement and migration? To date, not much. As an alternate regime to those previously examined, the climate regime was not developed with displacement in mind. Its formal structure was precipitated by meteorologists, climatologists, and technical advisors; in other words, no one with a social science background. The refugee regime, migration regime, and humanitarian regime all were established in response to refugees, migrants, and the otherwise displaced, respectively. The climate regime was established in response to high levels of GHGs in the atmosphere. Furthermore, most of the literature evaluating its current status and possibilities for future development focus solely on GHGs and the best ways to govern them. This work considers the implications of top-down and bottom-up agreements, if legally binding or flexible pledges are more apt to garner commitments and compliance, alternative designs, and equitable agreements (Barrett and Toman, 2010; Baumert et al., 2003; Keohane and Victor, 2010; Stokke et al., 2005; Thompson, 2010; Torvanger et al., 2005). And while an understanding of commitments, flexibility, and equity does also apply to governance considerations of migration and displacement in this context, it will be the commitments, flexibility, and equity of mitigation and adaptation agreements that will set the tone for any other subsequent issues.

Migration and displacement developments in the UNFCCC

Migration and displacement, as a long-term issue beyond or regardless of current mitigation and adaptation activities, has been articulated by island states for over a decade. But its discussion has been under an umbrella frame called "Loss and Damage". Loss and Damage was conceived of as the irreversible long-term harm faced by people and states which will occur because of historical fossil fuel emissions which has yet to be seen; it is based on the estimate of global warming beyond what any international mitigation or adaptation can tackle. Thus, many island nations are concerned about their long-term survival *even if* the UNFCCC negotiates a treaty to drastically lower emissions targets and fully funds global adaptation projects. The Small Island Developing States (SIDS) had been pushing for recognition for the ultimate "loss", their permanent forced displacement from their homes.

Loss and damage has been a priority of the Alliance of Small Island States (AOSIS) since 1991. In that year, they conceived of an international mechanism by which they could access funds immediately after a disaster. This way, beyond the current humanitarian disaster imperative to simply provide short-term assistance, they would be able to provide a better and quicker response to disasters that have been predicted to happen more frequently and to build back better. That same year, AOSIS originally proposed the establishment of an international insurance pool as a "collective loss-sharing scheme" to "compensate the most vulnerable small-island and low-lying coastal developing countries from loss and damage arising from sea level rise" (Mace and Schaeffer, 2013). The scheme was to be funded by mandatory contributions from the Annex I parties to the Convention, but it omits some highly antagonistic and complex practical issues concerning definitions of climate change, the standard of care, and the level of liability that should be applied to the culpable state(s) (Okereke et al., 2015). Thus, the proposal did not make it into the UNFCCC agreement in 1992.

Attention to loss and damage in the UNFCCC, then, did not begin in earnest until 2007 with the Bali Action Plan. It called for increased adaptation efforts, including strategies and means to address loss and damage in developing countries. Although the Bali Action Plan contained an entire section on (disaster) risk management and loss and damage associated with climate change, any association or mention of compensation or liability for such loss and damage was a cause for discomfort for industrialized countries (Warner and Zakieldeen, 2011). By 2008, the proposal developed into a multi-pronged instrument with provisions for disaster risk management, compensation, and rehabilitation for unavoidable and irreversible damage (Burkett, 2014). For many years, the idea and desire for an institutionalization of this concept had made little progress toward making it into any main negotiating document; again, this hung on the issue of "compensation" and "liability" – language which held up any progress at the COP 15 (Warner and Zakieldeen, 2011).

However, the COP 16 meetings in Cancun, Mexico, in late 2010 became turning point. It began as just another series of meetings in the continued discussion about prevention of further climate change and GHG emissions, but it eventually provided a step forward for those nations concerned with their eventual displacement. The text, Bolivia argued, was full of loopholes for polluters and reduced the obligation for developed countries to act (Solon, 2010). Bolivia also contended that the text replaced binding mechanisms for reducing greenhouse emissions with voluntary pledges. This would mean that the plight of the Carterets and the Maldives would not be isolated events, but would represent the beginning of a snowball effect which would threaten the homelands of many more peoples around the globe. Bolivia had reason to demand more from these negotiations as well. Many of its residents are already "climate migrants", as their country defines them (Bolivia Climate Summit: Climate Migrants, 2010). Adding to the list of nations becoming aware of how climate change is affecting its people is Ghana, whose Minister of Environment and Energy has admitted a concern for internal climate induced migration. The issue there has been drought and subsequent floods in their northern region. The minister noted that over 300,000 deaths were recorded annually due to climate change, while another 300 million people per year were affected by climate change (Ayittey, 2009). The Cancun Adaptation Framework noted that approaches to loss and damage should consider impacts, including sea level rise, increasing temperatures, and ocean acidification (Burkett, 2014). But the big win for AOSIS within the frame of loss and damage came in the negotiation over a particular subsection paragraph. Over the course of the meeting in 2010 at the COP 16, SIDS, along with their non-party advocates, negotiated paragraph 14 (f) into the Cancun draft decision. This subsection invites parties to take specific action nationally to enact: "Measures to enhance understanding, coordination and cooperation *with regard to climate change induced displacement, migration and planned relocation,*[4] where appropriate, at national, regional and international levels" (Draft decision -/CP.16). While this subsection, like the agreement, is not legally binding, it does ask individual nations to acknowledge the existence of climate change migration and displacement at several levels of governance. This decision launched the Work Programme on Loss and Damage, to be placed under the Subsidiary Body for Implementation (SBI) (McNamara, 2014), which would consider approaches to address this issue through workshops and expert meetings (Burkett, 2014).

The following year, at the COP 17 meeting in Durban, South Africa, no additions or changes were made to the previous text concerning migration or displacement, but Loss and Damage was elaborated on in terms of its overarching goals and its role in the Framework Convention

(Burkett, 2014). Negotiators reached a consensus on elements of the SBI Work Program (Decision -/CP.17); the decision requests the SBI to continue the implementation of the work program to make recommendations on Loss and Damage at the next COP session. It calls for stakeholders and experts to share the outcomes, lessons learned, and good practice related to the implementation of existing risk assessment and risk management approaches (Warner and Zakieldeen, 2011).

The decision emerging from the COP 18 meeting in Doha, Qatar, in 2012 represented a significant advance in the Loss and Damage discussion. It heightened the work stream's importance by calling for an advanced understanding of non-economic loss and damage, patterns of migration and displacement, and identifying the development of approaches to rehabilitation following climate-related loss and damage. Finally, the Doha Gateway mandated the formation of an institutional mechanism for the next COP (Burkett, 2014). This did, in a practical sense, recognize the particular threat posed by sea level rise (Bothe, 2014). It was an important point in that a group of small nations were able to bring to the forefront their most crucial issue.

At the COP 19 in Warsaw, parties convened to create the institutional mechanism mandated by the Doha Gateway. The WIM was established to address loss and damage, including both slow onset and extreme events. The parties created an executive committee which will report annually to the COP through both its subsidiary bodies, the Subsidiary Body for Scientific and Technological Advice (SBSTA) and SBI (Decision -/CP.19). The WIM was established under the Cancun Adaptation Framework and did not provide Loss and Damage with its own additional pillar; this made it a point of controversy (Simonelli, 2013). The G-77 countries and China argued that the new international mechanism should be housed as a separate entity under the Convention itself (McNamara, 2014), the reason being that "Loss and Damage" was originally intended to address long-term irreparable losses and damages beyond adaptation. Considering the minimal mitigation targets over the course of each COP meeting, many nations considered the LDM to be a measure of last resort, acknowledging that both mitigation and adaptation would not be enough to save certain vulnerable regions and countries. This opinion also weighed into the consideration of funding. AOSIS (2013) argued that LDM funding should be from a dedicated source and separate from that of adaptation funding. Again, if Loss and Damage were to be put under the Adaptation pillar of the UNFCCC, would any funds it may need get funneled into adaptation projects instead? How could it function to fix what adaptation and mitigation could not if it were possibly buried under them? The meeting became so intense that the G-77 bloc of developing countries walked out of the discussion during the second week of the meetings. The climax came at 4 am on Wednesday, November 20, when the lead negotiators from the G-77 and China walked out.[5] Bilateral discussions did resuscitate the talks, and compromise eventually won out in the creation of the mechanism – under the Cancun Adaptation Framework; this came as talks ran into overtime on Saturday (McNamara, 2014). The decision legitimizes the exploration of responses beyond mitigation and adaptation, but does not promise compensation (Burkett, 2014). Additionally, the committee is not empowered to decide any concrete claim (Bothe, 2014).

The initial meeting of the Executive Committee (ExCom) of the WIM was held in Bonn in March 2014, where it adopted a two-year work plan which includes the action areas of: (1) enhancing knowledge and understanding of comprehensive risk management approaches, including the identification of gaps or development of methodologies to be used by national governments, and (2) enhancing data knowledge and response measures concerning non-economic losses associated with climate change, including slow onset events such as sea level rise or glacial melt (Okereke et al., 2015). As the work plan currently stands from the September 2014 Adaptation Committee meeting,[6] it consists of eight points (and subpoints), with general time lines for each, and culminating in an additional five-year rolling work plan for consideration by COP 22, building on the results of the present work plan. Its main action items include: (1) enhancing the understanding of how loss and damage affects vulnerable developing countries, (2) enhancing understanding of and promoting comprehensive risk management, (3) enhancing data and knowledge on the risks of slow onset events and their impacts, (4) enhancing data and knowledge on non-economic losses, (5) enhancing understanding of capacity and coordination needed to prepare for and respond to loss, (6) enhancing understanding and expertise on how climate change affects patterns of migration and displacement, (7) encouraging comprehensive risk management through financial instruments, and (8) complementing the work of the existing bodies and expert groups under the Convention. The COP 20 in Lima, Peru, in 2014, confirmed this work plan of the ExCom and outlined its reporting and operational procedures. The next major step will be the negotiations of the WIM for Loss and Damage into the Paris talks (COP 21) in 2015. In early February 2015, the Adaptation Committee meeting in Geneva produced the first version of the full negotiating text, which, if accepted, includes several specific leaps forward for Loss and Damage. Under *Loss and Damage, Option 1*, 33.3 (a), it suggests "provisions for establishing a climate change displacement coordination facility that:

• Provides support for emergency relief;
• Assists in providing organized migration and planned relocation;
• Undertakes compensation measures." (ADP 2–8 Agenda item 3)

The text also leaves an opening for discussion as to which governing body the international mechanism on loss and damage should be subject to. This could reopen the dispute over which pillar (if any) the mechanism falls under, a subject that has already been contentious during previous negotiations. While almost the entirety of this text that came out of the Adaptation Committee is bracketed, meaning that it is not formally agreed to and is up for debate, these items are the most concrete options thus far to deal with climate induced displacement.

Institutional and political analysis

Regimes

In the previous chapter, three IGOs, and their preceding regimes, were introduced as potential conduits to assist with climate induced migration and displacement. The refugee regime, although ancient in sentiment, was institutionalized for a very specific purpose; to assist those displaced due to WWII. Because UNHCR was developed to be the solution to a singular problem, it had to expand in order for it to remain relevant, thus acquiescing to a broadening of protections in the 1967 Protocol. It was a critical acknowledgement that the 1951 Convention was too shortsighted; the need for protection was far broader than previous conceptions, and persecution would continue far beyond the Nazi or Communist regimes. The institutionalization of the refugee regime was only meant to be temporary. This is not to say that is was not a tremendous achievement, even as a short-term fix. It was not necessarily in the monetary or national interest for nations to agree to take in WWII refugees. However, the loss of sovereignty ceded to the terms of the 1951 Convention had allowed for a great leap forward in human rights. It is also important to note that the Convention and Protocol had overwhelming support from UN member states, but without clear compliance mechanisms. This was accomplished through a treaty which was integrated into national laws, thus keeping sovereignty intact.

The migration regime, on the other hand, is still decentralized, bereft of any binding legal treaties, and is unceremoniously reactive not only to migration flows, but to displacement as well. This regime remains tied much closer to the issue of sovereignty; there has not been any one migration crisis large enough to relinquish immigration policy to a larger body. While the history of humanity is that of migration, in modern times the migration regime seeks to regulate a phenomenon which continuously finds it way around rules and borders. The borders of the world are not distinguished by elaborate fences, and yet many nations continue to develop policies which could only be effective *if* this were the case. Individual states guard their sovereignty knowing full well that they cannot control their borders unilaterally. And instead of using this fact to orchestrate clear, binding regional treaties, politicians use nationalist rhetoric with xenophobic undertones to criminalize those who seek to subvert their ill-thought-out policies. Globalization of trade without a restructuring of immigration and border controls has proven inefficient, but governments are still wary of any sort of hard law in this area. Development has necessitated inflows of migrants to create modernity, and yet the modern state cannot cede control of its borders to keep up with the times.

The humanitarian regime, like the refugee regime, is highly connected to the perils of war. Extending medical treatment to the "other" or even to the enemy rejects the notion of separateness that borders ultimately create. Additionally, refugees come from both sides of a conflict. Humanitarianism comes from a deep-seated connection to the suffering of all people and a view that, as humans, everyone deserves minimal standards. These fall apart during wartime and often after natural disasters. However, the divide between Dunantans and Wilsonians does demonstrate a rift between the active and reactive forces within this regime. The Dunantist sect holds a close parallel to both the refugee and migration response in that these are seen as imperative only after a situation arises; Wilsonians see that one disaster can lead to changes which can prevent or at least assuage the next. However, when states began to invest in humanitarian projects they did so with political concerns far from either of these veins and placed conditionality on needs-based relief. One could argue that states could see the opportunity to be proactive if there was the chance of gaining stature or influence, but were reactive when no one could decide what to do otherwise. In this book, this discussion has focused on the concept of aid as relief, which could be seen as different from aid as redevelopment. Without a current focus on building back better, temporary aid projects do facilitate movement, but mostly to a shelter.

The climate regime is a very different from the previous three in many regards. This regime has slowly expanded as the consequences of climate change have become more apparent, but its focus has been more on gasses than people. UNHCR and UN OCHA have directly sought to deal with the human consequences of war and disaster, while IOM has sought to facilitate human mobility; but the UNFCCC's focus has been atmospheric gas percentages and their effects on the environment – not people. The human element had to evolve into this regime, which has taken a long time and is still unclear. The climate regime is an environmental regime; between carbon credits, emissions reporting, and clean development mechanisms, it can also be considered a partial development regime. Sovereignty, under this lens, is a sensitive issue. The divide between the developed and developing world is contentious regarding the rights of the sovereign state to develop (or emit) to certain levels. This reflects tensions over human as well as industrial development. Smaller nations (economically) resent that they are being asked to curb the industrial growth they need to elevate the wellbeing of their people, while the larger nations resent that the smaller nations want them to reduce their impacts significantly, fearing that it will hurt their strong economies. Finally, this regime has the shortest span of norm development. While all four regimes were formalized in the twentieth century, issues such as humanitarian relief and migration have been around since the dawn of man. Climate change was not acknowledged as a global threat until the late 1900s.

Institutionalization

The institutionalization of these regimes has allowed a much larger reach for collective action to assist those currently migrating or those who have been displaced. While international governance structures have greater capabilities, these structures are not always conducive to effective outcomes. IGOs only institutionalize out of compromise, and thus their structures will vary based on multiple dimensions. In this chapter, all four institutions will be evaluated according to the following variables: Organizational Structure, Origination of Research, Primary Sources of Funding, Legal Frameworks, Scope of Responsibility, Compliance Mechanisms, and Number of Member States. These particular variables provide a clear outline of the basic structural differences between each IGO.

Organizational structure is a necessary starting point. Structure allows one to see and understand how an organization functions. Institutions can be centralized or decentralized, which reflects either a hierarchical or a lateral structure. Both the UN-based organizations (UNHCR and OCHA) have a clear hierarchical flow, while IOM's structure is lateral. In UNHCR's Office of the High Commissioner, a Deputy High

Commissioner and two Assistant High Commissioners report directly to the High Commissioner. The responsibility of each subordinate Commissioner is clearly delineated; the only overlap concerns a connection between the regional bureaus and the Assistant High Commissioner for Operations and the Assistant High Commissioner for Protection. For OCHA, the Corporate Programme Division, the Coordination and Response Division, and the Geneva Office report to the under secretary general and emergency response coordinator, with the Strategic Planning Unit as an additional offshoot; there is no overlap across subunits. With IOM, there is a stark difference. The director general and deputy general have eight offices reporting to them directly: the Office of the Inspector General, Office of Legal Affairs, Senior Regional Advisors, Spokesperson, Staff Security Unit, Ombudsperson, Gender Coordination Unit, and the Occupational Health Unit. Directly under these is the Office of the Chief of Staff, which has the Department of Operations and Emergencies, Department of Migration Management, Department of International Coordination and Partnerships, Department of Resources Management, and the Administrative Centres of Manila and Panama reporting to it. Under the Chief of Staff is an assortment of nine regional offices, then two Special Liaison offices, and finally Country offices. Clearly, each main office of IOM directly handles more horizontal units, while UNHCR and OCHA are structured in a vertical fashion. Conversely, the structure of the WIM under the Loss and Damage work stream is developing in a consensual and democratic fashion. Its governing ExCom is to be composed of 10 members of the Annex I countries and 10 non-Annex I countries providing two representatives from the African, Asia-Pacific, Latin American, and Caribbean states, one from the LDCs, and an additional two from the non-Annex I states (Decision -/CP.19). The COP 20 decision elaborates on this; it explains the length of term and how many terms each member can serve, that it should internally elect co-chairs – one from the Annex-I and one from the non-Annex I countries, and finally, that all ExCom decisions must be taken by consensus (Decision -2/CP.20). However, what comes out of the ExCom must then be reported through both the SBSTA and the SBI to make recommendations. Thus whatever governance suggestions come out of the WIM need to go through both the SBSTA and SBI and can then come for a discussion on the main floor to be deliberated on by all of the parties. The WIM is a mechanism nested within an IGO – the other three are IGOs in their own right – and thus its structure is ultimately vertical, because what it recommends cannot be implemented unless also agreed on by the UNFCCC COP as a whole.

The origination of research is also important. It demonstrates openness to new ideas and willingness to investigate how a particular issue will affect the IGO. While all four get their scientific information from the IPCC, among other sources, when it comes to research about climate change and their operations, UNHCR, IOM, and UN OCHA invest in internal research, either directly sponsored by the institution or contracted out to other academics, which is still published internally. The WIM has access to expert working papers that come out of the SBSTA, the ability to establish its own expert groups to develop inputs and recommendations, and invites relevant outside actors to develop specific analyses to assist with activity numbers 5 and 6 of the ExCom's initial two-year work plan. Ultimately, each IGO/mechanism produces original research relating to its specific goals and mandates.

Sources of funding for all four IGOs are somewhat similar, coming from their member states and a handful of other outside sources consisting of the European Commission, the private sector, and individual donors. Most important to note is the desire of donors to have their money spent as they prefer. This point was previously noted with humanitarian donors. Money can arguably be a form of soft or hard power, depending on one's interpretation. Most of these IGOs are financially tied to their biggest donors, as these large donors do have influence in these organizations and the IGOs have to survive and fulfill their mandates. Member states are not pure Dunantists seeking to do charitable work; they seek to affect the areas in which they chose to participate/donate. This is especially apparent within IOM; the majority of its budget is allocated by donors for specific and time-bound projects only leaving a comparatively limited core administrative budget (Ionesco and Traore Chazalnoël, 2015). Additionally, neither UNHCR nor IOM has a balanced set of donations among its contributors; some pay much more than others and thus have a louder voice inside the institution. These particular donors are also the same ones in each organization. According to their 2010 financial reports, the USA is the top contributor by a significant amount. In this year the USA accounted for 31% of the total budget of UNHCR and 31.8% of the IOM budget. The WIM shares this similarity, but in a different way. While *all* parties to the UNFCCC pay into the system on a sliding scale, the vast majority (again) comes from the USA. The scale mandates that no party should pay less than 0.001% or more than 0.25% of the Convention budget (Decision 17/CP.4), with the lowest percentage paid being 0.2145% in 2010. However, as a nested mechanism within the UNFCCC, the WIM is only indirectly funded. The disproportionate contributions go into a main fund and the operating expenditures for each Convention-related activity are then divided out.

Thus, unlike UNHCR and IOM, no donor has a disproportionate say in the ExCom's direct undertakings and recommendations. When it comes to OCHA, the US' contribution only accounts for 11%; its highest contributing member state in 2010 was Sweden, and its other receipts are more evenly distributed at the top through Western Europe. The total funds to both UNHCR and IOM hover around 1.8 and 1.3 billion (USD) respectively, while OCHA only saw 186 million in 2010 and had to take out a Central Emergency Response Fund (CERF) loan from the General Assembly against its outstanding pledges. At the low end is the UNFCCC at 44,200,099 (Euro) in 2010.

Another point of comparison across the IGOs is their legal frameworks. This determines the flexibility of their mandates if backed up by some sort of hard or soft law. For UNHCR, the 1951 Convention and its update, the 1967 Protocol, are international law. For IOM and OCHA this is not so clear. IOM's International Migration Law Unit has compiled various sorts of migration-related legal instruments, but migration law is derived from state sovereignty and the human rights of those migrating; it is not independent and/or binding. For OCHA, humanitarian law relates to actions taken in and during war as it relates to armies – the Geneva Convention. It does not regulate the way in which those responsible deal with crises within their territory or the actions of outside responders. With respect to the WIM as it has developed, there is the potential for formal legalization, but this depends on how well its recommendations are taken by the UNFCCC as a whole. Currently, the WIM only has authority to bring proposals to the negotiating floor, but does not have the ability to make its own international law. However, if its suggestions are accepted by the parties and integrated into a treaty, the WIM has the potential to legalize its decisions. Legal frameworks also define the scope of responsibility which these IGOs have. Legal frames specifically define who the institution is responsible for. Refugee law includes a specific definition which qualifies a certain group within a set of particular circumstances. Migrants and those in humanitarian need vary; the mandates of both IOM and OCHA consider the breadth of these needs and vulnerabilities and thus take a broader view of their responsibilities. The current work plan endorsed by the WIM ExCom does so as well. Additionally, a legal regime should have some sort of compliance mechanisms to enforce such laws. However, in the realm of international law there are very few methods for this, as it creates pressure on state independence and sovereignty. In recent years, Belgium was sanctioned by the European Court of Human Rights (ECHR) for sending refugees back to Greece when it knew that Greece did not have the means to adequately support them.[7] But not every country gives authority to such a body, and only three regional bodies exist: the ECHR, the Inter-American Court of Human Rights (IACHR), and the African Court on Human and Peoples' Rights (ACHPR). Thus even a legal mandate can only provide a thin level of compliance.

An assessment of member states is also necessary for all four organizations. If IGOs are an extension of the desires of their member states, it is important to understand which member states belong to each organization. The numbers are also essential to know. As branches of the UN, UNHCR and OCHA's member states are the official UN members, which now consist of 193 individual nations. There have been suggestions to bring IOM into the UN Secretariat, but this has never happened. Its member state count stands at 157.[8] What becomes apparent is that many countries that are already experiencing displacement due to climate processes or are most vulnerable to it are not members of IOM, or have only become members in the last few years (this includes the Maldives, PNG, and the Marshall Islands). As for the UNFCCC, its member states count consists of each UN member and additionally offers the same status to Niue, the Cook Islands, and the EU, bringing its total to 196 (Table 7.1).

Table 7.1 Overview of case study structural variables

	Structural IGO variables			
	Intergovernmental organization			
Variables	UNHCR	IOM	UN OCHA	UNFCCC Warsaw International Mechanism
Organizational Structure	Vertical	Horizontal	Vertical	Vertical
Origination of Research	Original	Original	Original	Original
Primary Sources of Funding	USA	USA	Western Europe	USA (indirectly)
Legal Frameworks	Yes	No	No	No
Scope of Responsibility	Specific	Broad	Broad	Broad
Compliance Mechanisms	ECHR, IACHR, ACHPR	National courts	None	None
Member States	193	157	193	196

Implications

The structural components of these IGOs do not exist in a vacuum. Thus the political environments that they face also provide a context through which it is necessary to understand their potential suitability to govern this developing human security problem. First, it is understandable why many first look to UNHCR for guidance. Since its inception, it has assisted millions of people fleeing the most desperate of situations; the vast majority of nations have signed its treaties and have acknowledged their responsibility to refugees. As a regime it is far-reaching, and as an institution it has specific compliance mechanisms underscored by international law. Additionally, the bureaucratic label of "refugee" is universally recognized, if not for its correct legal meaning, at least for the implication of need and vulnerability that comes with the label. However, if those displaced by climate change are not being persecuted, do they need protection? This is the essential question to ask when analyzing this governance structure.

The feature of refugee law that is most powerful is the principle of non-refoulement; it is critical when administering protection to keep the affected individuals out from under the threat of harm. Although certain climate processes will hinder and in some cases prevent human existence in some areas, does returning them to these areas equate to an imminent threat? The threat normally under consideration when establishing refugee status is that from other humans, not the environment and its larger processes. Thus even refoulement as a protectionary measure does not exactly fit the circumstances of those displaced by climate change. Additionally, the non-entrée regime now guides how refugees are treated, administered, and processed. Non-entrée refers to the ways in which Western governments have made it more difficult for asylum seekers to enter and become refugees. Chimni (1998) outlines the actions which make this possible, such as the strict scholarship in this area of the positivist tradition of refugee law interpretation and the justification of different treatment of African refugees than of European refugees due to the different reasons they have had for flight. There has also been the growth of detention centers, external border processing centers, and European policies that only allow refugee petitions from one member state (such as the country of first arrival). This process is happening in other areas of the globe as well. Australia has faced harsh criticism in recent years for their Christmas Island processing center and its tighter restrictions. It is clear that the refugee regime is attempting to shrink, not expand. This is due to pressure from member states.

The previous chapter explains UNHCR's expansion into soft law instruments for IDPs. This juxtaposition of the tightening of the regime by member states while the bureaucracy of the institution is expanding through non-binding soft law demonstrates an operational rift. The UNHCR bureaucracy acknowledges the continued pressure it feels to assist more and more people in desperate situations and is offering solutions that are less demanding on the member states than a formal treaty. Soft law and formal recommendations offer a way to introduce member states to additional ideas without forcing them to formally act on them. But with the need for action, will the guidance of this institution accomplish much? With the increasing resistance toward accepting refugees from member states of the global North, UNHCR is already in a precarious position to simply execute its mandate, much less expand it. And while the previous chapter demonstrated that UNHCR's mandate has expanded, this has only happened once, through the 1967 Protocol. Its help with any other sort of displacement has been based on suggestions for new soft law instruments or reinterpretations of current soft law to emphasize avenues that are already available. Those being displaced by climate change are not "refugees" in any traditional sense, and with its current challenges in executing protections in a growingly securitized world, UNHCR may be able to offer expert advice in this situation, but this is all. In the strict legal sense, UNHCR has no obligation to accept a new group of displacees that are not being persecuted in its mandate, and although UNHCR continues to be implicated each time the media or other lay people use the term "climate refugee", this IGO will not be swayed simply by the misuse of a label.

IOM faces different challenges, but is also no better prepared to assist those displaced by climate change. The main drawback for IOM is its member states; not just that it is comprised of a lesser number of states than UNHCR, OCHA, and the WIM under the UNFCCC, but it is *who* is missing that counts. As outlined above, almost all of the nations which are the first ones to feel the damaging effects of climate change are not members of IOM. And of those island nations used as examples in Chapter 2, only PNG and the Maldives are *recent* members, with Tuvalu and Kiribati on the outside. This is highly problematic because an IGO that provides expertise toward migration governance (and seeks to facilitate it) is relatively confined in its actions if it does not include all nations in which migration occurs. Without the involvement of the most vulnerable nations, the institution has little impetus to act; projects proposed will solely relate to the needs of its member states directly. It would be prudent to further investigate why so many vulnerable states are not IOM members, additionally considering the potential barriers to membership.

Furthermore, if IOM were to attempt to govern this new form of movement, this would place those needing relocation under the migration regime, which is largely considered voluntary. Adding to the discussion of forced versus voluntary migration, climate change migration is considerably different from climate change displacement, and yet they are treated as one and the same. Choice here is key; to choose to migrate infers that one can also choose not to migrate. Many voluntary migrants choose to do so from a myriad of equally undesirable options; it is not a choice in so much as that if conditions are only going to deteriorate further, it is a matter of go now or go later – but one still has to go. Again, referring to this conceptualization back in Chapter 4, choice also equates to responsibility in that if seen through this lens, it may become more difficult to procure money to assist relocations if done early; regular voluntary migrants pay for their own journeys. Ultimately, climate processes will continue to degrade the ability of many areas to sustain human life, and even early migration can be equated to displacement in that hotspots will not regain their viability. Because of this, it is necessary to equate all movement due to changes in climate with displacement.

This is why labels and definitions matter. Choice of words precipitates governance; without the most accurate conceptions, the development of governance will be inadequate. Even though it has become obvious that globalization of trade and manufacturing has provided uneven community growth, many who have had to move to keep up have been at the losing end of these forces and yet they are considered "voluntary migrants". International business decisions are far from the control of the many that are affected by them. Economic migration, in this sense, can be very much a form of displacement if one's place of residence is negatively affected by larger economic forces. But referring to this phenomenon as "migration" instead of "displacement" has shifted the responsibility for such movement from the companies which have changed the economic landscape to the migrants themselves. This, of course, separates consequences from causation, and in doing do assumes that it is the individual's job to adjust *as if* macroeconomic changes were a natural phenomenon. And this is the hurdle with the terms "climate migrant" or "climate migration". The nomenclature will associate those being displaced by the climatic effects of GHG emissions with those whose agency has not been compromised by outside forces. It is the difference between the supposed push and pull factor dichotomy which has prevailed in the case of voluntary migration. By using the displacement label, it is less likely that climate displacees will be equated to voluntary economic migrants and the migration regime. But again, since the majority of those currently being displaced are not members of IOM, much of the previous discussion on conceptualization is moot without a broader membership.

Finally, while IOM has done quite a bit of research on the climate change and migration nexus, much of the work comes from outside academics. Publications written by the IOM bureaucracy tend to focus on how to extend the time people have in their communities, rather than providing migration alternatives. However, with a limited administrative budget for research and a precarious political position, its bureaucracy is much more responsive to the values and desires of its member states than to outside issues. A proactive migration program that takes into account the current needs of those already under climate induced stress may likely come out of IOM's newly minted Migration, Environment and Climate Change Division, but its implementation must be sponsored and paid for by direct project-related funds. This IGO functions specifically for its member states' needs, not for an established goal, as do UNHCR, OCHA, or the WIM under the UNFCCC. Because of this, it is less likely that IOM's projects will move toward assisting migrants due to climate change until its member states believe on their own that investing in projects to govern this type of migration is worthy project.

UN OCHA has a better chance of being the international governance structure that could include those displaced by climate change processes. While OCHA uses soft law, it can still foster compliance through learning and self-regulation. While hard law with working compliance mechanisms is ideal, even refugee law lacks complete compliance mechanisms and ample buy-in into comprehensive international courts for full compliance. Another feature which is helpful, however, is OCHA's broad mandate of responsibility. It necessitates no updates to be able to assist those who are labeled "migrants" or "displacees". Many of those whom it currently assists are at least temporarily living outside of their habitual residence due to generalized violence or the effects of a natural disaster, and thus OCHA is no stranger to dealing with many forms of migration and displacement. If the adverse effects of climate change on human populations are considered a humanitarian issue, this will not be a problem.

OCHA's primary source of funding is also different from the other three. While UNHCR, IOM, and the UNFCCC proper are primarily paid for by the USA, OCHA is primarily funded by Sweden. A consideration of the national culture of Sweden can tell much about the value it places in humanitarian relief. Sweden and Norway (from 2008–2010, Norway was OCHA's second, fifth, and third leading contributor, respectively) have the most advanced and comprehensive welfare systems across Europe. It is assumed that certain member states often exercise a disproportionate influence over IGOs; additionally, money is often the means of such power. In the case of UNHCR and IOM, it can be argued that this influence comes from the USA, due to its proportional contributions to these organizations; in the case of OCHA, it is the nations of Western Europe, including the European Commission. While the US' contributions have risen proportionally in the past few consecutive years (it was OCHA's fourth leading donor in 2008, third in 2009, and second in 2010), Western Europe still contributes much more as a whole (OCHA, 2010). It is also in Western Europe where interest in this topic is growing. In the summer of 2011, the Norwegian government and foreign ministry held the Nansen Conference on Climate Change and Displacement in the 21st Century. The event included practitioners, policy makers, and high level representatives from each of these IGOs. The initiative has produced regional dialogues around the world to assess the needs related to cross-border displacement and is funded by both Norway and Switzerland. The interest and influence that Western Europe has in OCHA is promising; these countries are less resistant to welfare spending and already have a deep sense of cooperation between them. However, they are also countries which have had significant challenges with immigration and, like the USA, tend to spend money overseas to fix certain foreign policy challenges as a way to prevent migration.

Bridging the gap

Out of the four governance mechanisms evaluated, the one with the most potential to bridge the governance gap is the WIM under the UNFCCC. While the climate regime would seem, at first, unamenable to taking on displacement due to its physical science focus, the development of the adaptation pillar has opened it up to assessing impacts on humanity. Had mitigation efforts been successful, there would be no need to delve into adaptation measures which considered disaster risk reduction and, thus long-term damage. Of the IGOs under consideration, the WIM is the only one which has gone beyond general discussions, expert papers, and broad-based recommendations to outline specific measures to be taken and within what time frame they should be accomplished with any hope that these can be formally institutionalized. To reiterate, the WIM cannot turn its proposed work plan into hard law by itself, but the fact that it exists in a formal negotiating space with this potential makes it much more capable of being able to do so than the others. The UNFCCC is in essence a treaty-making body, and thus the WIM, as a nested mechanism within such a body, has great potential to address climate induced displacement with formal and binding consequences. While the other three IGO bodies currently manage some forms of migration and displacement, they do so in specific contexts – the WIM has been developed to begin to manage migration and displacement within a context nexus. Additionally, it is able to comprehensively work on this governance without being directly overrun by one large donor. This is not to say that the UNFCCC's negotiating blocs do not function within this context for their own (sometimes moneyed) interests, but that the WIM has been designed with some balance of representation early on, and this will allow it produce recommendations without being completely stifled. These recommendations will eventually face tough criticism when reported back to the negotiating floor as a whole, but the multi-step nature of the process allows for positive movement within the mechanism first. The UNFCCC, as a workspace for the WIM, also allows for civil society involvement – something the others do not do. Civil society (through NGO workers, students, academics, and activists) has the ability to lobby nations about what should go into negotiating documents and put pressure on nations that are not living up to their commitments. While the WIM does not currently have a compliance mechanism, it is open to public opinion/debate/scrutiny as a function of non-formal compliance. An explanation of the full and formal role that civil society plays at the UNFCCC would necessitate a separate book, but because civil society tends to emphasize concerns before they are fully integrated into the formal negotiations, it is a persuasive sector which has had opportunities to push for strong language regarding safeguards, human rights, and equity.

While the WIM has several advantages over UNHCR, IOM, and UN OCHA, it is only its infancy. It may be a little uneven to compare three fully funded and institutionalized organizations to a newly embedded mechanism, but the WIM would be wholly unnecessary if any of the afore-mentioned IGOs had sought to expand their mandates into the area of climate induced displacement. Therefore, inaction on their part has brought about the development of an alternative.

Where the WIM goes from here will be debated at the COP 21 in Paris, but, minimally, the WIM has a mandate and proposed action items that, if adopted, are poised to actually govern climate induced displacement.

8
Conclusion

Not only is climate induced displacement a real and concerning phenomena, but the future of this migration is also poised to increase. While most movement is still situated in the future, what is happening now needs to be addressed. Whether it be the long struggle of the people of the Carterets for permanent relocation or the most recent cyclone activity in the Pacific, there is an increasing need for the international community to intervene and assist climate induced displacement as it happens. While the leaders of the nations most affected by the latest events (Vanuatu for Cyclone Pam and the Philippines for Typhoon Haiyan) have credited climate change with their severe effects, some atmospheric scientists have been able to substantiate this. Running several models after Cyclone Pam, Dr. Kerry Emanuel, a professor of Atmospheric Science at MIT, was able to assess that, even with some shorter-term data, the results suggest that all of this is consistent with the consensus that the frequency of high level tropical cyclones should increase as the planet gets warmer (Emanuel, 2015). Haiyan hit the Philippines just before the COP 19 in Warsaw. During the talks, the Philippine negotiator, Yeb Sano, began a fast which quickly spread among the civil society participants – especially the youth. "Fast for the climate" became a movement with a meaning that was more than symbolic – if the people of the Philippines were suffering, the participants would not eat until the meeting produced an agreement which provided justice for those harmed and displaced. Cyclone Pam displaced the *entire* island of Tuvalu in March 2015, forcing the relocation of a total of 40% of the whole population, contaminating freshwater supplies, destroying graveyards and the local septic systems (Radio New Zealand, 2015). Civil society leaders are already becoming engaged on the issue, arguing that a range of compensation is needed to handle these growing disasters. They have also argued that the Paris UNFCCC talks need to make real progress on this issue, because Cyclone Pam is another reminder that the world has reached the era of "Loss and Damage" (Singh, 2015). This language definitely demonstrates a shift from demands on UNHCR to those on the UNFCCC.

As demonstrated in Chapter 3, hyperbole can evoke vivid imagery and fantastical drama, but is not necessarily helpful. In the digital age, these ideas can move speedily through both print and online media, creating an echo chamber which reinforces what it produces – ineffectual language and labels. While it can be argued that drama demands attention until many more people know what is going on and call for action on the matter, "scary" imagery can also lead to the consideration of scary consequences. Those being displaced by climate effects are not necessarily "scary" nor do they pose a threat. Using hyperbolic scenarios to sell the problem infers that the process of displacement will occur much faster than in reality and can create unnecessary alarmism. "Where will all those people go?" is often a question posed as to what will happen if an island "sinks". However, since EUICs are not sinking, and there is time to negotiate long-term relocation plans, there may never be the need to ask such a question in these circumstances. Additionally, if people assume that everyone from an island nation will need to move at once, it can spark the fear of invasion. With the Maldives' population estimated at around 400,000, relocating them all appears to be an impossible task; where can the international community put that many people? Adding to the growing list of difficulties is the fact that the Maldives is an Islamic state and much of the developed world is fighting xenophobia toward Muslims from many places. Creating fear in this way can bolster calls for international security measures to prevent "them" from coming "here". However, this view is purely alarmist. Climate displacees should only pose a traditional security threat if their plight is ignored to the extent where it becomes desperate and they blame the international community for purposely abandoning them. First and foremost are the human security concerns of those left in poor conditions time and time again. There is literature that suggests that environmental scarcity can lead to violence, but the communities being displaced have not yet turned on each other, they are banding together – especially in the Pacific. While there is no way to see into the future to determine exactly how all of these factors will play out, alarmist rhetoric can force a self-fulfilling prophecy. Even some of the academic research cited in this book has actively used terminology which is alarmist and disempowering, but the majority of research does not. It tends to mention hyperbolic imagery in passing to acknowledge that it exists or to quickly dismiss it.

Academic research has been able to embrace the conceptual overlap that persists on this issue, emphasizing climate change as a compounding factor on top of economic and social circumstances. Additionally, it has avoided the issue of direct causation, leaning primarily on the IPCC and on other scientific sources to substantiate any statements that directly connect current displacement caused by large-scale events to climate change. Academics tend to understand that what they say matters and it is important to be overly cautious about the thorniness of certain issues. However, although they attempt to explain such complications, this does not always translate well into the policy arena. Policy makers are not academics and tend to need issues simplified in order to make decisions and facilitate political cooperation. Unfortunately, information in the academic arena can often be so watered down for policy that it leaves much to be desired. This book has analyzed how the "climate refugee" is not possible through two veins of analysis: the opinions of additional academics, and statements directly from UNHCR. And yet, this did not stop the US secretary of state, John Kerry, from warning his fellow ambassadors that "There'll be climate refugees that all of you will be coping with at some point. If not now, in the not-too-distant future . . . It is a national security threat, it is a health threat, it's an environmental threat, it's an economic threat" (Fitzgerald, 2015).

If there are this many threats, as per Kerry, then an IGO will be necessary to address them. It is not that the potential for bilateral or multilateral agreements should be ignored, but that thus far they have not proven to be enough in this area. The governments of Tuvalu and New Zealand have had a migration scheme for years which has allowed a small quota of migrants from Fiji, Tuvalu, Tonga, and Kiribati to settle in New Zealand in order to work. This is often discussed as an environmental migration agreement – although it is not (Gemenne and Shen, 2009). While some people have made their way through some strict conditions, New Zealand is not actively allowing Tuvaluans to stay past their visa expiry date due to the environmental degradation in the homelands. When this happened to Ioane Teitiota, a Tuvaluan living in New Zealand, he made a case of indirect persecution by the industrial nations and attempted to claim "climate refugee" status. His claim was rejected by the migration tribunal, the High Court, and the High Court for appeals. And he is not the first; refugee law scholar Jane McAdam argues that both Australia and New Zealand have refuted 17 such claims in the last 20 years (O'Brien, 2015). Legal challenges are not working, and individual nations are not willing to offer protections along the lines of "refugeehood" on their own. If not an IGO, what is the alternative?

Reevaluation of expansion theories

Considering the IGO evolution described earlier in the book, structure has been much less of a hindrance to expansion than political concerns. Each fully developed IGO (UNHCR, IOM, and OCHA) expanded geographically, from handling a specific locale to eventually reaching around the globe. Whether this expansion was done under a vertical or horizontal management structure appears to be unimportant. Each has also originated their own research, with the WIM in the development stage of this. Only UNHCR has a binding legal framework, but has been pulled into helping beyond its mandate geographically and in situations of generalized violence. It is also the only IGO with a specific mandate versus a broad one, and yet it has broadened its activities over the years. Finally, the increasing displacement due to tensions, either during or after the Cold War, has demonstrated that there has been no threat to the survival of UNHCR, IOM, or OCHA. Thus it has been outside forces which have demanded expansion of all three and then, additionally, the fourth.

UNHCR has experienced the biggest pull toward a theoretical "spillover". Nations came together on a mutually important and pressing issue and were so successful in addressing it that there has been extensive outside pressure for it to address additional displacees and expand its mandate. However, it has resisted expanding beyond its only formal expansion (the 1967 Protocol) at every turn. While its one-time expansion geographically did reinforce its dominance with displacement as a whole, its next expansion (into IDP assistance) was only a limited effort. This is not to say that UNHCR does not care about those internally displaced (their mandate was built on the ideal of protection), but in offering a soft law instrument instead of offering to expand its legal protections, it has created a hierarchy of displacement. Even staff (and scholars) have bought into the "special" status of those formally persecuted and see other forms of displacement as lesser – the debate in Chapter 4 over whether refugee studies should be housed under forced migration studies in reinforces this. The bureaucratic mythology of the refugee is real, and it has created a protective environment which has continued to be restrictive.

IOM, on the other hand, has also seen a demand to expand, but in a political manner. It can be more clearly related to firm theory in that its activities are a demand by "clients" and in this case, the "clients" are the member states. Its response, as a logistical facilitator, is guided by member state demands and its consistent ethos of management reflects this.

As receiving states prefer to manage migration flows to keep them consistent with their immigration policies, management *is* key. One way to manage flows is to demonstrate what can be done to keep people from migrating; this was seen in the IOM Mauritius report, emphasizing adaptation instead of planned relocations. IOM is also where organizational funding matters most; a firm is responsive to its clients and gives best service to its biggest clients. IOM is not UN affiliated, and as an outside organization it does not have a directive to protect human rights or assist suffering; it is a logistically based IGO and assists other agencies (UNHCR and OCHA) with their more altruistic missions.

An aid agency such as OCHA is tough to evaluate using either neofunctionalism/spillover or firm theory. While OCHA, as a UN agency, does have a macro bias and has been successful in temporarily assisting displacees, there has been no pressure for it to expand in any way to tackle anything other than what it currently does, and so European expansion theories fall flat. There is an awkward nexus between aid and development; temporary assistance and building back better would be the logical step forward for OCHA expansion to assist climate displacement. However, aid and development work in different silos and do not often cross paths. As firm theory explains, it takes major resources to be able to expand. While OCHA can operate around the globe, its operating budget is often small and the IGO can barely respond when it is needed. This is a structural instead of a political constraint, but an important one. As mentioned previously, aid agencies often are flush with contributions right after a disaster, but have a hard time keeping themselves funded between events. Additionally, there are many aid agencies across the globe which all fight for the same funds. Where things become complicated is when individuals are displaced after a disaster and go to the first place they know for help. While OCHA may not be the best or most appropriate IGO to offer assistance, it may not be able to stop being considered a resource of immediate help.[1]

Neofunctionalism and spillover is a much more clear fit to describe the expansion of the UNFCCC and to include a function such as a Loss and Damage mechanism. The UNFCCC institutionalized around a common issue that was not controversial in and of itself, but in terms of how it could be fixed. However, it did produce the Kyoto Protocol, a real binding agreement to reduce GHG emissions. While compliance has been less than impressive, it did develop further to tackle adaptation and has expanded impressively faster than the previous three IGOs and into various additional issue areas, as explained in Chapter 7. Furthermore, it has a macro bias and actors have mattered. Entrepreneurs such as AOSIS and its members and supporters have worked tirelessly to move the Loss and Damage mechanism forward, which has included reconstitution, in some manner, for the eventual displacement of many people and nations. Finally, civil society has a voice here, and like the EU, it is a venue in which individuals get a vote as to its business; NGO observers, academics, and activists alike are able to participate in the process, echoing the demands of the actors in the game. This additional level of involvement has been a successful driving force toward further spillover.

Political time horizons

An additional theoretical frame which has not been considered is the political time horizon. Political actors make decisions based on how they will affect their political careers. Thus, acting on climate change (or climate induced displacement) would assume that the project being proposed can be sold to their constituents in a manner that makes the actor look good. This does not necessarily assume that the constituents benefit directly from the actions of the politician, but that they believe the issue is important to them. This issue is also reflected in literature based on human nature as related to how people handle the future. What is considered the conventional wisdom in this area says that if left to our own devices, humans will satisfy their most urgent urges today and leave the future for another day. However, even biology argues that we humans are capable of short- and long-term thinking. This is based on our connection with our older generations and their tendency to think about their old age and death. The capacity even increases as people age (Princen, 2009). This suggests that political office holders, who are usually at least in their middle age, have the ability to look to the future; the only thing holding them back is a focus on their personal career, which necessitates shifting to a short-term interpretation of policies and politics.

However, the main subject of inquiry in this book has been IGOs, and while their member states are made of political office holders, they do not have this problem directly. While climate displacement may not be on the national agenda in many states, the IGO, as an instrument of international relations, has the ability to provide a space to tackle issues such as these without necessarily being relevant in every nation involved. Additionally, the one IGO where this would be most salient but yet has made the least difference is in the UNFCCC and its subsequent WIM. Its decisions most directly affect countries because it is developing treaties. Nevertheless, most climate migration is a problem situated in the future, which will affect the political careers of those negotiators and ministers not even in office yet. The leaders who have been arguing that something must be done are the leaders of the nations who will be directly affected. But even some of these leaders do not have to push the issue as of yet, as their tenure will be over when the time comes to move. In places like Tuvalu, the problem is not even salient with the public as a whole; they do not want to leave and have a religious belief system which supports a sense of spiritual optimism when it comes to their plight. They are a highly Christian nation and believe in the promise that God made after the great flood, that he would never do such a thing again (Morris, 2009; Patel, 2006). The leaders in the Pacific are looking to the international community to assist in a solution even when their people hold out religious hope, which suggests that long-term planning beyond political time lines is possible. Even in non-affected nations, the Nansen Conference on Climate Change and Displacement held in Oslo in 2011 brought together many foreign ministries to begin to plot out a direction on this emerging issue. Hosted by Norway, which is not going to face displacement in this way, the conference was a step in the direction of making the issue politically relevant. Since its initiation in 2011, the Nansen Initiative has conducted regional consultations around the globe to assess the needs and vulnerabilities concerning cross-border displacement due to climate processes. The results have been presented as side events at the UNFCCC COP and Intersessional meetings. Political time horizons may play a role in nations where inaction is still the norm, but the focus of this inquiry is the intergovernmental realm. Time horizons are not as relevant at the international level in this case; but considering the nations who are making progress regardless of them, it is not a frame which can explain the lack of expansion at this level of governance.

Implications

Overall, UNHCR and IOM have already conceded this issue to the UNFCCC. With their contributions to the COP 19 related to mobility in the context of loss and damage (2013) and to the Nairobi Work Programme (2014) as part of the Advisory Group on Climate Change and Mobility, it is clear that both of these IGOs have chosen their path forward on the issue. They will advise another IGO's nested mechanism and not expand. OCHA will continue to do what it can as a first responder, but it could not accommodate climate displacement if it wanted to and has not been pressured to do so. As a consequence, the WIM has grown to fill the governance gap that the others have left open, and it should do so. Within a treaty-making body with membership including all nations, a few territories, the EU, and civil society participation, the WIM is poised to create policy in this area unlike any other. But this does not mean that progress from here on will be easy. It took 22 years from the inception of Loss and Damage until the institutionalization of an actual mechanism. The commitment phase for the next Kyoto (K2) commences in 2015 at the COP 21 in Paris, but will not be implemented until 2020. Therefore, more damage will be done. And this is all the reason for those involved in pushing the agenda of the WIM forward to ensure some solid and binding policies, because climate events will still worsen. Additionally, the issue of "compensation", which aided the 22-year gap between the need for a mechanism to address irreparable loss and the establishment of the WIM, is back on the table in the latest negotiating document. Until this draft, the word had been taboo in the UNFCCC and with those who either negotiate or advocate at the COPs. It is unclear if this will again hold up the development of the WIM or the larger negotiation as a whole. Also in this latest negotiating text are two other suggested action items of the WIM which will institutionally overlap with IOM and OCHA: (1) to provide support for emergency relief and (2) to assist in providing organized migration and planned relocation. Again, it is unclear whether these functions will be accepted and how they could be implemented. Definition and clarification are necessary to determine whether this mechanism, IOM, or OCHA will be the lead. These items do overlap, and negotiators may create a new mechanism that is doing exactly what others were already built for. However, without direct expansion and participation by these other entities, some overlap in operations will be unavoidable. But there is the possibility that the WIM could evolve into something bigger and more comprehensive. Some individuals in the process of being displaced may choose to move as their livelihoods begin to collapse and may not wait until they are desperate and have suffered. If the WIM can identify communities at risk (by vulnerabilities, hotspots, and similar indicators) and allow for such people to be able to have a fast track through receiving nations' immigration lines, this can be an effective tool. In choosing to leave before the worst damage or suffering occurs, this subset of people will use their own resources to fund their own migration, which would be much more acceptable to receiving nations. In this way, an immigration status as a proposed outcome of the WIM can be based on the continued deterioration of one's homeland, but may not offer any protections or assistance. This function can be a starting point that will allow many nations to experience resettlement due to climate change without the fear of waves of "refugees" imposing on their nations. It also allows for the individual choice of the displacee; one can stay until it is impossible to sustain one's self or choose to leave before hardship occurs. Those who migrate sooner will be individuals and families which the governments of nations with high emissions do not yet have to fund. Without the WIM, this could only be a direct treaty between nations which are already under climate stress and others willing to take in those who wish to leave; within the WIM, there is more possibility.

However, if the WIM does stall and incidences like Cyclone Pam continue to displace those in the Pacific more often, there will be little that can be done without some sort of emergency measure. If the world is to be serious about this phenomenon, it needs to come together in pre-planning by offering an immigration scheme, as suggested above, and thoroughly developing the WIM's planned relocation arm with consideration for temporary housing, job training – basically a replication of IOM's procedures with OCHA's concerns and consideration to the human security and rights of those being displaced. With conditions likely to worsen, the stakes are high for such preparation. If the WIM is stalled, halted, gutted, or killed, the quote from Kerry mentioned earlier *may* be prophetic. Enabling the WIM to fully plan and prepare for inevitable movements and reconstitution can prevent traditional security concerns in this area. If individuals, communities, and nations are left in limbo with dwindling resources, they could become security threats to neighboring individuals, communities, and nations, thereby causing more global upheaval. This does not have to be the case, but can be. There is no immediate reason to expect that those being displaced will be an existential threat to nearby states. The success of UNHCR (in resettlement), IOM (in logistics and projects), and OCHA (in short-term dislocation) demonstrate this. But it will be vitally important to respect the human security and rights concerns of the displaced. In doing so, not only will the UNFCCC's WIM show a sense of global goodwill toward the most vulnerable, but prevent a difficult situation from becoming worse.

Further research

This investigation has only begun to evaluate all of the governance structures which have been implicated in assisting this new group of migrants in the past few years. The literature that surrounds humanitarianism is often paired with development. One major criticism of impartial humanitarian efforts is that although they alleviate suffering, they do nothing to prevent situations that cause suffering in the first place – it is a short-term fix. This comes from the Wilsonian view which seeks to leave those assisted better off than they were before the incident in which the humanitarian aid was needed. This view ties easily into the development literature. Lautze (1996) explains that often the goal of humanitarian assistance is to put the lives of those affected by some sort of disaster back to the levels they were before the incident. However, because climate change will eventually make rebuilding more and more difficult, it will be imperative to investigate the extent to which development mechanisms can be used to rebuild communities in alternate sites or with alternative materials and designs. The literature that connects humanitarian assistance and development is also light and will need to be developed on conceptual and theoretical levels before it will support the addition of climate change. This connection will be an important next step in the advancement of this field. Additionally, development policy may be included in the WIM pre-planning; it will be imperative to evaluate this if it does happen.

On the structural side, further research needs to delve deeper into IGO funding. For the purposes of this project, aggregate budget expenditures and the proportion of funding by certain nations were used. In the future development of governance, earmarked expenditures need to be addressed. Many member states earmark their funds for only specific expenditures – this is especially influential for IOM, as it is directly responsive to member state requests, and this structural component imposes particular constraints. Continuing research needs to take into consideration these earmarked funds and their influence over time on the IGO in a grander international relations context.

A reevaluation of land rights also needs to occur in this context. In a world where every inch of land has been claimed, purchased, or taken by force, resettlement is made very difficult. As the climate continues to change, private property rights may now protect historically illegitimate claims. But it is not only sovereign national land which is vulnerable to new demands; individual land owners and indigenous holdings will also be implicated. The changing climate means that the areas that were once fertile will shift to spaces that are currently used for other purposes; agriculture, coastlines, and wastelands will all realign. Unfortunately, property laws are not necessarily flexible, and many of those who currently own valuable land will see its value drop and yet need to purchase new land. However, falling property values will impoverish many, leaving them unable to afford this. Additionally, changes in land ownership will mean changes in the ownership of newly valued natural resources, including aquifers and minerals. Land disputes are arguably the biggest cause of war the world has ever seen. Without a new conception of ownership which coincides with a changing public good, more disputes may be on the horizon. This area poses the most concern for traditional security threats due to climate change. It is not the displacement of individuals but the displacement of economic value which should cause alarm.

Another area of research that needs development is integrating the concerns of those becoming displaced. In this book, many elites' voices are used to describe the needs of their people. The former president of the Maldives made his country's eventual displacement a major priority by using his office to make speeches and prepare media appearances (such as his underwater cabinet meeting) to bring attention to his nation's plight. Chapter 2 referenced the views of the former prime minister of Tuvalu, Ielemia, and his desire to sue large carbon emitting nations; even the Nansen Initiative's regional consultations are conducted through discussions with local elites, such as NGO representatives and local leadership. But there is little in the literature about the desires of the individual. One of the first surveys to ask those who will eventually be displaced about their thoughts on the matter was conducted in 2013 in the Maldives (Simonelli, 2014a, 2014b). Semi-structured interviews asked individuals on K. Guraidhoo and Dhuvaafaru (in the South Male and Raa Atoll respectively) about if/where they would go if climate processes necessitated that they leave. The majority – 11 out of 16 respondents on Guraidhoo and 15 out of 18 on Dhuvaafaru – said that they would move if need be, and most of those preferred to do so with their entire communities intact. The preferences of the majority were to move internally to either the capital or its redeveloped neighbor, Hulhulmale, but if prompted to think about it, some did have preferences for international relocation, if necessary.[2] Thus while many had not yet thought about the need to internationally relocate, their preferences were far from the proposal made by their outspoken previous president. The preferences of the displacees may be explicitly different than those of their national representatives. Ioane Teitiota, in his fight to stay in New Zealand, calls his bid for refugeehood "migrating with dignity" (Weiss, 2015). This phrase has also been used by other civil society groups in the Pacific (Randall, 2014) and the Kiribati government (Reuters, 2014). There is not enough evidence to assume that the view of the EUI leaders is fully representative of the opinions of their people, and more research is necessary to include the preferences of the displaced in displacement policy.

Finally, this research needs to collaborate with that in the field of physical science. Cooperation in this area can lead to the foundation of a better time line for displacement and the identification of hotspots. Accurate schedules for action can create a frame for appropriate responses which needs to be applied to the most vulnerable areas. Right now those academics who work on migration/displacement issues do not talk across disciplines to those in hydrology, climatology, and so forth. A strong connection will provide a sturdy bridge from science to policy and will connect the knowledge of environmental activity to how it will affect the earth's human inhabitants.

Notes

2 Current State of Affairs

1 https://www.cia.gov/library/publications/the-world-factbook/geos/tv.html
2 CIA World Fact Book.
3 CIA World Fact Book.
4 Added by Leckie in an interview for the *Financial Times*.

3 Hyperbole versus Fact

1 Search preformed March 5, 2015.
2 Plato mentions that he is using names that are references from his own society to describe similar affiliations in Atlantis.
3 Search performed March 6, 2015.
4 Search performed March 7, 2015.
5 Simonelli's (2014) interviews in the Maldives provide personal experience of many years of discussing rising tides on the island of Kandholhudhoo. One elderly gentleman had to be taken to the hospital on a neighboring island after one particularly bad incident where he passed out trying to bail the thigh-high water out of his house.
6 This frame has the potential to be expanded to other areas vulnerable to the same consequences through different climate processes (such as desertification), but for the purposes of this discussion about poor island descriptors, it will remain as is.

4 Academically Understood Context

1 Travel and tourism are also a prominent features of voluntary migration. I have omitted a discussion of them here, as it does not add any theoretical insight to the study at hand. A larger discussion about circular/seasonal migration has also been excluded for the same reasons.
2 Whether deserving of the status or not.
3 This makes it different from the way in which the refugee faces a threat; theirs is a right to security and liberty.
4 Article 1, Section B1, Subsection F.

5 Institutional Expansion

1 A more detailed explanation of this divide will be provided in chapter 7, in a discussion about the development of the climate regime.

6   Lack of Expansion

1   These nations signed during the years 1967 and 1968.
2   In order, these are: Food and Agriculture Organization (FAO), UN office for the Coordination of Humanitarian Affairs (OCHA), United Nations Development Program (UNDP), United Nations Population Fund (UNPFA), United Nations Human Settlements Programme (UNHABITAT), United Nations High Commissioner for Refugees (UNHCR), United Nations Children's Fund (UNICEF), World Food Programme (WFP), and the World Health Organization (WHO). Standing invitees include: International Committee of the Red Cross (ICRC), International Council of Voluntary Agencies (ICVA), International Federation of the Red Cross and Red Crescent Societies (IFRC), InterAction, International Organization for Migration (IOM), Office of the High Commission for Human Rights (OHCHR), Steering Committee for Humanitarian Response (SCHR), Office of the Special Rapporteur on the Human Rights of Internally Displaced Persons (SR on HR of IDPs), and the World Bank (WB) (IASC.org).
3   An IASC typology for climate change-related drivers of migration. Others include: environmental degradation and/or slow onset extreme hazard events, significant permanent losses in state territory as a result of sea level rise, and armed conflict/violence over shrinking natural resources.
4   In the week of October 10, 2011, several islands in the South Pacific, including Tuvalu, ran out of clean water due to the compounded effects of sea level rise and La Nina. The UN and Australia were in talks as to how to handle the matter as a humanitarian emergency.

7   Filling the Governance Gap

1   A negotiating group consisting of Japan, the USA, Switzerland, Canada, Australia, Norway, and New Zealand. Iceland, Mexico, and the Republic of Korea may also attend JUSSCANNZ meetings.
2   This means that there was no formal agreement on this text; it is draft language to be formally agreed upon during the COP session.
3   A poor joke coming out of commentaries about this COP (Roberts, 2013).
4   Emphasis mine.
5   The walkout was tentatively planned when this bloc went into negotiations that evening, as told to me on the evening of the 19th of November 2013 by an anonymous negotiator.
6   Version 18, September 11:00.
7   The case was presented to me in short form by Nicole DeMoor, a lawyer and refugee scholar at the University of Ghent.
8   The states in the UN but not members of IOM are: Andorra, Bahrain, Barbados, Bhutan, Brunei, China, Cuba, Democratic People's Republic of Korea, Dominica, Equatorial Guinea, Ethiopia, Grenada, Indonesia, Iraq, Kiribati, Kuwait, Lao People's Democratic Republic, Lebanon, Liechtenstein, Macedonia, Malaysia, Monaco, Oman, Palau, Qatar, Russian Federation, Saint Kitts and Nevis, Saint Lucia, San Marino, Sao Tome and Principe, Saudi Arabia, Singapore, Solomon Islands, South Sudan, Syria, Tonga, Tuvalu, United Arab Emirates, and Uzbekistan.

8 Conclusion

1 During the 2011 Nansen Conference in Oslo, one of OCHA's African Bureau Chiefs mentioned that it is often the agency of first response regardless of the situation and voiced the concern that those affected by climate-related disasters would swamp her office and other offices and they would not be able to keep up.
2 These included: France, the UK, India, the USA, Sri Lanka, Australia, Macau, and Saudi Arabia.

References

Carterets Integrated Relocation Program Guinea. Papua New Guinea: Tulele Peisa.

1948. *The Universal Declaration of Human Rights.* Geneva: United Nations.

1951. *Constitution of the International Organization for Migration.* Geneva: IOM.

1951. *Protocol Relating to the Status of Refugees.* Geneva: UNHCR.

1967. *Handbook on Procedures and Criteria for Determining Refugee Status under the 1951 Convention and 1967 Protocol Relating to the Status of Refugees.* Geneva: UNHCR.

1969. "Convention on the Status of Refugees." *The American Journal of International Law* 63: 389–407.

1998. "Kyoto Protocol to the United Nations Framework Convention on Climate Change." Geneva: United Nations.

2003. "Bangladesh: The Next Atlantis?" *Environment* 45 (5): 8.

2005. *Report of the Global Commission on International Migration.* Population and Development Review: GCIM.

2005. *Report of the International Meeting to Review the Implementation of the Programme of Action for the Sustainable Development of Small Island States.* Geneva: United Nations.

2008. *About* [Online]. Tulele Peisa. http://www.tulele-peisa.org/about/

2008. *The Bougainville Resettlement Initiative Meeting Report.* Displacement Solutions, Canberra.

2008. "Maldives Saving to Buy New Homeland in Face of Climate Change." *The Telegraph*, November 10.

2009. *Alliance of Small Island States (AOSIS) Declaration on Climate Change.* AOSIS.

2009. "Australia Coastal Living at Risk." *BBC*, October 27.

2009. *Climate Change and Humanitarian Action: Key Emerging Trends and Challenges.* Geneva: OCHA.

2009. "Climate Change Fight shouldn't Hit Development." *Business Standard*, October 23.

2009. "Climate Change, a Major Disaster." *Modern Ghana News*, October 21.

2009. *Compendium of IOM's Activities in Migration, Climate Change and the Environment.* Geneva: IOM.

2009. *Declaration on Climate Change.* New York: Organization of American States.

2009. "Developed Nations Should Lead in Combating Climate Change." *The Hindu*, October 18.

2009. "Factbox: The World's Water and Climate Change." *Reuters*, March 9.

2009. "Islands Fight the Tide." *Television New Zealand*, January 22.

2009. "Money Is the Key to Success of Copenhagen." *The Independent*, November 2.

2009. "Monitoring Disaster Displacement in the Context of Climate Change." In *Internal Displacement.* Geneva: OCHA and Norwegian Refugee Council.

2009. "A New (under) Class of Travelers." *The Economist*, June 25.

2009. "Protection of and Assistance to Internally Displaced Persons: Note by the Secretary-General." In *August 3.* Geneva: United Nations.

2010. *2010 Annual Report OCHA.* Geneva: United Nations.

2010. *Disaster Risk Reduction, Climate Change Adaptation and Environmental Migration*. Geneva: IOM.

2010. *Financial Report for the Year Ended 31 December*. International Organization for Migration.

2010. *Forced Migration Research and Policy Overview of Current Trends and Future Directions*. Oxford: Refugee Studies Centre, University of Oxford.

2010. *Funding UNHCR's Programs*. Geneva: UNHCR.

2010. "Human Rights and Humanitarian Law-Conflict or Convergence." *Case Western Reserve University School of Law*.

2010. *IASC Framework on Durable Solutions for Internally Displaced Persons*. Washington, DC: The Brookings Institution-University of Bern Project on Internal Displacement.

2010. Internal Displacement. *OCHA on Message*.

2010. *Sun Come Up* [Online]. Big Red Barn Films. http://www.suncomeup.com/film/Home.html

2011. "Climate Change, Environmental Degradation and Migration-Background Paper." In *International Dialogue on Migration*. Geneva: IOM.

2011. *Compendium of IOM's Activities on Migration, Climate Change, and the Environment*. Geneva: IOM.

2011. Decision-/CP.17. United Nations Framework Convention on Climate Change.

2011. "Intercessional Workshop on Climate Change, Environmental Degradation and Migration." In *International Dialogue on Migration 2011*. Geneva: IOM.

2011. *Maldives* [Online]. Climate Lab. Accessed March 29. http://climatelab.org/Maldives

2011. "OCHA and Slow Onset Emergencies." In *OCHA Occasional Policy Briefing Series*. Geneva: OCHA.

2011. "Statement by Antonio Guterres United Nations High Commissioner for Refugees." In *Nansen Conference on Climate Change and Displacement*. Oslo: UNHCR.

2012. *Addressing Climate Change and Migration in Asia and the Pacific*. Manila: Asian Development Bank.

2013. *Contribution and Political Elements Related to Human Mobility in the Context of a Warsaw COP 19 Decision on Loss and Damage*. Advisory Group on Climate Change and Mobility.

2013. "Decision-/CP.19." In *Warsaw International Mechanism for Loss and Damage Associated with Climate Impacts*. Geneva: UNFCCC.

2014. "2013 Survey on Environmental Migration." In *Migration, Environment, and Climate Change*. Geneva: IOM.

2014. "Capacity-Building Activities on Migration, Environment, and Climate Change." In *Migration, Environment and Climate Change*. Geneva: IOM.

2014. "Decision 2/CP.20." In *Report of the Conference of the Parties on Its Twentieth Session*, held in Lima Peru from 1 to 14 December 2014. UNFCCC.

2014. "Disaster Risk Reduction and Climate Change Adaptation in IOM's Response to Environmental Migration." In *Info Sheet*. Geneva: IOM.

2014. "Initial 2-Year Work Plan of the Executive Committee of the Warsaw International. Mechanism for Loss and Damage Associated with Climate Change Impacts (ExCom) in Accordance with Decisions 3/CP.18 and 2/CP.19." In *Second Meeting of the Warsaw International Mechanism (WIM) Executive Committee (ExCom)*, September 16–18, Bonn, Germany.

2014. "IOM Outlook on Migration, Environment, and Climate Change." In *Migration, Environment, and Climate Change*. Geneva: IOM.

2014. "IOM Perspectives on Migration, Environment and Climate Change." In *Migration, Environment and Climate Change*. Geneva: IOM.

2014. *Joint Submission on Activities for the Nairobi Work Programme*. Advisory Group on Climate Change and Mobility.

2014. "A New Climate Change Agreement Must Include Human Rights Protections." In *An Open Letter from Special Procedures Mandate-Holders of the Human Rights Council to the State Parties to the UN Framework Convention on Climate Change on the Occasion of the Meeting of the Ad Hoc Working Group on the Durban Platform for Enhanced Action in Bonn (20–25 October 2014)*. Geneva: Office of the High Commissioner for Human Rights.

2014. "Note on International Protection." In *Reports on the Work of the Standing Committee*. Geneva: United Nations General Assembly.

2014. "Push for Late Bougainville Referendum." *Radio New Zealand*, June 30.

2014. *Refugee Protection and International Migration: Trends August 2013–July 2014*. Geneva: UNHCR Division of International Protection.

2014. *UNHCR, the Environment & Climate Change*. Geneva: UNHCR.

2015. "One Tuvalu Island Evacuated after Flooding from Pam." *Radio New Zealand International*, March 18.

Abbott, K., R. O. Keohane, A. Moravcsik, A. M. Slaughter, and D. Snidal. 2000. "The Concept of Legalization." *International Organization* 54 (3): 401–419.

Abbott, K., and D. Snidal. 2000. "Hard and Soft Law in International Governance." *International Organization* 54 (3): 421–456.

Abbott, K. W., and D. Snidal. 1998. "Why States Act through Formal International Organizations." *Journal of Conflict Resolution* 42 (1): 3–32.

Abu-Sahleih, S. A. 1996. "The Islamic Conception of Migration." *International Migration Review* 30 (1): 37–57.

Adelman, H. 2001. "From Refugees to Forced Migration: The UNHCR and Human Security." *International Migration Review* 35 (1): 7–32.

Adelman, H., and S. McGrath. 2007. "To Date or To Marry: That Is the Question." *Journal of Refugee Studies* 20: 376–380.

Agrawala, S. 1999. "Early Science–Policy Interactions in Climate Change: Lessons from the Advisory Group on Greenhouse Gases." *Global Environmental Change* 9: 157–169.

Albala-Bertrand, J. M. 2000. "Responses to Complex Humanitarian Emergencies and Natural Disasters: An Analytical Comparison." *Third World Quarterly* 21 (2): 215–227.

Allen, L. 2004. "Will Tuvalu Disappear Beneath the Sea?" *Smithsonian* 35 (5): 44–52.

Andresen, S. 1998. *The Development of the Climate Regime: Positions, Evaluation and Lessons*. California: University of Southern California, Center for International Studies.

Andresen, Steinar, and Shardul Agrawala. 2002. "Leaders, Pushers and Laggards in the Making of the Climate Regime." *Global Environmental Change* 12: 41–51.

Andrijasevic, R., and W. Walters 2010. "The International Organization for Migration and the International Government of Borders." *Environment and Planning: Society and Space* 28: 977–999.

Anthony, D. W. 1990. "Migration in Archeology: The Baby and the Bathwater." *American Anthropologist* 92: 895–914.

AOSIS. 2013. *Submission of Nauru on Behalf of the Alliance of Small Island States: Views and Information on Elements of an International Mechanism to Address Loss and Damage from the Adverse Effects of Climate Change*. November 12.

Archer, Kevin. 2012. "Rescaling Global Governance: Imaging the Demise of the Nation-State." *Globalizations* 9 (2): 214–256.

Arifin, Z. 1997. "A Clearer Picture of Global Warming." *New Strait Times*, June 26.

Ashutosh, I., and A. Mountz. 2011. "Migration Management for the Benefit of Whom? Interrogating the Work of the International Organization for Migration." *Citizenship Studies* 15: 21–38.

Athanasiou, T., and P. Bear. 2002. *Dead Heat: Global Justice and Global Warming*. Westminster, MD: Seven Stories Press, Canada.

Axelrod, R. 1984. *The Evolution of Cooperation*. New York: Basic Books.

Ayers, Jessica, and Tim Forsyth. 2009. "Community-Based Adaptation to Climate Change." *Environment Magazine*, July/August.

Ayittey. 2009. "Climate Change, a Major Disaster." *Modern Ghana News*, October 21.

Babiker, Mustafa H., Henry D. Jacoby, John M. Reilly, and David M. Reiner. 2002. "The Evolution of a Climate Regime: Kyoto to Marrakesh." In *Joint Program on the Science and Policy of Global Change*, edited by MIT. Cambridge, MA: MIT Press.

Bailey, Adrian J. 2010. "Population Geographies and Climate Change." *Progress in Human Geography* 35 (5): 686–695.

Balter, Michael. 2011. "Human Evolutionary Genetics. Genes Confirm Europeans' Blow to Native Americans." *Science* 334 (6061): 1335.

Banjeree, S., R. Black, and D. Kniveton. 2012. *Migration as an Effective Mode of Adaptation to Climate Change*. Brussels: European Commission.

Barber, R. 2009. "Facilitating Humanitarian Assistance in International and Humanitarian Law." *International Review of the Red Cross* 91 (874): 371–397.

Barnett, M. 2005. "Humanitarianism Transformed." *Perspectives on politics* 3: 723–740.

Barnett, M. 2011. Humanitarianism, Paternalism, and the UNHCR. In *Refugees in International Relations*, edited by A. Betts and G. Loescher. Oxford, UK: Oxford University Press.

Barnett, M. N., and M. Finnemore. 1999. "The Politics, Power, and Pathologies of International Organizations." *International Organization* 53 (4): 703–733.

Barrett, Scott, and Michael Toman. 2010. "Contrasting Future Paths for an Evolving Global Climate Regime." In *Policy Research Working Paper*. The World Bank.

Baumert, Kevin A., James F. Perkaus, and Nancy Kete. 2003. "Great Expectations: Can International Emissions Trading Deliver an Equitable Climate Regime?" *Climate Policy* 3: 137–148.

Bearce, D. H., and S. Bondanella. 2007. "Intergovernmental Organizations, Socialization, and Member-State Interest Convergence." *International Organization* 61 (4): 703–733.

Beckfield, J. 2003. "Inequality in the World Polity: The Structure of International Organization." *American Sociological Review* 68 (3): 401–424.

Belton, K. A., and W. Q. Morales. 2009. "The Multi-Faceted Debate on Human Migration." *The Latin Americanist* 53 (1): 187–210.

Bernauer, T., A. Kalbhenn, V. Koubi, and G. Spilker. 2010. "A Comparison of International and Domestic Sources of Global Governance Dynamics." *British Journal of Political Science* 40: 509–538.

Berringer, A. C. 2010. "Are 'Climate Change Refugees' Really Refugees? An Analysis of the Convention and Appropriate Labels." *Western Political Science Association Annual Meeting*, San Francisco, CA.

Berringer, A. C. 2013. "Migration and Climate Change: Global Governance Regimes and the Incorporation of Climate Change Displacement." In *Climate Change and Global Policy Regimes: Towards Institutional Legitimacy*, edited by Tim Cadman. Hampshire: Palgrave Macmillan.

Betts, A. 2010. "Survival Migration: A New Protection Framework." *Global Governance* 16: 361–382.

Betts, A., and G. Loescher. 2011. *Refugees in International Relations*. Oxford: Oxford University Press.

Beyer, G. 1981. "The Political Refugee: 35 Years Later." *International Migration Review* 15 (1/2): 26–34.

Bharali, Gita. 2007. "Development-Induced Displacement: A History of Transition to Impoverishment and Environmental Degradation." In *Ecology and Environment in North East India: Past and Present*. Assam, India: Dibrugarh University.

Biermann, F., and I. Boas. 2008. "Protecting Climate Refugees." *Environment* (November/December): 8–17.

Biermann, F., P. Pattberg, H. van Asselt, and F. Zelli. 2009. "The Fragmentation of Global Governance Architectures: A Framework Analysis." *Global Environmental Politics* 9 (4): 57–78.

Black, R. 2001a. *Environmental Refugees: Myth or Reality?* Geneva: UNHCR.

Black, R. 2001b. "Fifty Years of Refugee Studies: From Theory to Policy." *International Migration Review* 35 (1): 57–78.

Blitz, Brad K. 2011. "Statelessness and Environmental-Induced Displacement: Future Scenarios of Deterritorialization, Rescue and Recovery Examined." *Mobilities* 6 (3): 433–450.

Boano, C. 2008. "Guide on Climate Change and Displacement." *Forced Migration Online*, August. http://www.forcedmigration.org/guides/fmo046

Boano, C., R. Zetter, and T. Morris. 2008. *Environmentally Displaced People: Understanding the Linkages between Environmental Change, Livelihoods, and Forced Migration*. Oxford: Refugee Studies Centre.

Boatright, R. G. 2009. "Cross-Border Interest Group Learning in Canada and the United States." *American Review of Canadian Studies* 39 (4): 418–437.

Bock, J. G. 2011. "Humanitarian Aid and the Struggle for Peace and Justice: Organizational Innovation after a Blind Date." *Journal of Sociology Social Welfare* 38 (2): 282–309.

Boehmer, C., and T. Nordstrom. 2008. "Intergovernmental Organization Memberships: Examining Political Community and the Attributes of International Organizations." *International Interactions* 34: 282–309.

Bohmer, C., and A. Shuman. 2008. *Rejecting Refugees: Political Asylum in the 21st Century*. New York: Routledge.

Borovnik, M. 2005. "Seaferers' 'Maritime Culture' and the 'I-Kiribati Way of life': The Formation of Flexible Identities." *Singapore Journal of Tropical Geography* 26 (2): 132–150.

Borovnik, M. 2006. "Working Overseas: Seaferers' Remittances and their Distribution in Kiribati." *Asia Pacific Review* 47 (1): 151–161.

Boswell, C. 2008. "Combining Economics and Sociology in Migration Theory." *Journal of Ethnic and Migration Studies* 34 (4): 549–566.

Boswell, C., and P. R. Mueser. 2008. "Introduction: Economics and Interdisciplinary Approaches in Migration Research." *Journal of Ethnic and Migration Studies* 34 (4): 519–529.

Bothe, Michael. 2014. "Doha and Warsaw: Reflections on Climate Law and Policy." *Climate Law* 4: 5–20.

Brachthauser, Christine. 2011. "Explaining Global Governance – A Complexity Problem." *Cambridge Review of International Affairs* 24 (2): 221–244.

Braman, L. M., P. Suarez, and M. K. van Aalst. 2010. "Climate Change Adaptation: Integrating Climate Science into Humanitarian Work." *International Review of the Red Cross Migration* 92 (879): 15–25.

Bronen, Robin. 2013. "Climate-Induced Displacement of Alaska Native Communities." In *Brookings-LSE Project on Internal Displacement*. Washington, DC: Brookings.

Brown, O. 2008. *Migration and Climate Change*. Geneva: IOM.

Brun, C. 2001. "Reterritorializing the Relationship between People and Place in Refugee Studies." *Geografiska Annaler* 83 (1): 15–25.

Buncombe, A. 2009. "Rising Sea Levels Inspire Maldives' Underwater Cabinet Meeting." *New Zealand Herald*, October 7.

Burkett, Maxine. 2014. "Loss and Damage." *Climate Law* 4: 119–130.

Cadman, Tim. 2012. "Evaluating the Quality and Legitimacy of Global Governance: A Theoretical and Analytical Approach." *International Journal of Social Quality* 2 (1): 4–23.

Cameron, F. R. 2011. "Saving the 'Disappearing Islands': Climate Change Governance, Pacific Island States and Cosmopolitan Dispositions." *Continuum: Journal of Media and Cultural Studies* 25 (6): 873–886.

Cardy, W. F. 1994. "Environment and Forced Migration." In *Fourth International Research and Advisory Panel Conference*, Sommerville College, University of Oxford.

Castles, S. 2003. "The International Politics of Forced Migration." *Development* 46 (93): 11–20.

Castles, S. 2007. "Twenty-First-Century Migration as a Challenge to Sociology." *Journal of Ethnic and Migration Studies* 33 (3): 351–371.

Castles, S. 2009. "Development and Migration – Migration and Development: What Comes First? Global Perspectives and African Experiences." *Theoria: A Journal of Social and Political Theory* 56 (121): 1–31.

Cavazos, Tereza. 2012. "Challenges of Mexico to Face Climate Change." In *Experimental and Theoretical Advances in Fluid Dynamics*. Heidelberg: Springer, Berlin.

Cernea, M. 1997. "The Risks and Reconstruction Model for Resettling Displaced Populations." *World Development* 25 (10): 1569–1588.

Chimni, B. 1998. "The Geopolitics of Refugee Studies: From the South." *Journal of Refugee Studies* 11 (4): 350–374.

Clark, W. A. V. 1986. *Human Migration*. London: Sage Publications.

Clarke, S. 2009. "Five Years to Save the World from Climate Change, Says WWF." *ABC*, October 19.

Clemens, E. S., and J. M. Cook. 1999. "Politics and Institutionalism: Explaining Durability and Change." *Annual Review of Sociology* 25: 441–466.

Cohen, R. 2007. "Response to Hathaway." *Journal of Refugee Studies* 20 (3): 370–376.

Colomy, P. 1998. "Neofunctionalism and Neoinstitutionalism: Human Agency and Interest in Institutional Change." *Sociological Forum* 13 (2): 265–300.

Connell, J. 2003. "Losing Ground? Tuvalu, the Greenhouse Effect and the Garbage Can." *Asia Pacific Viewpoint* 44 (2): 89–107.

Considine, M. 2009. "Sea Level Rise: The View from Ground Zero." *ECOS* (April/May).

Cordes-Holland, Owen. 2008. "The Sinking of the Strait: The Implications of Climate Change For Torres Strait Islanders' Human Rights Protected by the ICCPR." *Melbourne Journal of International Law* 9 (2).

Corson, C. 2010. "Shifting Environmental Governance in a Neoliberal World: US Aid for Conservation." *Antipode* 42 (3): 576–602.

Cottier, T. 2009. "Multilayered Governance, Pluralism, and Moral Conflict." *Indiana Journal of Global Legal Studies* 16 (2): 647–679.

Crisp, J. 2003. *A New Asylum Paradigm? Globalization, Migration and the Uncertain Future of the International Refugee Regime.* Geneva: UNHCR.

Crush, J., and S. Ramachandran. 2010. "Xenophobia, International Migration and Development." *Journal of Humanitarian Development and Capabilities* 11 (2): 209–228.

Davie, M. R. 1957. "Review." *Annals of the American Academy of Political Science* 296.

de Haas, H. 2006. "Turning the Tide? Why 'Development Instead of Migration' Policies are Bound to Fail." In *Working Papers.* Oxford: International Migration Institute.

de Waal, A. 2010. "The Humanitarians' Tragedy: Escapable and Inescapable Cruelties." *Disasters* 34 (s2): 130–137.

DeWind, J. 2007. "Response to Hathaway." *Journal of Refugee Studies* 20: 381–385.

Domroes, M. 2001. "Conceptualizing State-Controlled Resort Islands for an Environment-Friendly Development of Tourism: The Maldivian Experience." *Singapore Journal of Tropical Geography* 22 (2): 122–137.

Donnelly, J. 1981. "Recent Trends in the UN Human Rights Activity: Description and Polemic." *International Organization* 35 (4): 633–655.

Donnelly, J. 1986. "International Human Rights: A Regime Analysis." *International Organization* 40 (3): 599–642.

Dorussen, H., and H. Ward. 2008. "Intergovernmental Organizations and the Kantian Peace." *Journal of Conflict Resolution* 52 (2): 189–212.

Dowie, M. 2011. *Conservation Refugees.* Cambridge: MIT Press.

Downing, T. E. 2002. *Avoiding New Poverty: Mining-Induced Displacement and Resettlement.* London: International Institute for Environment and Development and World Business Council for Sustainable Development.

Dun, O., and F. Gemenne. 2008. "Defining 'Environmental Migration'." *Forced Migration Review* 31: 10–11.

Duvell, F. 2005. "Globalization of Migration Control." *Crossing Over: Comparing Recent Migration in the United States and Europe* 23 Lexington MA: Lexington Books, 23–46.

Dwivedi, R. 2002. "Models and Methods in Development-Induced Displacement (Review Article)." *Development and Change* 33 (4): 709–732.

Eberlein, B., and A. L. Newman. 2008. "Escaping the International Governance Dilemma? Incorporated Transgovernmental Networks in the European Union." *Governance: An International Journal of Policy, Administration, and Institutions* 21 (1): 25–52.

Ede, P. M. 2002/2003. "That Sinking Feeling." *Earth Science Journal* 17 (4): 39–40.

El-Hinnawi, E. 1985. *Environmental Refugees.* Nairobi, Kenya: UNEP.

Elie, J. 2010. "The Historical Roots of Cooperation between the Commissioner for Refugees and the International Organization for Migration." *Global Governance* 16: 345–360.

Elliott, Lorraine. 2012. "Climate Change and Migration in Southeast Asia: Responding to a New Human Security Challenge." In *Asia Security Policy Series*. Singapore: Centre for Non-Traditional Security Studies.

Emanuel, Kerry. 2015. "Severe Tropical Cyclone Pam and Climate Change." *Real Climate: Climate Science from Climate Scientists*, March 31.

Evans, J. 2008. "Climate Change Drives Maldives to Buy Land. *Financial Times*, November 10.

Eveland, W. P., A. F. Hayes, D. V. Shah, and N. Kwak. 2005. "Understanding the Relationship between Communication and Political Knowledge: Comparison Approach Using Panel Data." *Political Communication* 22: 423–446.

Farbotko, C. 2005. "Tuvalu and Climate Change: Constructions of Environmental Displacement in the Sydney Morning Herald." *Geografiska Annaler: Series B, Human Geography* 87: 279–293.

Farbotko, C. 2010. "The Global Warming Clock Is Ticking to See These Places While You Can': Voyeuristic Tourism and Model Environmental Citizens on Tuvalu's Disappearing Islands." *Singapore Journal of Tropical Geography* 31: 224–238.

Farbotko, C., and H. V. McGregor. 2010. "Copenhagen, Climate Science and the Emotional Geographies of Climate Change." *Australian Geographer* 41 (2): 159–166.

Farrag, M. 1997. "Managing International Migration in Developing Countries." *International Migration (Geneva, Switzerland)* 35 (3): 315–336.

Fatoric, Sandra. 2014. "Migration as a Climate Adaptation Strategy in Developed Nations." In *Briefer*. Washington, DC: Center for Climate and Security.

Felts, T. 2005. "US Firms Return to Expansion, Opening Branches Internationally." *Of Counsel: The Legal and Management Report* 24 (1).

Ferris, Elizabeth. 2012. "Protection and Planned Relocations in the Context of Climate Change." In *Legal and Protection Policy Research Series*. Geneva: UNHCR.

Finnemore, M., and K. Sikkink. 1998. "International Norm Dynamics and Political Change." *International Organization* 52 (4): 887–917.

Finucane, Melissa L. 2009. "Why Science Alone Won't Solve the Climate Crisis: Managing Climate Risks in the Pacific." In *Asia Pacific Issues*. Hawaii: East West Center.

Fitzgerald, Sandy. 2015. "John Kerry Warns Ambassadors about 'Climate Refugee' Threat." *News Max*, March 31.

Fligstein, N., and R. Freeland. 1995. "Theoretical and Comparative Perspectives on Corporate Organization." *Annual Review of Sociology* 21: 21–43.

Fulu, E. 2007. "Gender, Vulnerability, and the Experts: Responding to the Maldives Tsunami." *Development and Change* 38 (5): 843–864.

Funk, M. 2009. "Come Hell or High Water." *World Policy Journal* Summer 26 (2): 93–94.

Gallagher, D. 1989. "The Evolution of the International Refugee System." *The International Migration Review* 23 (3): 579–598.

Gelsdorf, K. 2010. *Global Challenges and Their Impact on International Humanitarian Action*. Geneva: OCHA.

Gemenne, F., and A. Magnan. 2011. *The Other Migrants Preparing for Change: Environmental Changes and Migration in the Republic of Mauritius*. Geneva: IOM.

Gemenne, F., and Shawn Shen. 2009. "Tuvalu and New Zealand D 2.3.2.3." In *Scientific Support to Policies*. Brussels: EACH-FOR.

Goodwin, G. 1960. "The Expanding United Nations." *International Affairs* 36 (2): 174–187.

Goodwin-Gill, G. S. 1989. "International Law and Human Rights: Trends Concerning International Migrants and Refugees." *International Migration Review* 23 (3): 526–546.

Goodwin-Gill, G., and J. McAdam. 2007. *The Refugee in International Law*. Oxford, UK: Oxford University Press.

Graff, Van de, and De Ville. 2013. "Regime Complexes and Interplay Management." *International Studies Review* 5: 568–571.

Grahl-Madsen, A. 1966. "The European Tradition of Asylum and the Development of Refugee Law." *Journal of Peace Research* 3 (3): 278–289.

Grahl-Madsen, A. 1983. "Identifying the World's Refugees." *Annals of the American Academy of Political Science* 467: 11–23.

Gramlich, E. M. 2002. "The Methodology of Cost Benefit Analysis: Avoiding Politics in Decision Making." Remarks by Governor Edward M. Gramlich at the Werner Sichel Economics Lecture-Seminar Series, Western Michigan University, Kalamazoo, Michigan, October 16.

Greenwood, C. 2010. "Human Rights and Humanitarian Law-Conflict or Convergence." *Case Western Reserve Journal of International Law* 43: 491.

Guterres, A. 2011. "Statement by Antonio Guterres, United Nations High Commissioner for Refugees." Nansen Conference on Climate Change and Displacement, Oslo, Norway.

Haas, E. B. 1953. "The Balance of Power: Prescription, Concept, or Propaganda?" *World Politics* 5: 442–477.

Haas, E. B. 1958. *The Uniting of Europe: Political, Social, and Economic Forces, 1950–1957*. Indiana: University of Notre Dame Press.

Hafner-Burton, E. M. 2005. "Trading Human Rights: How Preferential Trade Agreements Influence Government Repression." *International Organization* 59 (3): 593–629.

Hafner-Burton, E. M., and Kiyoteru Tsutsui. 2007. "Justice Lost! The Failure of International Human Rights Law to Matter Where Needed Most." *Journal of Peace Research* 44 (4): 407–425.

Hakovirta, H. 1993. "The Global Refugee Problem: A Model and Its Applications." *International Political Science Review* 14 (1): 35–57.

Hall, S. 1999. "Thinking the Diaspora: Home Thoughts from Abroad." *Small Axe* 6: 1–18.

Hammar, A. 2008. "In the Name of Sovereignty: Displacement and State Making in Post-Independence Zimbabwe." *Journal of Contemporary African Studies* 26 (4): 417–434.

Hammerstad, A. 2011. "UNHCR and the Securitization of Forced Migration." In *Refugees in International Relations*, edited by A. Betts and G. Loescher. Oxford, UK: Oxford University Press.

Hardin, G. 1968. "The Tragedy of the Commons." *Science* 162 (3859): 1243–1248.

Hare, William, Claire Stockwell, Christian Flachsland, and Sebastian Oberthur. 2010. "The Architecture of the Global Climate Regime: A Top-Down Perspective." *Climate Policy* 10: 600–614.

Hathaway, J. C. 1984. "The Evolution of Refugee Status in International Law." *The International and Comparative Law Quarterly* 33 (2): 1920–1950.

Hathaway, J. C. 2007. "Forced Migration Studies: Could We Just Agree to Date?" *Journal of Refugee Studies* 20 (3): 349–369.

Haug, S. 2008. "Migration Networks and Migration Decision-Making." *Journal of Ethnic and Migration Studies* 34: 585–605.

Haviland, C. 2009. "Maldives Anger at Climate Inertia." *BBC*, November 9.

Hawkins, D. 2004. "Explaining Costly International Institutions: Persuasion and Enforceable Human Rights Norms." *International Studies Quarterly* 48 (4): 779–804.

Hayes-Brown, L. 2004. "Climate Change Its Effect Explored." *Jamaica Information Service*, March 4.

Heinmiller, B. T. 2007. "Do Intergovernmental Institutions Matter? The Case of Water Diversion Regulation in the Great Lakes Region." *Governance: An International Journal of Policy Administration and Institutions* 20 (4): 655–674.

Held, D. 1995. *Democracy and the Global Order: From the Modern State to Cosmopolitan Governance*. Stanford, CA: Stanford University Press.

Helton, A. C. 2001. "Bureaucracy and the Quality of Mercy." *International Migration Review* 35 (1): 192–225.

Helton, A. C. 2002. "Review: Refugee Rights and Realities: Evolving International Concepts and Regimes by Frances Nicholson; Patrick Twomey." *Human Rights Quarterly* 24 (2): 561–563.

Heming, L., P. Waley, and P. Rees. 2001. "Reservoir Resettlement in China: Past Experience and the Three Gorges Dam." *The Geographical Journal* 167: 195–212.

Hilhorst, D., and B. J. Jansen. 2010. "Humanitarian Space as Arena: On the Everyday Politics of Aid." *Development and Change* 41 (6): 1117–1139.

Horst, H. A. 2007. "You Can't Be in Two Places at Once': Rethinking Transnationalism through Jamaican Return Migration." *Identities: Global Studies in Culture and Power* 14: 63–83.

Hu, W., L. J. Cox, J. Wright, and T. R. Harris. 2008. "Understanding Firms' Relocation Decisions Using Self-Reported Factor Importance Rating." *The Review of Regional Studies* 38 (1): 67–88.

Huq, S., and Y. Sokona. 2001. "Climate Change Negotiations. A View from the South." *Opinion: World Summit on Sustainable Development*.

Huyck, E. E., and L. F. Bouvier. 1983. "The Demography of Refugees." *The Annals of the American Academy of Political and Social Science* 467: 39–61.

Hyman, G., and D. Gleave. 1978. "A Reasonable Theory of Migration." *Transactions* 3 (2): 179–201.

Hyndman, J. 2009. "Acts of Aid: Neoliberalism in a War Zone." *Antipode* 41 (5): 867–889.

Ielemia, A. 2007. "A Threat to Our Human Rights, Tuvalu's Perspective on Climate Change." *UN Chronicle* 44 (2), June.

Ikenberry, G. J. 2010. "A Crisis of Global Governance?" *Current History* 109 (730): 315–321.

Ingram, P., J. Robinson, and M. L. Busch. 2005. "The Intergovernmental Network of World Trade: IGO Connectedness, Governance, and Embeddedness." *American Journal of Sociology* 3: 824–858.

IO. 1947a. "General International Organizations." *International Organization* 1 (1): 144–145.

IO. 1947b. "International Refugee Organization." *International Organization* 1 (1): 137–138.

IO. 1947c. "VI. Other Functional Agencies." *International Organization* 1 (3): 556.

IO. 1947d. "Constitution for the International Refugee Organization, as Approved by the General Assembly, December 15, 1946." *International Organization* 1: 577–590.

IO. 1949. "Specialized Agencies." *International Organization* 3 (1): 156–158.

IO. 1950. "Specialized Agencies." *International Organization* 4 (3): 492–493.

IO. 1952. "International Refugee Organization." *International Organization* 6 (3): 447–448.

IO. 1952. "Specialized Agencies." *International Organization* 6 (2): 306–310.

IO. 1952. "Specialized Agencies." *International Organization* 5 (3): 125–127.

Ionesco, D., and M. Traore Chazalnoël. 2015. "The Role of the International Organization for Migration (IOM) in the International Governance of Environmental Migration." In *Organizational Perspectives on Environmental Migration*, edited by K. Rosenow-Williams and F. Gemenne. London: Routledge.

IPCC. 2007. "Summary for Policymakers." In *Climate Change 2007: The Physical Science Basis. Contribution of Working Group I to the Fourth Assessment Report of the Intergovernmental Panel on Climate Change*, edited by S. Solomon, D. Qin, M. Manning, Z. Chen, M. Marquis, K. B. Averyt, M. Tignor, and H. L. Miller. Cambridge, UK and New York: Cambridge University Press.

IPCC. 2013. "Summary for Policymakers." In *Climate Change 2013: The Physical Science Basis. Contribution of Working Group I to the Fifth Assessment Report of the Intergovernmental Panel on Climate Change*, edited by T. F. Stocker, D. Qin, G.-K. Plattner, M. Tignon, S. K. Allen, J. Boschung, A. Nauels, Y. Xia, V. Bex, and P. M. Midgley. Cambridge, UK and New York: Cambridge University Press.

Jacobsen, K. 2002. "Can Refugees Benefit the State? Refugee Resources and African State Building." *The Journal of Modern African Studies* 40 (4): 577–596.

Jain, Abhimanyu George. 2014. "The 21st Century Atlantis: The International Law of Statehood and Climate Change-Induced Loss of Territory." *Stanford Journal of International Law* 50 (1).

James, S. 2009. "Carteret Islands Evacuation: Climate Change Refugees in the Pacific Retrieved from Blue Living Ideas." Accessed May 25, 2009. http://blue-livingideas.com/topics/climate-change/carteret-islands-evacuation-climate-change-refugees-pacific/

Kahler, Miles. 2013. "Rising Powers and Global Governance: Negotiating Change in a Resilient Status Quo." *International Affairs* 89 (3): 711–729.

Kakahel, Shafqat. 2012. "Financing of Climate Change-Related Actions: Recent Developments." In *Peace and Sustainable Development in South Asia*, edited by Sarah S. Aneel, Uzma T. Haroon, and Imrana Niazi. Lahore: Sang-E-Meel Publications.

Karatani, R. 2005. "How History Separated Refugee and Migrant Regimes: In Search of their Institutional Origins." *International Journal of Refugee Law* 23 (4): 517–541.

Kartiki, Katha. 2011. "Climate Change and Migration: A Case Study from Rural Bangladesh." *Gender & Development* 19 (1): 23–38.

Keck, M. E., and K. Sikkink. 1999. "Advocacy Networks in International and Regional Politics." *International Social Science Journal* 51 (159): 89–101.

Kench, P. S., and R. W. Brander. 2006. "Wave Processes on Coral Reef Flats: Implications for Reef Geomorphology Using Australian Case Studies." *Journal of Coastal Research* 22: 209–233.

Kench, P. S., R. F. McLean, and S. L. Nichol. 2005. "New Model of Reef-Island Revolution: Maldives, Indian Ocean." *Geology* 33 (2): 145–148.

Kennedy, D. 1986. "International Refugee Protection." *Human Rights Quarterly* 8 (1): 1–69.

Kennedy, W. 2008. "Dance of Dangerous Sea." *Canadian Geographic* 128 (5); 851–869.

Kent, R. C. 2004. "International Humanitarian Crises: Two Decades Before and Two Decades Beyond." *International Affairs* 80 (5): 851–869.

Keohane, R. 1984. *After Hegemony: Cooperation and Discord in World Politics.* Princeton, NJ: Princeton University Press.

Keohane, R., and L. Martin. 1995. "The Promise of Institutionalist Theory." *International Security* 20 (1): 629–652.

Keohane, R., and D. Victor. 2011. "The Regime Complex for Climate Change." *The Harvard Project on International Climate Agreements.* Cambridge, MA: John F. Kennedy School of Government.

Kinsella, J., and E. Brehony. 2009. "Are Current Aid Strategies Marginalizing the Already Marginalized? Cases from Tanzania." *Development in Practice* 19 (1): 383–396.

Kneebone, S. 2010. "The Governance of Labor Migration in Southeast Asia." *Global Governance: A Review of Multilateralism and International Organizations* 16: 383–396.

Kohut, A., and R. Wike. 2008. "Xenophobia on the Continent." *The National Interest*, November/December 46–52.

Kolers, Avery. 2012. "Floating Provisos and Sinking Islands." *Journal of Applied Philosophy* 29 (4): 333–343.

Kolmannskog, Vikram. 2012. "Climate Change, Environmental Displacement and International Law." *Journal of International Development* 24: 1071–1081.

Korememos, B., C. Lipson, and D. Snidal. 2001. "The Rational Design on International Institutions." *International Organization* 55 (4): 761–799.

Koser, K. 2008. "Gaps in IDP Protection." *Forced Migration Review* (October).

Koser, K. 2010. "Introduction: International Migration and Global Governance." *Global Governance* 16 (3): 301–315.

Krasner, S. D. 1993. "Global Communications and National Power: Life on the Pareto Frontier." In *Neorealism and Neoliberalism, The Contemporary Debate*, edited by D. A. Baldwin, 116–140. New York: Columbia University Press.

Krenz, F. E. 1966. "The Refugee as a Subject of International Law." *The International and Comparative Law Quarterly* 15 (1): 90–116.

Kreps, G. 1984. "Sociological Inquiry and Disaster Research." *Annual Review of Sociology* 10: 309–330.

Kuhl, S. 2009. "Capacity Development as the Model for Development Aid Organizations." *Development and Change* 40 (3): 551–577.

Kuhlman, T. 1991. *Towards a Definition of Refugees.* Oxford: Refugee Studies Centre.

Kumar, S., and U. Waheed. 2007. "Foreign Market Entry and Expansion – Directions for Strategic Organizations Growth Based on a Global System Perspective. *Information Knowledge Systems Management* 6: 177–196.

Laczko, F., and K. Warner. 2008. "Migration, Environment and Development: New Directions for Research." *International Migration and Development, Continuing the Dialogue: Legal and Policy Perspectives.* New York and Geneva: International Organization for Migration and Center for Migration Studies (CMS).

Laczko, F., and B. Wijkstrom. 2010. *Enhancing the Contribution of Migration Research to Policy Making: Intergovernmental Workshop.* Geneva: IOM.

Lake, D. A. 2010. "Rightful Rules: Authority, Order, and the Foundations of Global Governance." *International Studies Quarterly*: 587–613.

Lane, J.-E., and S. O. Ersson. 2000. *The New Institutional Politics: Performance and Outcomes.* London: Routledge.

Lasas, A. 2008. "Restituting Victims: EU and NATO Enlargements through the Lenses of Collective Guilt." *Journal of European Public Policy* 15 (1): 98–116.

Lateu, J. 2008. "That Sinking Feeling." *New Internationalist*, March 1.

Laucci, C. 2009. "Customary International Humanitarian Law Study: Fundamental Guarantees." *Slovenian Law Review* 6: 191.

Lautze, S. 1996. "Lives versus Livelihoods How to Foster Self Sufficiency and Productivity of Disaster Victims." In *Occasional paper Number One.* Washington, DC: Office of US Foreign Disaster Assistance Bureau for Humanitarian Agency for International Development.

Leckie, S. 2008. "Human Rights Implications." *Forced Migration Review* 31.

Leebaw, B. 2007. "The Politics of Impartial Activism: Humanitarianism and Human Rights." *Perspectives on Politics* 5 (2): 223–239.

Lentz, E. C., and C. B. Barrett. 2008. "Improving Food Aid: What Reforms Would Yield the Highest Payoff?" *World Development* 36 (7): 1152–1172.

Lewis, G. C. 1982. *Human Migration: A Geographical Perspective.* London: Croom Helm.

Linde, T. 2009. "Humanitarian Assistance to Migrants Irrespective of Their Status – Towards a Non-Categorical Approach." *International Review of the Red Cross* 91 (875): 171–180.

Locke, J. T. 2009. "Climate Change-Induced Migration in the Pacific Region: Sudden Crisis and Long-Term Developments." *The Geographical Journal* 175: 171–180.

Loughry, M. 2009. "Stressed Islands No Longer Pacific." *Eureka Street* 19: 42.

Loughry, M., and J. McAdams. 2008. "Kiribati – Relocation and Adaptation." *Forced Migration Review* 31.

Lynas, M. 2004. "Warning from a Warming World." *Geographical* 76 (5): 51.

Mace, M. J., and M. Schaeffer. 2013. "Loss and Damage under the UNFCCC: What Relationship to the Hyogo Framework?" *Climate Analytics* (http://www.lossand-damage. net).

Malin, P. M. 1947. "The Refugee: A Problem for International Organization." *International Organization* 1 (3): 443–459.

Malkki, L. H. 1995. "Refugees and Exile: From 'Refugee Studies' to the National Order of Things." *Annual Review of Anthropology* 24: 495–523.

Malla, Katak. 2012. "The EU and Climate Treaty Negotiations after the Durban Conference." *European Policy Analysis* March: 2epa.

Mangalam, J. J., and H. K. Schwartzweller. 1968. "General Theory in the Study of Migration: Current Needs and Difficulties." *International Migration Review* 3 (1): 3–18.

Mansfield, E., H. V. Milner, and E. Reinhardt. 2002. "Why Democracies Cooperate More: Electoral Control and International Trade Agreements." *International Organization* 56 (3): 323–329.

Marchi, S. 2010. "Global Governance: Migration's Next Frontier." *Global Governance* 16: 301–318.

Martin, S. 2004. "Making the UN Work: Forced Migration and Institutional Reform." *Journal of Refugee Studies* 17 (3): 107–150.

McCarthy, J., O. Canziani, N. Leary, D. Dokken, and K. White. 2001. "Climate Change 2001: Impacts, Adaptation and Vulnerability." In *Contribution of Working Group II to the Third Assessment Report of the Intergovernmental Panel on Climate Change*. Cambridge: Cambridge University Press.

McCormick, J. M. 1980. "Intergovernmental Organizations and Cooperation among Nations." *International Studies Quarterly* 24: 75–98.

McLeman, R. 2011. "Climate Change, Migration and Critical International Security Considerations." *IOM Migration Research Series*. Geneva: IOM.

McLeman, R., and B. Smit. 2006. "Migration as Adaptation to Climate Change." *Climatic Change* 76: 31–53.

McNamara, D. 2006. "Humanitarian Reform and New Institutional Responses." *Forced Migration Review*, 9–10.

McNamara, Karen Elizabeth. 2014. "Exploring Loss and Damage at the International Climate Talks." *International Journal of Disaster Risk Science* 5 (3): 242–246.

McNeill, W. H. 1978. "Human Migration." *Bulletin of the American Academy of Arts and Sciences* 31 (8): 8–17.

McNeill, W. H. 1984. "Human Migration in Historical Perspective." *Population and Development* 10 (1): 1–18.

Mearsheimer, J. J. 1995. "The False Promise of International Relations." *International Security* 19 (3): 340–363.

Messina, C. 2007. "Strengthening the Humanitarian Coordinator System." *Forced Migration Review* 29: 23.

Meyer, J. W., and B. Rowan. 1977. "International Organizations: Formal Structure as Myth and Ceremony." *American Journal of Sociology* 83: 340–363.

Milner, H. V. 1997. *Interests, Institutions, and Information: Domestic Politics and International Relations*. Princeton, NJ: Princeton University Press.

Minnich, D. J. 2005. "Veto Players, Electoral Incentives, and International Commitments: The Impact of Domestic Institutions on Intergovernmental Organization Membership." *European Journal of Political Research* 44: 295–325.

Mitchell, Ronald B. 2005. "Flexibility, Compliance and Norm Development in the Climate Regime." In *Implementing the Climate Regime*, edited by Olav Schram, Jon Hovi Stokke, and Geir Ulfstein. London: Earthscan.

Mitrany, D. 1948. "The Functional Approach to World Organization." *International Affairs* 24: 350–363.

Mittelman, J. H. 2010. "Crisis and Global Governance: Money. Discourses, and Institutions." *Globalizations* 7 (1/2): 157–172.

Monbiot, G. 2009. "Climate Change Displacement Has Begun – But Hardly Anyone has Noticed." *The Guardian*, May 7.

Moravcsik, A. 2000. "The Origins of Human Rights Regimes: Democratic Delegation in Post War Europe." *International Organization* 54 (2): 217–252.

Morgenthau, H. J. 1985. "A Realist Theory of International Politics." In *Politics among Nations: The Struggle for Power and Peace*, edited by Kenneth W. Thompson and W. David Clinton. New York: McGraw Hill.

Morris, R. 2009. "To the Lifeboats." *Mother Jones* 34 (6).

Morrow, J. D. 1999. "The Strategic Setting of Choices: Signaling, Commitment, and Negotiation in International Politics." In *Strategic Choice and International Relations*, edited by D. A. Lake and R. Powell, 77–114. Princeton, NJ: Princeton University Press.

Morton, A. 2009. "Climate Change must be 'a Factor' in Deciding Whether to Rebuild." *The Age*, February 11.

Mosley, P. 2008. "Aid Volatility, Policy and Development." *World Development* 36 (10): 2082–2102.

Mullikin, T. 2007. *Global Solutions: Demanding Total Accountability for Climate Change*. Charlotte: Vox Populi Publishers.

Najam, Adil, Saleem Huq, and Youba Sokona. 2003. "Climate Negotiations beyond Kyoto: Developing Countries Concerns and Interests." *Climate Policy* 3: 221–231.

Nassari, J. 2008. "Refugees and Forced Migrants at the Crossroads: Forced Migration in a Changing World." *Journal of Refugee Studies* 22 (1): 1–10.

Ness, G. D., and S. R. Brechin. 1988. "Bridging the Gap: International Organizations as Organizations." *International Organization* 42 (2): 245–273.

Nettleford, R. 2010. "Migration, Transmission and Maintenance of the Intangible Heritage." *Museum International* 56: 1–2.

Newland, K. 2010. "The Governance of International Migration: Mechanisms, Processes, and Institutions." *Global Governance: A Review of Multilateralism and International Organizations* 16: 331–343.

Nickel, J. M. 2002. "Human Rights as Global Governance." *The Journal of Ethics* 6 (4): 353–371.

Nine, C. 2010. "Ecological Refugees, States Borders, and the Lockean Proviso." *Journal of Applied Philosophy* 27 (4): 359–379.

O'Brien, E. 2015. "An Islander's Bid to be the World's First Climate Refugee." *Bloomberg Business*, March 30.

O'Brien, P. 1996. "Migration and Its Risks." *International Migration Review* 30: 1067–1077.

Okereke, C., P. Baral, and Y. Dagnet. 2014. "Options for Adaptation and Loss and Damage in a 2015 Climate Agreement." In *Working Paper*. Washington, DC: Agreement for Climate Transformation 2015 (ACT 2015).

Oliver-Smith, A. 2002. "Displacement, Resistance, and the Critique of Development: From the Grass-Roots to the Global." In *RSC Working Paper*. Oxford: Refugee Studies Centre, University of Oxford.

Olson, M. 1965. *The Logic of Collective Action*. Cambridge: Harvard University Press.

Omidi, M. 2009. "Preview-Maldives Goes Underwater for Climate Change." *Reuters*, Accessed October 15, 2009. http://www.reuters.com/article/idUSCOL357547

Orsini, Amandine, Jean-Frederic Morin, and Oran Young. 2013. "Regime Complexes: A Buzz, a Boom, or a Boost for Global Governance?" *Global Governance* 19: 27–39.

Ostrom, E., R. Gardner, and J. Walker. 1994. *Rules, Games, and Common-Pool Resources*. Ann Arbor: University of Michigan Press.

Overbeek, H., K. Dingwerth, and D. Compagnon. 2010. "Forum: Global Governance: Decline or Maturation of an Academic Concept?" *International Studies Review* 12: 696–719.

Papademetriou, D. 1997/1998. "Migration." *Foreign Policy* 109: 15–31.

Papademetriou, D. 2003. "Managing Rapid and Deep Change in the Newest Age of Migration." *The Political Quarterly* 74: 39–58.

Parks, B. C., and J. T. Roberts. 2006. "Globalization, Vulnerability to Climate Change, and Perceived Injustice." *Society and Natural Resources* 19: 337–355.

Parks, Bradley C., and Timmons Roberts. 2008. "Inequality and the Global Climate Regime: Breaking the North-South Impasse." *Cambridge Journal of International Affairs* 21 (4): 621–648.

Patel, Samir S. 2006. "Climate Science: A Sinking Feeling." *Nature* 440 (7085): 734–736.

Pecoud, A., and P. de Guchteneire. 1998. "Migration without Borders: An Investigation into Free Movement of People." *Third World Quarterly* 19 (1): 7–27.

Petersen, W. 1978. "International Migration." *Annual Review of Sociology* 4: 533–575.

Phillips, T. 2010. "Awarding of Brazilian Dam Contract Prompts Warning of Bloodshed." *The Guardian*, April 21.

Plato. *Critias*. http://ascendingpassage.com/plato-atlantis-critias.htm

Plato. *Timaeus*. http://ascendingpassage.com/plato-atlantis-timaeus.htm

Prasad, Biman Chand, and Paresh Kumar Narayan. 2008. "Reviving Growth in the Fiji Islands: Are We Swimming or Sinking?" *Pacific Economic Bulletin* 23 (2): 5–26.

Price, T. 2003. "High Tide in Tuvalu." *Sierra* July/August.

Princen, S. 2009. "Agenda-Setting in the European Union." *Journal of Common Market Studies* 48 (1): 176–177.

Pronk, J. 1993. "Migration: The Nomad in Each of Us." *Population and Development Review* 19: 323–327.

Puchala, D. 1970. "International Transactions and Regional Integration." *International Organization* 24: 732–763.

Putnam, R. 1988. "Diplomacy and Domestic Politics: The Logic of the Two-Level Game." *International Organization* 42 (3): 341–351.

Rakova, Ursula. 2014. *The Ocean Is Taking Away My Island*. The Blog: Huffington Post.

Ramachandran, V., E. J. Rueda-Sabater, and R. Kraft. 2009. "Rethinking Fundamental Principles of Global Governance: How to Represent States and Populations in Multilateral Institutions." *An International Journal of Policy, Administration, and Institutions* 22: 314–351.

Ramesh, R. 2008. "Paradise Almost Lost: Maldives Seek to Buy a New Homeland." *The Guardian*, November 9.

Randall, A. 2014. "Don't Call Them 'Refugees': Why Climate Change Victims Need a Different Label." *Guardian*, September 18.

Rees, E. 1960. "The Refugee Problem: Joint Responsibility." *Annals of the Academy of Political and Social Science* 329: 15–22.

Reuters, 2014. "Township in Solomon Islands Is 1st in Pacific to Relocate Due to Climate Change." *Scientific American*, August 15.

Rey, D., and J. Barkdull. 2005. "Why Do Some Democratic Countries Join More Intergovernmental Organizations Than Others?" *Social Science Quarterly* 86 (2): 291–309.

Risse, T. 2005. "Neofunctionalism, European Identity, and the Puzzles of European Integration." *Journal of European Public Policy* 12: 291–309.

Robbins, R. 1956. "The Refugee Status: Challenges and Response." *Law and Contemporary Problems* 21 (2): 311–333.

Roberts, Debra. 2013. "Cities OPT in While Nations COP Out: Reflections of COP18." *South African Journal of Science* 109 (5/6).

Roberts, J. Timmons. 2011. "Multipolarity and the New World (Dis)Order: US Hegemonic Decline and the Fragmentation of the Global Climate Regime." *Global Environmental Change* 21: 776–784.

Roberts, J. T., and B. C. Parks. 2007. *A Climate of Injustice: Global Inequality, North-South Politics, and Climate Policy*. Cambridge/MA-London.

Robinson, W. Courtland. 2003. "Risks and Rights: The Causes, Consequences, and Challenges of Development-Induced Displacement." In *An Occasional Paper*. Washington, DC: Brookings Institution – SAIS Project on Internal Displacement.

Rogers, R. 1992. "The Future of Refugee Flows and Policies." *The International Migration Review* 26 (4): 1112–1143.

Rosamond, B. 2005. "The Uniting of Europe and the Foundation of EU Studies: Revisiting the Neofunctionalism of Haas." *Journal of European Public Policy* 12 (2): 237–254.

Rubenstein, M. J. 1936. "The Refugee Problem." *International Affairs* 15 (5): 716–734.

Rubin, S. J., and A. P. Schwartz. 1951. "Refugees and Reparations." *Law and Contemporary Problems* 16 (3): 377–394.

Rucker, S. A. 1949. "The Work of the International Refugee Organization." *International Affairs* 25 (1): 66–73.

Ruggie, John Gerard. 2014. "Global Governance and 'New Governance Theory': Lessons from Business and Human Rights." *Global Governance* 20: 5–17.

Sassen, S. 1999. *Guests and Aliens*. New York: The New Press.

Schmitter, P. C. 2005. "Haas and the Legacy of Neofunctionalism." *Journal of European Public Policy* 12 (2): 255–272.

Scholte, J. A. 2001. "Civil Society and Democracy in Global Governance." *CSGR Working Paper No. 65/01*. Coventry: Centre for the Study of Globalisation and Regionalisation, University of Warwick.

Scholte, J. A. 2004. "Globalisation and Governance. From Statism to Polycentrism." *CSGR Working Paper No. 130/04*. Coventry: Centre for the Study of Globalisation and Regionalisation, University of Warwick.

Scholz, J. T., R. Berardo, and B. Kile. 2008. "Do Networks Solve Collective Action Problems? Credibility, Search, and Collaboration." *The Journal of Politics* 70 (2): 393–406.

Seaman, J., and J. Rivers. 1988. "Strategies for the Distribution of Relief Food." *Journal of the Royal Statistical Society* 151 (3): 464–472.

Seo, M. G., and W. D. Creed. 2002. "Institutional Contradictions, Praxis, and Institutional Change: A Dialectical Perspective." *The Academy of Management Review* 27 (2): 222–247.

Shacknove, A. E. 1985. "Who Is a Refugee?" *Ethics* 95 (2): 274–284.

Shaffer, G. C., and M. A. Pollack. 2010. "Hard v Soft Law: Alternatives Complements and Antagonists in International Governance." *Minnesota Law Review* 94: 706–799.

Shanks, C., H. K. Jacobson, and J. H. Kaplan. 1996. "Inertia and Change in the Constellation of Intergovernmental Organizations, 1981–1992." *International Organization* 50 (4): 593–627.

Shanmugaratnam, N., R. Lund, and K. A. Stolen. 2003. "In the Maze of Displacement." In *Conflict, Migration and Change*. Oslo: Norwegian Academic Press.

Shaw, R. P. 1975. *Migration Theory and Fact: A Review and Bibliography of Current Literature*. Philadelphia, PA: Regional Science Research Institute.

Sheehan, G. 2002. "Tuvalu Little, Tuvalu Late." *Harvard International Review* Spring: 11–12.

Shen, Shawn, and F. Gemenne. 2011. "Contrasted Views on Environmental Change and Migration: The Case of Tuvaluan Migration to New Zealand." *International Migration* 49 (1): 224–242.

Shenk, Jon. 2012. *The Island President*. USA: Samuel Goldwyn Films.

Simmons, B. 2000. "International Law and State Behavior: Commitment and Compliance in International Monetary Affairs." *American Political Science Review* 94: 819–835.

Simonelli, Andrea C. 2013. "Summary of Rights-Related Developments at COP19." In *Loss and Damage*. Human Rights and Climate Change Working Group.

Simonelli, Andrea C. 2014a. *Perceptions and Understandings of Climate Change and Migration: Conceptualizing and Contextualizing for Lakshadweep and the Maldives*, Field Report to the Norwegian Research Council.

Simonelli, Andrea C. 2014b. *Perceptions and Understandings of Climate Change and Migration: Conceptualizing and Contextualizing for Lakshadweep and the Maldives*, Field Report to the Norwegian Research Council.

Simonelli Berringer, Andrea C. 2012. "Climate Change Displacement and Global Governance: A Case Study of Three Intergovernmental Organizations and the Conflict between Member States and Bureaucracy." Dissertation Political Science, Louisiana State University.

Singh, Harjeet. 2015. "Vanuatu Needs Compensation, Not Charity." *Eco Business*, March 30.

Solon, P. 2010. "Why Bolivia Stood Alone in Opposing the Cancun Climate Agreement." *The Guardian*, December 21.

Sowell, T. 1996. *Migrations and Cultures: A World View*. New York: Basic Books.

Spencer, S. 2003. *The Politics of Migration: Managing Opportunity, Conflict and Change*. Oxford: Blackwell.

Sperling, L., and S. J. McGuire. 2010. "Persistent Myths about Emergency Seed Aid." *Food Policy* 35: 195–201.

Stanley, J., O. Stark, and J. E. Taylor. 1989. "Development Induced Displacement and Resettlement & Relative Deprivation and International Migration." *Demography* 26 (1): 1–14.

Stark, O., and J. E. Taylor. 1989. "Relative Deprivation and International Migration." *Demography* 26: 1–14.

Stein, B. N. 1981. "Refugee Research Bibliography." *The International Migration Review* 15 (1–2): 331–393.

Stein, B. N. 1983. "The Commitment to Refugee Resettlement." *Annals of the American Academy of Political and Social Sciences* May: 187–201.

Stein, J. V. 2010. "The International Law and Politics of Climate Change: Ratification of the United Nations Framework Convention and the Kyoto Protocol." *The Journal of Conflict Resolution* 52 (2): 209–229.

Stein, M. 1998. "The Three Gorges: The Unexamined Toll of Development-Induced Displacement." *Forced Migration Review* 1.

Stokke, O. S., J. Hovi, and G. Ulfstein. 2005. "Implementing the Climate Regime." *International Compliance*. London.

Storey, D., and S. Hunter. 2010. "Kiribati: An Environmental 'Perfect Storm'." *Australian Geographer* 41: 167–181.

Streimikiene, Dalia. 2013. "The 18th Session of the Conference of the Parties to the United Nations Convention on Climate Change (UNFCCC)." *Intellectual Economics* 7 (2): 254–259.

Swyngedouw, E. 2005. "Governance Innovation and the Citizen: The Janus Face of Governance-beyond-the-State." *Urban Studies* 42: 1991–2006.

Taran, Patrick A. 2000. "Human Rights of Migrants: Challenges of the New Decade." *International Migration* 38 (6): 7–51.

Taylor, S. T. 2005. "While Many Firms Stand Pat of Expansion, Notable Exceptions Enter New Markets." *Of Counsel: The Legal and Management Report*.

Thakur, Ramesh, Brian Job, Monica Serrano, and Diana Tussie. 2014. "The Next Phase in the Consolidation and Expansion of Global Governance." *Global Governance* 20: 1–4.

Thompson, Alexander. 2010. "Rational Design in Motion: Uncertainty and Flexibility in the Global Climate Regime." *European Journal of International Relations* 16 (2): 269–296.

Tol, Richard S. J., Maria Bohn, Thomas E. Downing, Marie-Laurie Guillerminet, Eva Hizsnyik, Roger Kasperson, Kate Lionsdale, Claire Mays, Robert J. Nicholls, Alexander A. Olsthoorn, Gabrielle Pfeifle, Marc Poumaldere, Ferenc L. Toth, Athanasios T. Fafedis, Peter E. Van Der Werff, and I. Hakan Yetkiner. 2006. "Adaptation to Five Meters of Sea Level Rise." *Journal of Risk Research* 9 (5): 467–482.

Torvanger, Asbjorn, Guri Bang, Hans H. Kolshus, and Jonas Vevatne. 2005. *Broadening the Climate Regime: Design and Feasibility and Multi-Stage Climate Agreements*. Oslo: Center for International Climate and Environmental Research (CICERO).

Trent, S. 2009. "Protecting Climate Change Refugees." *The Guardian*, November 2.

Tvedt, T. 2007. "International Development Aid and Its Impact on a Donor Country: A Case Study in Norway." *The European Journal of Development Research* 19 (4): 614–635.

UNHCR. 2005. 'An Introduction to International Protection: Protecting Persons of Concern to UNHCR." *Self-Study Module 1*. Geneva.

Vandergeest, P. 2003. "Land to Some Tillers: Development-Induced Displacement in Laos." *International Social Science Journal* 55 (175): 47–56.

Vanderheiden, S. 2008. *Political Theory and Global Climate Change*. Cambridge: MIT Press.

Vaux, T. 2006. "Humanitarian Trends and Dilemmas." *Development in Practice* 34: 240–254.

Volgy, T. J. 2008. "Identifying Formal Intergovernmental Organizations." *Journal of Peace Research* 45 (6): 837–850.

Von Stein, J. 2008. "The International Law and Politics of Climate Change: Ratification of the United Nations Framework Convention and the Kyoto Protocol." *Journal of Conflict Resolution* 52: 243–268.

Waltz, K. 1979. *Theory of International Politics*. New York: McGraw-Hill.

Walzer, M. (2011). "On Humanitarianism." *Foreign Affairs* 90 (4): 69–80.

Warne, K. 2008. "Dance of a Dangerous Sea." *Canadian Geographic* 128 (5).

Warner, Koko, and Sumaya Ahmed Zakieldeen. 2011. *Loss and Damage Due to Climate Change: An Overview of the UNFCCC Negotiations*. Oxford: European Capacity Building Initiative.

Webb, Arthur P., and P. S. Kench. 2010. "The Dynamic Response of Reef Islands to Sea-Level Rise: Evidence from Multi-Decade Analysis of Island Change in the Central Pacific." *Global and Planetary Change* 72 (3): 234–246.

Webber, F. 2011. "How Voluntary Are Voluntary Returns?" *Race & Class* 52 (4): 98–107.

Weidlich, W., and G. Haag. 1988. *Interregional Migration. Dynamic Theory and Comparative Analysis.* Berlin: Springer-Verlag.

Weiner, M. 1995. *The Global Migration Crisis: Challenge to States and to Human Rights.* New York: Harper Collins.

Weiss, K. R. 2015. "The Making of a Climate Refugee." Accessed January 28, 2015. http://foreignpolicy.com/2015/01/28/the-making-of-a-climate-refugee-kiribati-tarawa-teitiota/

Weiss, Thomas G., and Rorden Wilkinson. 2014a. "Global Governance to the Rescue: Saving International Relations?" *Global Governance* 20: 19–36.

Weiss, Thomas G., and Rorden Wilkinson. 2014b. "Rethinking Global Governance? Complexity, Authority, Power, Change." *International Studies Quarterly* 58: 207–215.

Wendt, A. 1992. "Anarchy Is What States Make of It: The Social Construction of Power Politics." *International Organization* 46 (2): 391–425.

Wenk, M. G. 1968. "The Refugee: A Search for Clarification." *International Migration Review* 2 (3): 62–69.

Whitty, J. 2003. "All the Disappearing Islands." *Mother Jones*, 23–43.

Winchie, D. B., and D. W. Carment. 1989. "Migration and Motivation: The Migrant's Perspective." *The International Migration Review* 23 (1): 96–104.

Woldemeskel, G. 1989. "The Consequences of Resettlement in Ethiopia." *African Affairs* 88: 359–374.

Wong, Derek. 2013. "Sovereignty Sunk? The Position of 'Sinking States' at International Law." *Melbourne Journal of International Law* 14.

Wood, W. B. 1994. Forced Migration: Local Conflicts and International Dilemmas. *Annals of the Association of American geographers* 84: 607–634.

Woodroffe, Colin D. 2008. "Reef-Island Topography and the Vulnerability of Atolls to Sea-Level Rise." *Global and Planetary Change* 62: 77–96.

Wright, J. 2009. "Climate Change to Kill Local Tourist Attractions." *Courier-Mail*, May 28.

Yamano, H., H. Kayanne, T. Yamaguchi, Y. Kuwahara, H. Yokoki, H. Shimazaki, and M. Chikamori. 2007. "Atoll Island Vulnerability to Flooding and Inundation Revealed by Historical Reconstruction: Congealable Islet, Funafuti Atoll, Tuvalu." *Global and Planetary Change* 57: 407–416.

Young, D. A. 2002. "Space-Time Computer Simulation Method for Human Migration." *American Anthropologist* 104: 138–158.

Zetter, R. 1991. "Labeling Refugees: Forming and Transforming a Bureaucratic Identity." *Journal of Refugee Studies* 4 (1): 39–62.

Zetter, R. 2007. "More Labels, Fewer Refugees: Remaking the Refugee Label in an Era of Globalization." *Journal of Refugee Studies* 20: 172–192.

Index

Norway, 77, 79, 83, 137, 145, 152
November, 19, 88, 115, 125, 152
NWP, 118

OAU, 78–79
objectives, 68, 108–109, 116
obligation, 6, 16, 33, 77, 80, 82, 84, 86, 100, 105, 116–117, 123, 134
observers, 144
occupation, 40, 78, 95
ocean, 9, 13, 15, 18, 23, 29, 32, 114, 123
oceanography, 114
OCHA, 6–7, 11, 73, 106–113, 128–132, 134, 136–138, 142–143, 145–147, 152–153
October, 152
ODPM, 94
OECD, 118
officials, 19, 36, 78, 109
OHCHR, 152
Okereke, 118, 122, 125
Oman, 152
Omidi, 20
operationalized, 117–118, 120
operations, 47, 60, 88–89, 102, 129–130, 146
opportunities, 8, 18, 35, 37, 52, 75, 91–92, 95, 138
organization, 6–7, 15, 22, 40, 55–57, 59–60, 63–65, 68–72, 75–78, 80–81, 83, 87–89, 91, 100–102, 104–105, 107–109, 112–114, 128, 130, 132, 137–138, 143, 152
organizational, 59, 61, 71, 81, 108–109, 128, 132, 143
origin, 40, 45, 76, 78, 80, 84, 90
originate, 47, 65
originated, 74, 77, 101, 113, 142
origination, 105, 128, 130, 132
Orsini, 59
Oslo, 83, 145, 153
Ostrom, 62
Ottawa, 95
Ottoman, 75, 87
outcome, 3, 42, 63, 66–67, 69, 93, 101, 120, 124, 128, 146
outline, 7, 9, 11, 14, 20–21, 39, 43, 49, 53, 65, 71, 74, 76, 80, 82–83, 86, 88,

93, 96, 98, 110, 112, 125, 128, 133–134, 137
outlook, 64, 97–98
outmigration, 20
overarching, 58, 123
Overbeek, 57
overdevelopment, 62
overpopulation, 17–18
oversimplifications, 10, 23
overview, 9, 11–13, 55, 80, 83, 132
Oxfam, 15, 102

Pacific, 3, 9, 13, 15, 19, 21, 23, 32, 139–140, 145, 147, 149, 152
Pakistan, 88–89, 98, 102
Palau, 152
Palestine, 75
Panama, 129
Papademetriou, 36, 87
Papua New Guinea, 9, 14
paradigm, 39, 51
paradise, 28–29
Paris, 5, 77, 115, 125, 138, 140, 146
participation, 25, 43, 55, 57, 59, 66, 146
Patel, 29, 145
PD, 62
perceptions, 43, 117
persecuted, 48–49, 83, 133–134, 142
persecuting, 49
persecution, 34, 39–40, 47–49, 79, 126, 141
personnel, 58, 69, 72, 109
persons, 6, 33, 40, 44–45, 48, 51, 75–76, 80, 82, 88, 91, 152
perspective, 43, 64, 74, 96
Peru, 75, 125
Petersen, 34–35
phenomenon, 5–7, 9–10, 17, 20, 23–24, 30, 33, 35, 38, 43–44, 55, 64, 86, 95, 99, 114, 127, 135, 139, 147
Philippine, 139
PICMME, 87
pilgrimage, 84
pillar, 84, 106, 118, 124, 126, 137
planet, 18, 139
Plato, 25–27, 151
plenary, 119
PM, 1–12, 54, 73–138